ORGANISM AND THE ORIGINS OF SELF

BOSTON STUDIES IN THE PHILOSOPHY OF SCIENCE

VOLUME 129

ORGANISM AND THE ORIGINS OF SELF

Edited by

ALFRED I. TAUBER
Boston University School of Medicine

SPRINGER SCIENCE+BUSINESS MEDIA, B.V.

Library of Congress Cataloging-in-Publication Data

Organism and the origins of self / edited by Alfred I. Tauber.
p. cm. -- (Boston studies in the philosophy of science ; v. 129)
Proceedings from a symposium held at Boston University, Apr. 3-4, 1990, under the auspices of the Boston University Center for Philosophy and History of Science.
ISBN 978-94-010-5507-9 ISBN 978-94-011-3406-4 (eBook)
DOI 10.1007/978-94-011-3406-4
1. Self (Philosophy)--Congresses. 2. Evolution--Congresses.
3. Immunology--Congresses. 4. Philosophy and science--Congresses.
I. Tauber, Alfred I. II. Boston University. Center for Philosophy and History of Science. III. Series.
Q174.B67 vol. 129
[BD438.5]
001'.01 s--dc20
[126] 91-10258

ISBN 978-94-010-5507-9

Printed on acid-free paper

To my mother, in loving memory

TABLE OF CONTENTS

evidence concerning the AIDS epidemic, completing our consideration of the Self as defined by a broadened formulation of immune identity, how it is constructed and its metaphysical origins.

The next group of essays may be considered as a group because of the explicit use of evolutionary problems to examine a definition of Self. The companion papers by Foster and Sarkar treat the issue of [illegible] [illegible] of Self [illegible] appeal to directed mutation. Sober discusses the problem of selection as [illegible] "Why are [illegible] adapted?", why [illegible] organisms, rather than [illegible] at some other level of organization, are the loci of adaptation. [illegible] presents a rather controversial theory of an evolutionary mechanism based on "amalgamation" (hybridization or [illegible] of species, [illegible]) which deals [illegible] and developmental biology [illegible] [illegible] represent [illegible] of defining the Self as ultimately a [illegible] Selves.

Although [illegible] theme is remarkable, there is no doubt that the anthology does not [illegible] a complete [illegible] of how the Self is defined by [illegible] [illegible] [illegible] [illegible] the Self is a metaphor and concept, clustered by its many presentations [illegible] a scientific [illegible] of endless variety. [illegible] for [illegible] criteria, [illegible] governing integrity and its [illegible] of biology. The problems presented [illegible] [illegible] [illegible] [illegible]

[illegible]

ALFRED I. TAUBER

[illegible]

FOREWORD

It is often said that the Copernican/Galilean revolution was one that dethroned the human species from its place at the center of the Universe and relegated human beings to their appropriate role as minor elements in the modern scientific view of nature. Indeed, the volume published by the U.S. National Academy of Science in 1973, at the 500th anniversary of Copernicus's birth, took the view that modern science is an outcome of the "Copernican Revolution" that destroyed and anthropocentric world. There is an extraordinary irony in this view because the entire ideological movement of European thought has been in precisely the opposite direction since Galileo's death in the middle of the 17th century. While physicists may have come to regard a human individual as a mere speck of jelly on a mole of dust circling a pinpoint of fire in a minor galaxy of a vast universe, the whole development of social theory in the last 300 years has been characterized by the progressive apotheosis of the individual as causally primary and as central to our concerns. Nor has scientific biology been formed differently. Nothing better characterizes bourgeois thought than its obsessive concentration on individuals.

We need to remind ourselves that it was not always so. The concentration on individual rights, properties and powers, whether in political or natural economy, is an outcome of the ideological shift that took place beginning in the 17th century with the rise of modern capitalism. In pre-capitalist Europe, customary relations between groups and institutions were definitive in the allocation of the rights and powers of individuals, where individuality was secondary and without independent power. Individual lives were the embodiments of collective properties not their causes. Developing capitalist relations demanded at the very least, that individuals could sell their labor power in the labor market and that people with money could engage in making more.

In France, until the revolution, to engage in trade was automatically to forfeit one's patent of nobility (How different from the nobility of the Second Empire!), and Russian serfs had an absolute right to remain on the land, although their owners often tried to ship them to cities as factory labor. The individual as social atom had to be freed from customary restraints; the individual, as individual, had to become the source of power and the repository of rights. So, social theory became, as C.B. MacPherson so convincingly argued, a theory of "possessive individualism". Individual rights and powers are at the center of modern political theory and "to secure these rights, Governments are instituted among men." The bourgeois theory of political economy, in the view of Adam Smith and the Scottish economists, is a theory of the individual as actor and as causally primal. All the collective properties of the economy arise as if by an invisible hand from the individualist behavior of social atoms, the individual producers, investors, buyers and sellers. The modern neo-classical economic theory that was built on that 18th century foundation is primarily a *behavioral* science, making strong claims about the psychology of individual preference. Indeed, the rise of psychology as a science, from its origins in experimental epistemology, is the manifestation of the central place that individual properties are seen to hold in schemes of social explanation. Psychology, in this sense, is the quintessential bourgeois science, because its central problematic is precisely the ontogeny of the *individual.*

It has long been a commonplace of the history of science that Darwin's theory of the natural economy of organisms was simply the political economy of the 18th and 19th centuries writ large. After all, we have Darwin's own word for it. And all agree that fundamental to Darwin's transfer of economic theory onto Nature was a concentration on the individual organism rather than the group, as the central unit of both analysis and cause. Nor are we surprised that the most recent extension of Darwinism as a mode of explanation, the theory of the evolution of behavior that calls itself 'Sociobiology', is, despite its name, explicitly a theory of the evolution of collective properties as a direct consequence of individual, selfish, fitness-maximizing action. But Mendelism, too, was a successful program because it focused attention on individuals rather than groups. It was by counting and classifying individual progeny, by taking account of the *differences* between individual offspring, rather than concentrating on the average properties of the ensemble that Mendel and his successors solved the problems of heredity. In fact, the relative failure of geneticists to this moment, to create a real predictive genetics of quantitative characters, has been a consequence of their forced reliance on ensemble properties like means and variances, and their inability to connect these group properties to individual differences. And, of course, embryology has as its central problematic the origin and development of the individual. In medicine, the ideopathic has triumphed over the sociopathic mode of explanation. Koch and Metchnikoff have vanquished Virchow who remains, nevertheless, a scientific model for the socialist left because of his emphasis on social causation of disease.

Because bourgeois economic and social ideology, and by extension, biological theory as well, had placed the individual at the causal and analytic center, the chief ideological preoccupation of the 19th and 20th centuries has been the problems of the *self*. How does the individual know itself, how has its psychic and material self-hood developed, how does it protect its privileged self-hood not only from the invasive attacks of other selves, but from the engulfing collective that threatens to consume and digest the self, assimilating it with other selves into an undifferentiated mass? These problems of self-hood are intertwined with two other ideological commitments that also arose with the restructuring of social relations consequent on the rise of modern capitalism: the alienation of the internal from the external, the organism from the environment, and the general commitment to reductive explanation. What is striking about these commitments is that they are in contradiction to the self and threaten to annihilate the privileged position of the individual.

The historical development of a modern mechanistic biology had depended critically on a successful separation of the internal from the external. The Darwinian theory of evolution regards the world external to organisms as an autonomous domain with its own dynamic, a dynamic that creates the "stage" on which the "evolutionary play" is enacted. Environments set the problems. Organisms whose inner nature allow them to solve the problems successfully survive and leave offspring. The others fail. The nature of the organism itself is a consequence of internal forces that are independent of the external world, that is, at random with respect to the problems created by the environment. The individual organism is then the locus of connection between the internal and external. It is called into being by the internal and disposed of by the external. It has, in this way, no separate existence, but is simply the nexus of autonomous internal and external forces. We then have the curious irony that although Darwinism is a theory of individual survival and reproduction, of individual adaptation, the organism as organism plays no role at all. There is no unique domain of forces that are seen at the level of the organism. So, the critical role of the individual is threatened and contradicted by its placement at the boundary between autonomous internal and external forces. The contradiction is a very deep one, for if there is no boundary between the internal and the external, if they flow continuously into each other as the pre-modern natural philosphers thought, then how do we locate the individual at all? "Self", by its own nature, demands the definition of its dialectical partner, "other". The concept of self is an alienating one, dividing the world into non-overlapping domains; yet once the individual has been placed at the boundary between internal and external, it seems to have a passive existence. Nor is it

is the 20th century's struggling and dissenting voice, while chemo-reductionism has grown to dominate modern molecular biology and biochemistry [2, pp. 103–106]. It is not self-apparent why this particular program has triumphed at the (exclusive) expense of other research programs when complex systems such as the immune or neurological seem so obviously to require organizational and hierarchical paradigms. This question appears to reflect complex issues of a broadly ideological nature, and whether we wish to analyze by one or another of various models, it is obvious that we must stand back and reconsider our metaphysical orientation in order to initiate an understanding of what we are doing, beyond accumulating data and solving "scientific" problems. The issue is quite simply, to what do we attribute meaning, i.e. personal knowledge—Polanyi's term—in modern biology? It is to this question that we address our effort.

My own orientation has focused on how Darwinism has influenced medicine, and in particular immunology, in defining the Self. On one level, Darwinism (its modern versions and alternatives) has had little direct impact on medicine's superficial consciousness. Neither physician, nor patient explicitly acknowledges his evolutionary debts, and while aware that profuse genetic variety exists for each of our proteins, sometimes manifest in clinical disease, the direct influence of evolutionary principles has had little, if any direct influence on modern medical thinking. This is actually quite extraordinary. For it was in the early reception of Darwin's natural selection theory that a profound alteration in the concept of organism took place: no longer was the organism a pre-set balance of forces (humors, elements of any kind), but was intrinsically a product of evolved, potentially imbalanced structures/functions that were harmonized by evolutionary forces and self needs, to yield functional units more adaptive in a competitive, hostile environment.

Chernyak and I have recently completed a study of how the explicit appreciation of this basic shift in the metaphysical structure of the organism was made by Elie Metchnikoff in the 1870's and 1880's (Fig. 1) [3]. He presented the phagocytosis theory, the basic conceptual notion of immunity, in response to how the organism was defined by such evolutionary challenges; in the process, he established a modern definition of self-hood. I have structured this discussion on Metchnikoff for three compelling reasons: First, the emergence of the modern concept of organism might begin with any number of events in the 18th or 19th centuries, and to a certain extent, I must plead a degree of arbitrariness in choosing to place Metchnikoff in central focus. I do so primarily because he so clearly illustrates the bridge from descriptive

Fig. 1. Elie Metchnikoff, 1907.

biology to the biochemical reductionism of the 20th century. In the process, he captured the essential centrality of organism and erected upon that crucial concept a new metaphysical domain for investigation. Second, and more specifically, I wish to emphasize the extraordinary importance of Metchnikoff's theoretical formulation for establishing the conceptual basis of modern immunology, namely the *active* host response to pathogens. This notion of self-definition, as an active process, was novel and had an enormous impact on biology. It will serve us well to begin our discussion of origin of Self with a careful exploration of his theory. The third interesting feature is that Metchnikoff was unabashedly philosophical about his science.

Perhaps the clearest path to trace Metchnikoff's scientific legacy as an evolutionist resides in his early polemic with Haeckel. Metchnikoff was successful in supplanting Haeckel's theory of ontogeny recapitulating phylogeny with a more complex appreciation of ontogenetic development, rejecting a single pattern of gastrulation extrapolated from vertebrates to primitive invertebrates. Whether gastrulation occurs by invagination, or by unipolar or multipolar introgression, the result is the same: a single bilayered, ciliated embryo incapable of further development is transformed into a two-layered organism capable of further movement via the ciliated ectoderm and of further development by the unciliated endoderm. As Leo Buss has argued, explaining this embryologic strategy is a central problem in modern evolutionary theories, where the modern synthesis (evolution and genetics) has omitted ontogeny (the synthetic theory of evolution is a "theory of adults") [5, p. 50]. A single-celled protist must simultaneously express specialized modes of locomotion, feeding and behavior, and yet retain the capacity for cell division. Metazoans have no such constraint, and have taken the strategy of differentiation and segregation of germ cells. At issue is the simultaneous need for an organism to move through fluid with cilia or flagella and to divide using a mitotic spindle. Unless a cell possesses microtubule organizing centers capable of performing both tasks, or possesses multiple microtubule organizing centers per cell, the cell's functional range will be constrained. In certain protist groups, cell division and locomotion can occur simultaneously; in others they cannot. While many protist taxa overcome the ciliation constraint to division, those protists giving rise to metazoans did not. Metazoans inherited the constraint limiting simultaneous mitosis and ciliation. Leaving ciliated cells on the surface for locomotion and feeding, the movement and subsequent proliferation of germ cells from the blastular surface into the center of the sphere is gastrulation, a metazoan solution to the requirement of simultaneous movement and development.

There is a familiar parallel to Metchnikoff's orientation in modern biology: Metazoan ontogeny is a sequence of cell lineages progressively denying their own capacity to increase for the collective interest of the individual (as above, Buss, and on a more basic level the same basic process is postulated to have occurred within cells by symbiosis, as argued by Lynn Margulis [6]). These notions of co-evolved microbial communities, where the eukaryotic cell is composed of several genomes from different sources—heterogenomic development (Margulis), or the strategies where the metazoan collective of parent and variant daughter cells imposes constraints on cell lineages for the collective interest of the individual (Buss), are each closely related to

Metchnikoff's original constructions, which were based on the conflict between the potentially opposing processes of various somatic elements and organismal integrity. The existence of harmonious function in favor of the individual (and total) organism in Metchnikoff's original terms is an active process. How have organisms evolved so that some cells have abandoned their own capacity to replicate? The strategies are complex, but generally, patterns in cleavage and regulation are adaptations that serve the function of imposing selection at the level of the cell lineage. What Metchnikoff recognized, albeit in a poorly, and from our point of view, unsophisticated manner, was that evolution must be understood by selective processes that operate on the interactions of cell lineages. "Evolutionary pattern has arisen not by selection on individuals alone, but by the interactive effects of selection operating at differing levels of biological organization". [5, p. 68] This is a dialectical position, which we will have occasion to explore below, but suffice to note here that Buss and Margulis share the same broad metaphysical construct as first proposed by Metchnikoff over 100 years ago. It is the re-emergence, sometimes in different guises, sometimes in exactly the same format, of the view that organisms are dynamic subjects of their own self-definition, and not simply objects of "evolutionary forces" or physico-chemical laws. For example, the application of this concept in the modern context, results in a formulation that generates a highly evocative model for examining the immune system. Normal development in ontogeny proceeds not with every detail of cell interaction programed, but metazoan cells must interact as a consequence of traits developed in the ancestral past; the genome encodes the relative competitive relationship of developing cell lineages [7]. Edelman's use of "competition" in this case is controversial, and might be viewed more accurately as "cooperative". The interaction of cell lineages then may be regarded as self-limitation selected through evolution, rather then competition, *per se* [8]. In any case, an epigenetic landscape proposed by Waddington, portrays the indeterminism of development: The undulating peaks and valleys, where a ball placed above the ridges may proceed down any one of several pathways, depicts each valley as the origin of a variant cell lineage in the course of ontogeny [9]. The graphic then depicts how a particular ontogenetic expression is but one of several potential pathways whose potential (i.e. mechanism of interaction) is programmed, but whose final declaration is determined by self-constructed community of cell lineages.

In this view, explanation of epigenesis must equally share with genetics an understanding of development, for "genes specify local rules, not global pattern. Above all, developmental events intervene between genotypic and

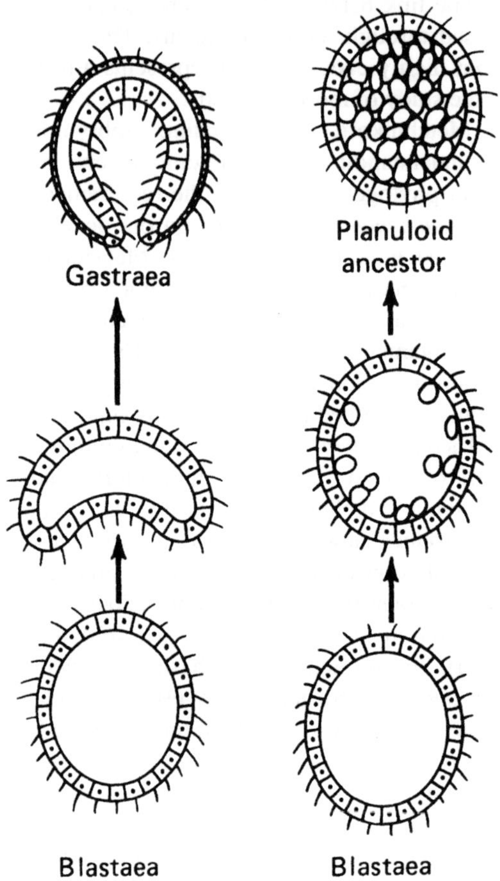

Fig. 4. Hypothetical stages in the evolution of early metazoans. A colonial flagellate as metazoan precursor was postulated in Haeckel's original theory (left column), as possessing a distinct anterior-posterior axis and differentiation of somatic and reproductive cells. This first stage, blastea (blastula was considered its recapitulation in existent metazoans) developed into the metazoan ancestor, gastraea (analogous to gastrula of metazoans), by invagination to form a double walled, sac-like organism. Metchnikoff, citing that cnidarians gastrulate by ingression, where cells proliferate from the blastula wall into the interior blastocoel, producing a solid gastrula, suggested that invagination may have arisen as a secondary mechanism of gastrulation. The planuloid ancestor (i.e., planula larva of cnidarians) was first named *parenchymella*, and then *phagocytella*. [Reproduced from ref. 3, Chernyak and Tauber, 1988, p. 229.]

which appear to enhance phagocytosis and may well be the most primitive humoral recognition factors of non-self markers. (These substances are distinct from aggregation factors which mediate self recognition in syngeneic and allogenic interactions (see below).) It is of interest that sponges tolerate an extensive interaction with commensals and symbionts, in stable relation that appears to form a dynamic mutual exchange of metabolites. In certain sponges, bacteria may occupy up to 40% of the mesohyl volume [13]. There is an implicit assumption that the sponge is able to control the density and composition of its internal symbiont community, but it is of interest from this perspective how Metchnikoff incorporated divergent data into his grand scheme, choosing to ignore the problem of distinguishing defensive from nutritive functions.

It was in the coelenterates that he found a more orthodox defensive phagocyte. Ameboid cells are found in all three cell layers—but Metchnikoff did not differentiate between the non-phagocytic interstitial cells of the *Hydrozoa* class, and the true phagocytes, called amebocytes in modern parlance, that appear in *Scyrhozoa* (jellyfish) and *Anthozoa* (sea anemones, corals) [14]. While interstial cells of hydra may participate in graft rejection and wound healing, they are not required. In contrast, the amebocyte participates in healing, tissue reorganization and phagocytosis of foreign tissue. True parasites are uncommon although coelenterates have not been found to generate agglutinins, bactericidins or antibody-like substances, they utilize antibiotics, the stolon armed with nematocytes (stinging cells), and mucus for non-cellular defense. The different classes have distinct phagocytic responses: Hydrozoans do not have an inflammatory reaction, since only stationary, endodermal cells phagocytose. In scyphozoans and anthozoans, infiltration occurs, but a basic difference is observed between the response to foreign invasion and feeding, involving two different cell types. Truly not much is known beyond Metchnikoff's first observations published in *Lectures* of how a splinter injected into a jellyfish is surrounded within 24 hours by numerous amebocytes, but that observation and similar descriptions in phyla extending to vertebrates, enabled Metchnikoff to erect a grand scheme of host defense. It was truly an extraordinary intuitive grasp of an underlying biological process.

Again, it is in primitive organisms that principles of allorecognition and the development of the immune system may be sought and possibly extrapolated to vertebrate immunity. Fusion between individuals with different commitments to somatic function results in parasitism, and mechanisms to prevent

support for the hypothesis arises from several levels of investigation: 1) The role innervation of spleen, thymus and lymph nodes plays is unclear, but the anatomy is striking [21]. 2) The embryological development of the thymus is closely integrated with neural crest interactions, showing a developmental basis for linkage between the immune and neurologic systems [22]. 3) In the case of mast cell activation, (the cell principally responsible for the manifestations of the acute allergic response) the reaction is initiated by binding of antigen to the surface membrane IgE-receptor; while representing the classical "immune" response, this pathway now appears as only one avenue of initiation of the hypersensitivity reaction. The selective release of distinct sensory neuropeptides from different subsets of nerves, the specificity of neuropeptide recognition by mast cells, and other target cells, and the diversity of direct and indirect activities of neuropeptides suggest that sensory nerves may initiate and modulate immediate hypersensitivity by heretofore undescribed mechanisms [23]. 4) There is compelling data showing bidirectional circuits of circulating hormones between the immune and nervous systems [24]. Thymosins, lymphokins and certain complement proteins affect neurologic function, as well as lymphocyte-elaborated opioid peptides, adrenocorticotropic hormone (ACTH) and thyroid stimulating hormone (classically described neuroendocrine hormones). In turn these neuroendocrine hormones and autonomic pathways directly affect thymus and bone marrow function, and modulate macrophage and lymphocyte behavior.

We are also intrigued with the likely evolutionary origins of neuropeptides, hormones, and receptors, on the one hand, and the super-immunoglobulin family that includes the cell adhesion molecules (CAM's) on the other [17]. Roth and co-workers have presented a compelling hypothesis that communicative substances have a long phylogenetic history, since unicellular organisms contain molecules resembling messenger peptides of vertebrates [25]. They argue that many hormones and their receptors originated with microbes, and later in evolution, anatomic and functional diversity led to the neuroendocrine and immune systems. This so-called paleocentric or unification theory, would then predict that a hormone may be synthesized not only in the specialized tissue, but by other cell types as well (e.g. ACTH and endorphins are normally produced in seven different tissues). Materials that resemble hormonal peptides and neuropeptides, until recently thought restricted to multicellular animals, are present in protozoa, bacteria and higher plants. There is also evidence for substances in microbes that bind hormones and other messengers, which resemble vertebrate receptors. "In extending by two-to-four-fold the evolutionary age of these molecules, we extend the range of biologic distribution by orders of magnitude [25]." Similar insight has been obtained for the immunoglobulin recognition family, where structural and genetic analysis has shown homology not only between immunoglobulin, MHC and T4/T8 lymphocyte receptors, but also with vertebrate embryonic cell adhesion molecules (CAM). The topobiological address system of vertebrate morphogenesis [7] then appears as a primitive recognition system employed in ontogenetic development, and not surprisingly, structural homology with neural CAM has been established with the cell adhesion protein (cs-A, contact site A) of the slime mold, *Dictyostelium* [17]. It appears that the immunoglobulin suprafamily originated from more ubiquitous cell recognition molecules which are common in all metazoan species.

The biochemical parallels of the immune and neuronal system are of increasing interest not only in speculative construction of their respective evolutionary and ontogenetic development [26], but in expanding the conceptual basis by which the immune system might be characterized. A series of papers first based on the concept of autopoiesis, and later developed into an "autonomous network" concept, have attempted to reformulate the intellectual orientation of immunology in a direction we consider both interesting and potentially productive [27]. Most saliently, the effort of what I will refer to as the Paris School, conforms to our endeavor of tracing the intellectual idea of immunity that posits the direction of its future growth. We have occasion to reformulate the construction of immune theory, but the Paris School perspective offers us the foundation for a revision [18].

The immune system asserts a molecular self during ontogeny, and for the entire lifetime of the individual, it keeps a memory of this molecular identity.

> It must be stressed that the self is in no way a well-defined (neither pre-defined) repertoire, a list of authorized molecules, but rather a set of viable states, of mutually compatible groupings, of dynamical patterns. In effect, a molecule is neither self nor anti-self, as a musical note does not belong more to a composer than to another one. The self is not just a static border in the shape space, delineating friend from foe. Moreover, the self is not a genetic constant. It bears the genetic make-up of the individual and of its past history, while shaping itself along an unforseen path [27, Varela *et al.*, p. 363].

When viewed primarily as a cognitive or informational system the immunological process becomes a system of self-definition, a self-referential process.

> ... from our vantage point, the only valid sense of immunological self is the one defined by the dynamics of the network itself. What does not enter into its cognitive domain is ignored (i.e., it is non-sense). This is in clear contrast to the traditional notion that immune system sets a boundary between self in contradistinction to a supposed non-self [*Ibid*, p. 365].

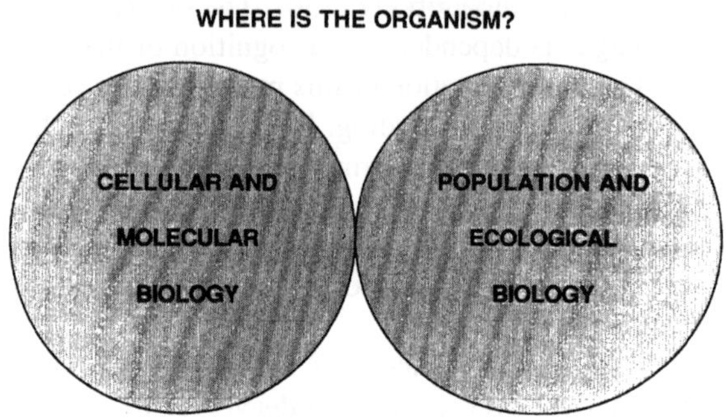

Fig. 5. Where is the organism?

respective points of view, each contribute an elucidating facet to illustrate the problem. But the concept of organism seems to fall into the precarious tangent between molecular and cellular biology on the one hand, and behavioral-ecological studies on the other (Fig. 5).

Organismic biology is as yet an undernourished orphan in biology, unwanted because it is not understood by biologists interested in levels above and below the central one of individual organisms [33].

Clearly the concept of organism contains an essential, but ever shifting metaphysical focus of the life sciences; when the problem of self-hood is added, our inquiry is somewhat contained by the borders of individuality, its genesis, evolution, and maintenance. A sympathetic forum for discussion addresses the *reactive*, self-defining, organism, as recently re-described by Levins and Lewontin in regards to the evolutionary process:

.... Cartesian reductionism is sometimes spoken of as the "Cartesian method," as a way of finding out about the world that entails cutting it up into bits and pieces (perhaps only conceptually) and reconstructing the properties of the system from the parts of the parts so produced. But Cartesianism is more than simply a method of investigation; it is a commitment to how things really are.

....The great success of Cartesian method and the Cartesian view of nature is in part a result of a historical path of least resistance. Those problems that yield to the attack are pursued most vigorously, precisely because the method works there. Other problems and other phenomena are left behind, walled off from understanding by the commitment to Cartesianism. The harder problems are not tackled.

.... One way to break out of the grip of Cartesianism is to look again at the concepts of part and whole. "Part" and "whole" have a special relationship to each other, in that one cannot exist without the other, any more than "up" can exist without "down". What constitutes the parts is defined by the whole that is being considered. Moreover, parts acquire properties by virtue of being parts of a particular whole, properties they do not have in isolation or as parts of another whole. It is not that the whole is more than the sum of its parts, but that the parts acquire new properties. But as the parts acquire properties by being together, they impart to the whole new properties, which are reflected in changes in the parts, and so on. Parts and wholes evolve in consequence of their relationship, and the relationship itself evolves. These are the properties of things that we call dialectical: that one thing cannot exist without the other, that one acquires its properties from its relation to the other, that the properties of both evolve as a consequence of their interpenetration [34].

Dialectical biology, at least by this definition, rests on the singled integration of part and whole as only existing together and thus characterizing the other. The parts and whole have a dialectical relationship with the other, conferring properties, back and forth, so that neither has its full meaning independent of its *relation* to the other. This philosophical attitude is obviously related to the pre-Darwinian scientific tradition, where the chain of being stressed the connectedness of each stratum of the natural order. Each being found its allotted place within the hierarchy of nature, each had its assigned responsibility, and each had its function relative to the larger whole. But by mid-19th century, the universe had lost its order:

...There have been during that period [1848-1865], in consequence of revelations by scientific research in this direction and in that, some most notable enlargements of our views of physical nature and of history—enlargements even to the breaking down of what had formerly been a wall in the minds of most, and the substitution on that side of a sheer vista of open space....Dykes have been burst; boundaries removed; we hardly know the old landmarks [35].

The metaphysical basis of the Darwinian revolution as interpreted by Metchnikoff was that the parts in fact strove against each other, within the organism. Integrity was not assigned or assumed, but accomplished as a result of an active process-harmonization. What the harmonizing force was could be argued—Metchnikoff assigned the responsibility to the phagocytes (broadly, immunity). As important as the immune system is for achieving organismic integration, surely each system must be responsive to others, and as Buss so elegantly argued, it is in fact a characteristic of all cell lineages to assert their hegemony and at the same time, their dependence on cooperative venture. This is obvious for eukaryotic life, but also applies to prokaryotes [36], and each is clearly dialectical in character.

Our focus on organism as dialectical may obviously be constructed on several levels, or perhaps better, by various vectors, each pointing towards a different aspect of organic function, interaction, or organization. Using Metchnikoff as our foil, so to speak, we have placed his concept of organism

purpose, and 2) a resultant ordering of their structure/function. Biologists working with such principles in conjunction with strong descriptive techniques (e.g. Johannes Müller and Karl von Baer) were instrumental in establishing physiology and embryology on a firm experimental basis. Parallel teleological constructions were made by Georges Cuvier in comparative anatomy and paleontology. It is not our purpose to further describe teleology's career in the mid- and late 19th century [42], beyond noting that in its first skirmishes with chemico-reductionism, the Kantian precept was drawn with clarity: teleology was to be used as a means to organize and establish criteria of organismic wholeness. It was not be contaminated with conflicting assignments i.e. causal purpose. The mechanical framework would "unravel the complex of effective causes" and teleology would posit modes of organization. These are mutual and complementary roles, not opposing. Each might optimally supplement the capability of the other, and thereby enhance the scientific enterprise. But in the ardor of the late 19th century, teleology was associated with vitalism, and the two were thrown out together. The vitalistic contamination created by von Baer, Müller and other second generation heirs to Kant, was not easily shed, especially under the power of effective advance made by the biochemists, whose reductionism served them well in their particular program. It is well to remember however, that the early reductionist program which focused its attack on vitalism did so with a conscious philosophical agenda: "Having failed to find a way of combating vitalism on empirical grounds, they attacked it on an epistemological basis" [37]. The great intellectual re-orientation ushered in by *Origin of Species*, not only presented a new paradigm for species transformation, but had enormous influence on general concepts concerning the nature of the organism. It offered a severe materialistic orientation, where blind struggle for survival determined form and function. The teleomechanical search for an organizing force or principle was quickly subsumed beneath the banner of chance. Search for an organizing principle to explain *telos* could not compete with the rapid successes of the materialistic program. Virchow, Bernard, Pasteur, trained in the pre-Darwinian period, were never fully satisfied with the radical reductionist program, but their cautionary notes were largely ignored.

Upon this stage, Elie Metchnikoff (in 1883) offered a new concept of organism, that both followed evolutionary principles and maintained the inner necessity of defining self-hood. Metchnikoff does not fit neatly into any of these ideological categories. His research program found no particular reason to resurrect the teleomechanical banner, which essentially expired with von

Baer, but his search for organismic "integrity", "wholeness" (and within the Darwinian currents), "harmonization" served as a dangerous alarm to the reductionists ever on the look-out for vitalistic/teleological sentiment. Metchnikoff was neither a von Baerian vitalist (a biological problem), nor teleologist (a metaphysical issue) (despite their professional and personal affinity), for in a radical fashion, on the edifice of the Darwinian program, he effectively changed the rules of the earlier debate, and erected a new definition of organismic integrity resulting in a novel science, immunology [3]. Metchnikoff reformulated the Kantian teleological framework, and to understand the uniqueness of this position, we must further outline the previous concept.

Kant had explored whether purpose (end, design) has any value for the investigation and explanation of nature. In the *Critique of Pure Reason*, he attempted to show that the fundamental principles of Newtonian physics were objectively valid for our experience. The unity of nature (required for science) and the unity of objective human experience are rooted in this common source, so that to move from the subjective order of our representations to an objective order of events, our representations must be connected in a certain sequence, whose order is governed by an objective law or rule. This causal principle, one of Kant's famous *a priori* concepts of synthesis, was however only applicable to mechanical and mathematical studies of nature, not the living. The idea of mechanism was concerned simply with those features capable of quantitative treatment, and thus only certain features of nature are relevant for mechanical explanation. From a mechanistic point of view the question *why* events happens is illegitimate, because it is in principle unanswerable; teleological explanation (*why*?) is quite different from a causal one (*how*?), although the former may be superimposed upon the latter. Kant first allowed teleology in the biological realm as a heuristic strategy to gain insight or direct our attention to novel questions, which would not be asked simply by addressing issues of *how*. He rigorously accepted only mechanical explanation as objective, for teleological explanation was not admissible as part of scientific knowledge. But Kant allows an approach to nature *as if* it were ultimately teleological to suggest avenues of investigation [41, p. 35]. Teleology then assumes a *regulative* validity in so far as it suggests how knowledge may be brought into systematic unity. And perhaps more fundamentally, he viewed organisms as possessing a different form of causality from that governing the inorganic realm. Linear modes of causation are inadequate for biological systems which transcend this form of reason and rely on different assumptions and strategies. While organisms may be viewed mechanically, there are fundamental differences. Kant viewed the organism as a "system", where parts exist for the whole and all parts for the sake of one.

It is here that Metchnikoff makes his radical departure, for he postulated a sub-system (the immune function) acting as the arbiter of organismic integrity, and almost as an autonomous center of activity, defined the Self. What regulated phagocyte definition of identity and how it was subsumed under the more encompassing requirements of the organism as a whole was largely dictated by both ontogenetic and phylogenetic constraints, individual history, and external challenges to the organism's self-aggrandizing function. This then is a teleology defined not from an encompassing Romantic morphotype or *Blauplan*, but from within, in a dynamic and dialectical fashion. It arose out of the metaphysical struggle of disharmonious parts harmonized by a self-regulating core activity. The extension of this position is presented in our accompanying essay [18], but suffice here to note that Metchnikoff's novelty and the implications of his theory were not only misunderstood, but dismissed under the rubric of an old teleological argument.

Despite Metchnikoff's departure from Kantian precepts, he implicitly understood his debt. Teleology might still be used as a fundamental guide in studying the organic realm. Biology could not be reduced to mechanical principles alone (*viz.* physics and chemistry), but began by recognizing organization as the beginning of any investigation of life processes. Biological research required the assumption of *zweckmassig*, or purposiveness, as a regulative concept. Remaining however is the fundamental paradox of Kant's view. Nature operates by blind mechanical laws, yet biological systems were by definition more than mere aggregates and operated with purpose. There is an irreconcilable dualism as long as Kant argued that the only objective principles we possess are mechanical and that organisms are by their very nature non-mechanical. Organic forms are not analogous to a watch, which functions by "external cause", for organism is both cause and effect of itself. If we are to remove the notion of external cause, then we must regard organisims as themselves producing their characteristic organization and unity.

> ... if parts are removed from the watch, it does not replace them on its own; nor if parts were missing ... does it compensate ... by having the other parts help out, let alone repair itself on its own when out of order: yet all of this we can expect organized nature to do. Hence an organized being is not a mere machine. For a machine has only *motive* force. But an organized being has within it formative force, ... that propagates itself [143, p. 253, No. 375].

Thus organisms produce their own organic unity and are defined by two primary characteristics: 1) constitutive parts function only in their relation to the whole, and 2) organic parts reciprocally produce each other. This self-organization was a mystery to Kant, for by rejecting the blind accident of organic function, order must be accounted for in terms of the order itself, i.e. a regulative, *a priori* principle. The Darwinian construct offered a fundamental re-orientation, and novel explanation. But ironically Kant offered the foundation of Metchnikoff's cardinal conceptual hypothesis: the organism is responsible for itself. In this sense, he is heir to Kant: the organism is both cause and effect of itself. The profound irony is that Metchnikoff resurrected a dynamic, assertive Self on the Darwinian scaffold of chance and blind materialism. That he was both misunderstood and vilified is lucid testament to the metaphysics of his opponents.

IV

The organism-mechanism debate continued into the 20th century, essentially oblivious to Metchnikoff's redefinition of the problem. Thus Jacques Loeb, in posing (in 1912) the reduction of organic processes solely to physico-chemical terms concluded that organic organization was derived "from the common source of all life phenomena, i.e. the chemical activity of the cell: [44, p. 103]; consistently then, psychology was "accessible to analysis by means of physical chemistry" (ibid, p. 61), our emotions are based on instincts, "whose analysis, from the mechanistic point of view, is only a question of time" (ibid, p. 32), and that such instincts are the root of our ethics and that the "instincts are just as hereditary as is the form our body" (ibid, p. 33). But at the same time E.S. Russell viewed biology rather differently:

> Complete uncertainty reigns as to the main principles of biology. Many of us think that the materialistic and simplistic method has proved a complete failure, and that the time has come to strike out on entirely different lines.... It may well be that the intransigent materialism of the 19th century is merely an episode, an aberration rather, in the history of biology [Russell, op. cit., note No. 39, p. 345].

Maybe J.H. Woodger presented the issue most succinctly:

> Only two types of theoretical biology have so far been devised, both involving using the analogy of a humanly constructed machine: 1) vitalism (with a mechanic), and 2) the 'machine theory' (without a mechanic). This provides no independent biological way of thinking, because machines presuppose organisms. The vitalist puts himself into the machine he has made. The

other type of theorist forgets he has made it. You obviously cannot escape from teleology in this way, because machines are teleological instruments made by men. Any explanation of teleology by analogy with machines simply attempts to explain internal teleology by means of external teleology and hence still remains teleology. The machine theory without a mechanic illustrates the way in which biology has turned either to one or the other of the two modes of thinking resulting from the Cartesian dualism. It only seems 'materialistic' because the psychological origin of the machine is easily forgotten and omitted from the analogy. Any one who wishes to spare himself further thought on this troublesome question can do so by adopting either of the current alternatives, but he will not thereby avoid 'teleology' [45, p. 441].

It is not our purpose to further explicate the debate, but only to note that neither Russell's dismay nor Woodger's skepticism did much to stem the positivistic tide of the reductionist program. But we must note, in reaction to the severe mechanical approach advocated by Wilhelm Roux (for embryology) and Jacques Loeb (for physiology) in the early 20th century, there arose a countervailing integrative approach which re-emphasized a holistic view of biological phenomena. The work of Sherrington, Henderson and Cannon focused on the interrelation of parts comprising complex physiological control, and for evolutionary theory, the synthesis constructed by R.A. Fisher, J.B.S. Haldane, and Sewall Wright, resulted from an appreciation that adaptation was the result of integrated processes (gene actions) that could only be understood if viewed as part of a whole: the gene pool was subject to the total environment (physical and biological) in which the organism "lived." Twentieth century biology has not resolved these competing scientific programs, or maybe more accurately, philosophical outlooks. The reductionists consider that all higher level interactions may be predicted (and ultimately confirmed) by knowledge of physico-chemical interactions (i.e. the whole is equal to the sum of its parts). To the holistic materialists [2], certain aspects of the whole emerge distinctly by integrative processes, i.e. properties are conferred on a component by the act of interaction itself; thus a complete description requires the study of interactions as well as the examination of parts, as separate. The extreme position of either school may easily be criticized, but the issue is one of dominance. The extraordinary recent advances in molecular biology might proclaim the gene as hegemonic, but the 19th conflict continues in new format.

Notwithstanding the lingering ideological issue, a science of *organism* has not developed, where the laws of organization and hierarchy have been sought as a *primary* focus of biology. This is not to say that the problem has been ignored; there is serious attention to the issue by both philosophers and scientists, e.g. the Eigen-Schuster hypercycle [46], Gatlin's game theory [47], macromelecular organization [48], and in several other disciplines [49], but the problem does not appear high on the agenda of our funding agencies. "Theoretical biology" has a highly limited audience and such discussion exerts little influence on the research strategies of most working biologists [50]. Part of the reason must reside in the mechanistic view that still guides our research. Clearly the impact of post-Newtonian physics has had little impact on our discernment of life processes. Quantum principles have recently been applied to biological systems in the areas of catalysis and proton transport [51], but these are mere extensions of current biochemical constructs. The impact of relativity and quantum mechanics on the biological orientation has at first glance been negligible, for the processes studied are at a level where such phenomena are not apparently relevant. But, if the history of science teaches us anything, it is how science directs epistemology, and implicit metaphysics governs scientific constructs. As Bouchelard said, "Science in effect creates philosophy" [52], but the converse is also true. In this regard, our choice of Metchnikoff as a paradigmatic thinker of 20th century biology may now become clearer.

As noted, Metchnikoff cannot be assigned the label, "teleomechanist." He implicitly invoked teleology, but highly radicalized from its Kantian formulation, and far closer to the 20th century neo-Kantian position than the concept understood by von Baer, Müller, or Leukhart. The traditional Kantian construct assumed that the organism was "given" as whole, that "integrity" (wholeness) was naturally provided or self-evident. Metchnikoff viewed integrity as actively attained—the natural state of the organism was disharmonious and wholeness (integrity) was actively pursued and sought by a harmonizing sub-system (the immune system), which is responsible for defining self-hood. It is an on-going activity, and one of a highly dynamic character. (Defending the Self may be viewed as a distinct property of the immune system [18].) The Self is ever-changing as it engages the environment, exposed to potential immune challenges, some of which are rejected and others incorporated into an ever newly synthesized whole. From this processed dynamic, self becomes Self and is never a static, given whole. As opposed to the classical scientific position, where the object is given and self-evident (an epistemological interpretation which emphasizes the "active" role of the intellect that "transforms" the object into knowledge about the object), in modern science to establish and define the object is essential to its development. Whether we view space in relativistic terms or the problem of defining atomic phenomena by an uncertainty principle, the object of our inquiry must be defined in its context and is never "given." It is almost uncanny how modern Metchnikoff's formulation of organismic integrity is in this ontolog-

nical role in Russian science [59], its influence on Western biology has been minimal; we would do well to seek the antecedents of dialecticism in our late 20th century context with organismic biologists such as E.S. Russell.

> ... the idea of an active creative organism is repugnant to the intelligence, and that we try by all means in our power to substitute for this some other conception. In so doing we instinctively fasten upon the relatively less living side of organisms—their routine habits and reflexes, their routine structure—and ignore the essential activity which they manifest both in behaviour and in form-change.
>
> We tend also to lay the causes of form-change, of evolution, as far as possible outside the living organism. With Darwin we seek the transforming factors in the environment rather than within the organism itself. We fight shy of the Lamarckian conception that the living thing obscurely works out its own salvation by blind and instinctive effort. We like to think of organisms as machines, as passive inventions gradually perfected from generation to generation by some external agency, by environment or by natural selection, or what you will. (Russell, E.S., op. cit., p. 307.)

I have chosen to use Metchnikoff as our initial model to illustrate a dialectical orientation toward biological inquiry that fulfills the criteria I have distilled as essential. We posit that Metchnikoff defined purpose in the definition of self-hood. He took Kantian organic teleology, from a question of organizational construction to a new dynamic process. Of course crucial to whether we are speaking of abstract philosophy or generative thinking to scientific advances, we can now assess the value of his philosophy: the product of his musings was the establishment of a novel biological idea, active host defense, self-definition, immunity in all its manifestations. In Metchnikoff's dynamic theory of immunity, the dialectical interaction serves as the basis of his concept of organism. Self arises out of an interactive process.

The key issues are that the definition of integrity arose from a dynamic *process*; never given, Self-hood arose from the striving self-definition of a special "immune" sub-system, whose on-going encounter with altered host elements or intrusion of external pathogens or molecules, required constant proclamation of self-integrity. We know that the history of the individual organism not only programs that encounter, but is altered by it, so that the process of self-definition is a dialectical one. In a most profound way, Metchnikoff's metaphysical vision of the organism is consistent with our modern ontological concern for definition of subject-object relations. As reflected in post-Newtonian physics, our object requires definition, and that very definition is subject to change. As Whitehead formulated the issue, it is *process* dynamics which forms our reality. In the organic realm, it is a novel formulation of teleological principles which must construct the axis of organismic self-actualization. It is a metaphysical problem that contains enormous scientific value for progress in understanding complex systems. Teleology does not impose upon the materialistic probe, but would ask non-competing questions of function, to integrate and synthesize. Metchnikoff was acutely aware of the organism in its entirety, of establishing the integrity of the whole. The parts were dialectical — competing and yet held by cooperative bonds, cell lineages constantly striving, but always within the context of the organism. His venture, as Lord Lister described it in 1896, truly was a "romantic chronicle", and I would argue romantic in the profound sense in which Whitehead read Wordsworth — to constantly seek *meaning* in the natural world by self-reference. The phagocyte was the vehicle of self-actualization, and as Whitehead may be coupled to Wordsworth, we see Samuel Taylor Coleridge as Metchnikoff's romantic prophet:

> I define life as *the principle of individuation*, or the power which unites a given *all* into a *whole* that is presupposed by its parts. The link that combines the two, and acts throughout both, will, of course, be defined by this *tendency* to *individuation* [57, p. 42].

Metchnikoff took to the extreme the applicability of the phagocytosis theory to restore health, prolong life, combat senility, but it was his overwhelming passion to confer an integrating principle to the life process. That he failed to direct immunology in the first half of this century bespeaks the enormous allure and effectiveness of chemo-mechano-reductionism, and the inability perhaps to apply his vision to problems amenable to study during that period [3]. Now that the organizational and systematic issues in immunology are becoming predominant, re-examination of Metchnikoff's original formulation may add an heuristic dimension to the way we think of self-hood.

We must continue to seek those principles that unify biological phenomena. Biology must remain biological, to wit, organism is purposeful (in that organizational insight and integrated function may be examined) and exists as process (Whiteheadian process is the metaphysic replacing atomized materialism, with functional activity having a history and integration-relation to other parts). The atomization of life processes can only lead to certain restrictive conclusions. Examination of systems as organizing has led to powerful paradigms in areas in which such principles are applied, i.e. ecology and immunology, but these are still in their infancy, and unifying principles for the neurosciences, developmental biology and evolutionary mechanisms remain obvious areas for creative and sustained attention [49].

Our inquiry is but the current expression of a dissonant view that has nagged mechano-reductionists since modern biology emerged in the 19th century, namely, how does one explain the unique biological features of organism? One avenue of investigation is to search for the origin of Self. Pondering the metaphysical basis of that question is an essential element of our quest. In my opinion, an exciting era in biology is emerging not in the sequencing of the genome, nor the further molecular definition of biological structure, but in the scientific inquiry as to what confers biological integrity to the immune system [18], the cooperation of the bacterial biosphere [36], the evolution of species beyond blind accumulation of mutations [60], and the epigenetic development of the embryo [7, 17]. To what end can organism be viewed as self-fulfilling, self-determined, self-identified, and then integrated and coordinated with other like units? Each of these issues pertains to problems of 1) organizational or hierarchical control, 2) stochastic versus directed choice, and finally, 3) whether purpose can be defined in current notions of dynamic process. The first of these three questions, has been under intense scientific and philosophical inquiry at least since the 1930's. The second issue, (viz. are there mechanisms for directing a responsive genome and novel modalities for macroevolutionary speciation?) is now becoming a legitimate area of scientific inquiry [61], and we might expect a revolutionary re-orientation in our concepts of organismal integrity if the first clues are in fact correct. As for the third area, we would probably err, as did von Baer and Müller, in attempting to apply teleology beyond its appropriate domain and assign it a role in elucidating "mechanism." Process and purpose must remain distinct in our characterization of the organism, for Self arises from each. Elucidating that synthesis must excite the biologist and serve as the foundation of his science.

Boston University School of Medicine

REFERENCES

1. Alvan Feinstein has eloquently argued that medicine requires its own unique science that does not rely on other disciplines, resists legitimatizing reductionism, and employs mathematical analysis suitable to its specific purposes. See *Clinical Judgement*, Baltimore: Williams and Wilkins, 1967 and a series of biometric papers: An additional basic science for clinical medicine. Ann. Int. Med. 99: 393–397, 544–550, 705–712, 843–848, 1983 and Clinical biostatistics. LVII. A glossary of neologisms in quantitative clinical science. Clin. Pharm. Therap. 30: 564–577, 1981.
2. Allen GE. *Life Science in the Twentieth Century*: Cambridge University Press, 1981.
3. Tauber AI and Chernyak L. *Metchnikoff and the Origins of Immunology*. New York: Oxford University Press, 1991. Portions of the study have been published separately: Chernyak L and Tauber AI. The birth of immunology: Metchnikoff, the embryologist. Cell. Immunol. 117: 218–233, 1988; Chernyak L and Tauber AI. The idea of immunity: Metchnikoff's

metaphysics and science. J. Hist. Biol., 23: 187–249, 1990; Tauber AI and Chernyak L. The birth of immunology: II. Metchnikoff and his critics. Cell. Immunol. 121: 447–473, 1989. Issues directly germane to modern interpretation have been explored in Tauber AI and Chernyak L. Metchnikoff and a theory of medicine. J. Royal Soc. Med., 82: 699–701, 1989; Tauber AI. Metchnikoff, the modern immunologist, J. Leuk. Biol. 47: 560–566, 1990 and Tauber AI. The immunological Self: A centenary perspective. Perspect. Biol. Med., In Press. Many elements of this Introduction are products of the exciting collaborative effort I have enjoyed with Chernyak, and have been explored with a different orientation in the above cited works.

4. Metchnikoff's mature position, although basically unchanged from 1882 to his death in 1916 is best summarized in his *Immunity in Infective Diseases* (1901) trans. F.G. Binnie, Cambridge, 1905. Reprinted by Johnston Reprint Corp., 1968.
5. Buss LW. *The Evolution of Individuality*. Princeton, 1987.
6. Margulis L. *Symbiosis in Cell Evolution*. San Francisco, Freeman, 1981.
7. Edelman GM. *Topobiology. An Introduction to Molecular Embryology*. New York: Basic Books, 1988.
8. Henry JJ, Amemiya S, Wray GA and Raff RA. Early inductive interactions are involved in restricting cell fates of mesomeres in sea urchin embryos. Dev. Biol. 136: 140–153, 1989.
9. CH Waddington. Evolutionary adaptation in *Evolution. After Darwin. Vol. 1. The Evolution of Life. Its Origin, History, and Future*. S. Tax, (ed.) Chicago: The University of Chicago, 1959, pp. 381–402. Reprinted in CH Waddington. *The Evolution of an Evolutionist*. Ithaca: Cornell University Press, 1975, pp. 36–59. See also the discussion by Bonner JT. *The Evolution of Development*. Cambridge University Press, 1958.
10. Lewontin R. *The Genetic Basis of Evolutionary Change*. New York: Columbia University Press, 1974, and further discussed by Eliott Sober in *The Nature of Selection. Evolutionary Theory in Philosophical Focus*. Cambridge, MIT Press, 1984.
11. Metchnikoff's intuitive proposition was experimentally proven by Chabry (Chabry, L. J. Anat. Physiol (Paris) 23: 167–319, 1887), who showed that destruction of one cell of a two-celled frog embryo resulted in half a tadpole; the later classical studies of Roux and Driesch based on this work directed the course of experimental embryology for decades.
12. Metchnikoff E. *Lectures on the Comparative Pathology of Inflammation*, trans. F.A. Starling and E.H. Starling. London: Kegan Paul, Trench, Trubner and Co., Ltd., 1893. Reprinted by Dover Publications Inc., NY, 1968.
13. Johnston IS and Hildemann WH. Cellular defense systems of the Porifera in N. Cohen and M.M. Sigel, editors, *The Reticuloendothelial System. A Comprehensive Treatise* Vol. 3, *Phylogeny and Ontogeny*, New York: Plenum Press, 1982, pp. 37–57.
14. Bigger CH and Hildemann WH. Cellular defense system of the coelenterata in N. Cohen and M.M. Sigel, op. cit., pp. 59–87.
15. The issue is well reviewed in Ratcliff, N.A. and Cooper EL. Invertebrate defense systems: An overview in N. Cohen and M.M. Sigel, op. cit., pp. 1–35 and Marchallionis JJ. *Immunity in Evolution*. Cambridge: Harvard University Press, 1977.
16. The vertebrate immune system may represent a convergent evolution of allorecognition, one which arose as a nonadaptive byproduct of sophisticated modes of xenorecognition. Alternatively, the immune systems of vertebrates and echinoderms may be homologous with those found in clonal invertebrates, only to have subsequently become adopted as a mechanism of xenorecognition. The latter hypothesis is supported by the fact that several primitive echinoderms (which are presumed ancestral stock for the chordates) were sedentary, potentially clonal, organisms. Thus, vertebrate ancestors may well have

encountered fusion as a naturally occurring event and developed allorecognition as a response to the threat of somatic cell parasitism following fusion. Further support is found in the fact that vertebrate recognition of foreign tissue still requires simultaneous self-recognition (i.e., antigens presented on macrophages result in the release of interleukins only to T-cells which match the antigen in the context of appropriate self markers). Hence the primitive system, though no longer required for fusion once mobility was acquired, was nevertheless required in xenorecognition and, accordingly, not lost in the course of evolution [Buss, op. cit., pp. 151–152].

See also Buss LW and Green DR. Histoincompatibility in vertebrates: The relict hypothesis. J. Comp. Dev. Immunol. 9: 191–201, 1985.

17. Edelman E. CAM'S and Igs: Cell adhesion and the evolutionary origins of immunity. Immunol. Rev. 100: 9–43, 1987; Rutishauser U, Acheson A, Hall AK, Mann DM and Sunshine J. Science 240: 53–57, 1988; Matsunaga T and Mori N. Globulin superfamily molecules are all related. Scand. J. Immunol. 25: 485–495, 1987.
18. Chernyak and I discuss these issues in our essay of this volume, The dialectical Self: Immunology's contribution.
19. See for general review: Adler R (ed) *Psychoneuroimmunology*. New York: Academic Press, 1981. Korneva EA, Klimenko VM and Shkhinek EI. *Neurohumoral Maintenance of Immune Homeostasis* (S.A. and E.O. Corson, translators and editors). Chicago: University of Chicago Press, 1985. Locke S, Ader R, Besedovsky H, Hall N, Solomon G and Strom T (ed). *Foundations of Psychoneuroimmunology*. New York: Aldine Publishing Co., 1985.
20. Metalnikov S and Chorine V. Role des reflexes conditionnels dans l'immunite. Ann. de l'Inst. Pasteur 40: 893–900, 1926. English translation and commentary in J.D. Bibel, *Milestones in Immunology*. Madison: Science Tech Publ., 1988. pp. 312–316.
21. Felten DL, Felten SY, Carlson SL, Olschowka JA and Livnat S. Noradrenergic and peptidergic innervation of lymphoid tissue. J. Immunol. 135: 755s–765s, 1985.
22. Bockman DE and Kirby ML. Neural crest interactions in the development of the immune system. J. Immunol. 135: 766s–772s, 1985.
23. Goetzl EJ, Chernov T, Renold F and Payan DG. Neuropeptide regulation of the expression of immediate hypersensitivity. J. Immunol. 135: 802s–805s, 1985 and Russell M, Dark KA, Cummins RW, Ellman G, Callaway E and Peeke HVS. Learned histamine release. Science 225: 733–734, 1984.
24. Hall NR, McGillis JP, Spangelo BL and Goldstein AL. Evidence that thymosins and other biologic response modifiers can function as neuroactive immunotransmitters. J. Immunol. 135: 806s–811s, 1985. Ruff MR, Sacerdote P, Wiedermann CJ and Pert CB. Neuropeptide receptors are shared components of nervous and immune systems. In: *Neuropeptides and Stress,* Hans Selye Symposia on Neuroendocrinology and Stress, Tache Y, Morely J and Brown M (Eds.). Springer-Verlag, New York, 1989, pp. 235–246. Subject well reviewed by J.E. Blalock. A molecular basis for bidirectional communication communication between the immune and neuroendocrine systems. Physiol. Reviews 69: 1–32, 1989.
25. Roth J, LeRoith D, Collier ES, Weaver NR, Watkinson A, Cleland CF and Glick SM. Evolutionary origins of neuropeptides, hormones, and receptors: Possible applications to immunology. J. Immunol. 816s–819s, 1985. Roth J, LeRoith D, Lesniak, MA, de Pablo F, Bassas L and Collier E. Molecules of intracellular communication in vertebrates, invertebrates, and microbes: do they share common origins? *Progress in Brain Reseach Vol. 68.* Hokfelt, T, K. Fuxe, and B. Pernow (ed.) pp. 71–79, 1986.

26. Edelman G. *Neural Darwinism. The Theory of Neuronal Group Selection.* New York: Basic Books, 1987.
27. See F.J. Varela, Organism — A meshwork of selfless selves in this volume, and earlier theoretical work of particular interest: Varela FJ, Coutinho A, Dupire B and Vaz NN. Cognitive Networks: Immune, neural and otherwise. in *Theoretical Immunology*, Part Two. A.S. Perelson (ed.). Redwood City: Addison-Wesley Publishing Co., 1988, pp. 359–375. Coutinho A, Forni L, Holmberg D, Ivars F and Vaz N. From an antigen-centered, clonal perspective of immune responses to an organism-centered network perspective of autonomous activity in a self-referential immune system. Immunol. Rev. 79: 151–168, 1984. Vaz NM and Varela FJ. Self and non-self: An organism-centered approach to immunology. Med. Hypoth. 4: 231–267, 1978. An experimental program has been generated by the Paris School: Stewart J and Varela FJ. Exploring the meaning of connectivity in the immune network. Immunol. Rev. 110: 37–61, 1989; Varela FJ and Coutinho A. Times and Trends. Immune networks: Getting on to the real thing. Res. Immunol. 140: 837–845, 1989; Stewart J, Varela FJ and Coutinho A. The relationship between connectivity and tolerance as revealed by computer simulation of the immune network: Some lessons for an understanding of autoimmunity. J. Autoimmun. 2: 15–23, 1989; Lundkvist I, Coutinho A, Varela F and Holmberg D. Evidence for a functional idiotypic network among natural antibodies in normal mice. Proc. Natl. Acad. Sci. 86: 5074–5078, 1989.
28. Jerne NK. The generative grammar of the immune system. EMBO. J. 4: 847–852, 1985. Jerne NK. Idiotypic networks and other preconceived ideas. Immunol. Rev. 79: 5–24, 1984. For an excellent history of modern immunology, see AM Silverstein, *A History of Immunology*. New York: Academic Press, 1989.
29. Benner R; van Oudenaren M, Bjorklund F, Ivars F and Holmberg D. "Background" immunoglobulin production measurement, biological significance and regulation. Immunol. Today 3: 243–249, 1982.
30. The macrophage appears to recognize antibody bound to newly exposed "senescent antigens", derived from band 3, the erythrocyte transporter. (Kay MMB, Bosman GJ, Johnson GJ and Beth AH. Band-3 polymers and aggregates, and hemoglobin precipitates in red cell aging. Blood Cells 14: 275–289, 1988.) Whether the neo-antigen is a proteolyzed or oxidized product, or a newly formed dimer of a native protein is not known. The sensecent antigen, in addition to being found on the erythrocyte, has been demonstrated on the surface of lymphocytes, neutrophils, platelets, embryonic kidney cells and adult liver cells. (Kay MMB. Isolation of the phagocytosis-inducing IgG-binding antigen on senescent somatic cells. Nature 289: 491–494, 1981.) In this case, the neo-antigen is recognized by immunoglobulin, however, altered physico-chemical plasma membrane structures may be recognized by phagocytes (Tanaka Y and Schroit AJ. Insertion of fluorescent phosphatidylserine into the plasma membrane or red blood cells. J. Biol. Chem. 258: 11335 –11343, 1983), thus the relative importance of immune (i.e. antibody) vs. non-immune recognition mechanisms of phagocyte targets is unsettled.
31. For a provocative hypothesis concerning autoimmunity see R.S. Root-Bernstein, Self, nonself and the paradoxes of autoimmunity in this volume, pp. 159–209.
32. Models recently proposed by Deboer, Perelson and Segel in A.S. Perelson (ed.) *Theoretical Immunology*, Vol. 1 and 2. Redwood City: Addison-Wesley, 1988; G. Parisi. A simple model for the immune network. Proc. Nat. Acad. Sci. USA 87: 429–433, 1990.

in biological systems, i.e. organic function could not be ultimately explained by reductive sciences. It is of note that by the early 1960's Bohr although asserting the crucial role of teleology in biology, had become more respectful of the reductionist approach.

Surely, as long as for practical or epistemological reasons one speaks of life, such teleological terms will be used in complementing the terminology of molecular biology. This circumstance, however, does not in itself imply any limitation in the application to biology of the well-established principles of atomic physics. [Niels Bohr, Light and Life revisited (address given June, 1962; unfinished manuscript, *Philosophical Writings*, Vol. III, op. cit., p. 26].

Delbrück's own musings on these issues are discussed in "A physicist's renewed look at Biology: Twenty years later". Science 168: 1312–1315, 1970; "Light and Life III". Carlsberg Research Communications. 41: 299–309, 1976, and *Mind from Matter. An Essay on Evolutionary Epistemology*. Palo Alto: Blackwell Scientific Publications, 1986.

51. Devault, D. *Quantum Mechanical Tunnelling in Biological Systems*, 2nd Edition. Cambridge University Press, 1984; Cha, Y., Murray, C.J. and Klinman, J.P. Hydrogen tunneling in enzyme rections, Nature 243: 1325–1330, 1989; Kuri, A. and Wolynes, P.G. Electron tunneling paths in proteins. Science 236: 1647–1652, 1987; Klinman, J.P. Quantum mechanical effects in enzyme-cataysed hydrogen transfer reactions. Trends Biochem. Sci. 14: 368–373, 1989. Recent discussion of the interface between quantum mechanics and biology has been speculatively explored by Muses and Yates. See C.A. Muses, Interfaces between quantum physics and bioenergetics, and F.E. Yates, Quantumstuff and biostuff. A view of patterns of convergence in contemporary science in *Self Organizing Systems*, ed. F.E. Yates, op. cit. pp. 489–497 and 617–644.
52. Bachelard, G. *The New Scientific Spirit*. Boston: Beacon Press, 1984, p. 3.
53. For example, the success of gene control and lambda phage molecular biology as a paradigm for eukaryote gene regulation is a particularly exciting chapter in the power of reductionist biology (see Mark Ptashne, *A Genetic Switch*. Cambridge: Cell Press, 1987), and for a summary of gene regulation at the mRNA level, see Marilyn Kozach, A profusion of controls, J. Cell. Biol. 107: 1–7, 1988.
54. As I have examined reductionism and the role it has played in science — in biology particularly — I have found that it has served primarily as a program or a policy; it has not served really, for the biologist, as a rule or law. The biologist might like to think of it as such, but more often than not he has used reductionism as a means of advocating a research strategy, of delineating what to him seems to be an acceptable type of explanation for the biological sciences. [Mendelsohn, E. Commentary, Part I. Conference on Explanation in Biology: Historical. Philosophical, and Scientific aspects. J. Hist. Biol. 2: 135–140, 1969, p. 139.]
55. Shimony, A. Some proposals concerning parts and wholes, in *Parts and Wholes*. Swedish Council for Planning and Coordination of Research. Stockholm, 1983, pp. 115–122.
56. Whitehead AN. *Science and the Modern World*. New York: Macmillan Company, 1925.
57. Coleridge ST. *Hints Towards the Formation of a More Comprehensive Theory of Life*. London: John Chruchill, 1848. Apropos this dialectical issue, there are some uncanningly pre-Whiteheadian statements in this monograph:

But to form the first conception of a real thing, we state both as one in the idea, *duration*.... or the oneness of space and time, is the predicate of all real being.... But as little can we conceive the oneness, except as the mid-point producing itself on each side; that is, manifesting itself on two opposite poles. Thus, from identity we derive duality, and from both together we obtain polarity, synthesis, indifference, predominance.... (p. 53).

That Coleridge can be so easily linked to Whitehead, simply re-emphasizes the affinity Whitehead had for the Romantic (as acknowledged in his debt to Wordsworth). What I find of particular interest is the strong dialectical component, the polarity and synthesis in Nature that so marked *Naturphilosophie* and so profoundly influenced Coleridge, and then re-emerged in Whitehead's metaphysics, albeit in a refracted form.

In a sense, Naturphilosophie can be viewed as one among many forms in which Romanticism manifested itself. It was perhaps, as one commentator put it, merely—one extreme systematization [Snelders, HAM Inorganic natural sciences 1797–1840: An introductory survey. Studies in Romanticism 9: 193–215, 1970, p. 195.] of Romantic ideas. [Raimonda Modiano, *Coleridge and the Concept of Nature.* Tallahassee; Florida State Univ. Press, 1985, p. 141.]

For analysis of *Theory of Life*, see T.H. Levere, *Poetry Realized in Nature, Samuel Taylor Coleridge and Early 19th Century Science.* Cambridge, 1981, pp. 42–45; 161–166; 215–219, and Coleridge's "reactionary" influence on medicine and biology see A. Desmond, *The Politics of Evolution.* Chicago: Univ. of Chicago Press, 1989.

58. David Guest in *A Textbook of Dialectical Materialism* (New York: International Publishers, 1939) cites several attempts (e.g. H. Levy, *Aspects of Dialectical Materialism,* London: Watts and Co., 1934, J.B.S. Haldane, *The Marxist Philosophy and the Sciences.* Freeport, N.Y. Books for Library, Press, 1939. Marcel Prenant *Biology and Marxism* translated by C.D. Greaves, New York: International Publishers, 1943) at showing the relevance of dialectical materialism for biology "when he is not afraid of the social conclusions to be drawn from it" (p. 102). In fact, these studies, like virtually every treatment of the subject, are politically motivated and either are passionately supportive or rejecting it out of hand (i.e. Sidney Hook's *Dialectical Materialism and Scientific Method,* Manchester: Committee of Science and Reason, 1955). A dispassionate assessment of dialectical thinking in biology, that would attempt to place dialectical materialism within its appropriate extra-Soviet context, is still required. In the context of our concern, dialectical is used in the sense used by E.S. Russell, without the embrace of Engel, whose political motivations were apparently overlooked by infatuated sympathizers like Joseph Needham:

[Engel's concept of nature as dialectical] was rightly directed against the static conceptions of the scientists of his time, who were unprepared for the mass of contradictions that science was about to have to deal with, and who did not appreciate that nature is full of apparently irreconcilable antagonisms and distinctions which are reconciled at higher organisational levels. [Needham, J. A Biologist's view of Whitehead, in *The Philosophy of Alfred North Whitehead. The Library of Living Philosophers.* ed. PA Schilpp. Menasha, Wisconsin: Banta Publishing Co., 1941, pp. 243–271.]

The sentiment is sympathetic, but the scientific contribution remains problematic.

59. Wetter, G.A. *Dialectical Materialism. A Historical and Systematic Survey of Philosophy in the Soviet Union.* Translated by P. Heath. New York: Praeger, 1958.
60. Gould, S.J. Is a new and general theory of evolution emerging? Paleobiology 6: 119–130, 1980.
61. See the respective essays of Patricia Foster and Sahotra Sarkar in this volume.

EDITOR'S COMMENTS TO LÖWY

Beginning with an analysis of Ludwik Fleck's constructionist concepts, Löwy has provided a historical synopsis of immunology's notions of the Self. The power of Fleck's argument that a scientific fact derives meaning from multiple sources and context, and is essentially derivative from a "thought collective" of divergent and heterogenous resources, remains a central concern for epistemologists of scientific knowledge. We find it particularly conductive to a "dialectical" sensitivity, which Löwy has illustrated for the case of immunology, a discipline clearly wedged between the thought styles of biological (holistic) and chemical (reductionist) approaches. She concludes that it is the dialogical processes demanded by interaction of these alternate thought styles that has yielded both expansion of each and their ever growing synthesis. It is in the "indeterminacy" of such key concept as organism and the Self that demands interactive investigation and receptivity to various modes of inquiry. Precisely in this refracted vision of the Self is dynamic dialogue demanded. This rationale has served to unite the divergent approaches presented in this anthology under such a constructionist principle.

ILANA LÖWY

THE IMMUNOLOGICAL CONSTRUCTION OF THE SELF

I. INTRODUCTION

> An organism can no longer be construed as a self-contained, independent unit with fixed boundaries, as it was still considered according to the theory of materialism. That concept became more abstract and fictitious, and *its particular meaning depended upon the purpose of the investigation.* For the morphologist it has changed into the concept of genotype as the abstract and fictitious results of hereditary factors. In physiology we find the concept of "harmonious life unit", according to Gradmann, "characterized by the notion that the activities of the parts are mutually complementary, mutually dependant upon each other, and form a viable whole through their cooperation". Morphological organisms of the type which are self-contained units do not have this ability. But a lichen, for instance, whose constituents are of completely different origins, one part an alga, another a fungus, constitutes such a harmonious life unit. The constituents are closely interdependent and on their own are usually not viable. All symbioses, for instance between nitrogen fixing bacteria and beans, between mycorrhiza and certain forest trees, between animals and photogenic bacteria, and between some wood beetles and fungi, form "harmonious life units", as do animals such as the ant colony and ecological units such as a forest. A whole scale of complexes exists, which, depending on the purpose of the investigation, are regarded as biological individuals. For some investigators the cell is considered the individual, for others it is the syncytium, for still others a symbiosis or lately even an ecological complex. "It is therefore a prejudice to stress the idea of organism", in the old sense of the word, "as a special kind of life unit, a prejudice which is unbecoming to modern biology" [Gradmann] [1].

The affirmation that the definition of what an organism is depends on the goal of the research that deals with this organism was made in 1935 by Ludwik Fleck, in his main epistemological study, *Genesis and Development of a Scientific Fact.* Fleck lived in Lwow (then in Poland), famous between the two wars among other things for its school of philosophy and logic (the Warsaw-Lwow school: Adjukiewicz, Kotarbinski, Lukasiewicz, Tarski). Fleck was not, however, a professional philosopher: he was a physician who specialized in bacteriology and immunology. In the years 1927–1939 he published a number of epistemological works. The most important among them is *Genesis*, in which the analysis of the origins and fate of the Wassermann reaction for the detection of syphilis led Fleck to general conclusions concerning the genesis of scientific knowledge. Fleck's book, published in German, obtained a small number of favorable reviews mainly in microbiology and pathology journals, but it failed to reach a wider

Alfred I. Tauber (ed.), Organism and the Origins of Self, 43–75.

it is very doubtful whether an invasion in the old sense is possible, involving as it does an interference by completely foreign organisms in natural conditions. A completely foreign organism could find no receptors capable of reaction and thus could not generate a biological process. It is therefore better to speak of a complicated revolution within a complex life-unit than of an invasion of it [8].

Fleck's understanding of infectious disease, informed by his experience as a medical bacteriologist and immunologist, was the origin of his epistemological reflections. Already in his first theoretical work, "Some specific features of the medical way of thinking", Fleck argued in favor of a holistic conception of diseases. Diseases, claimed Fleck, cannot be explained by a simplistic, reductionist view, such as the widespread belief that infectious diseases had a single etiological cause, namely a pathogenic microorganism. For him, an infectious disease was a highly complex process, the result of multifactorial and multidimensional interactions between microorganisms and their host. In order to describe such complexity one should observe it from several distinct and incommensurable points of view:

... it is impossible to get in medicine an idea that will embrace the entirety, like atomism in chemistry or energetics in physics. Neither cellular nor humoral theory, nor the functional understanding of diseases alone, nor their "psychogenic" conditioning, by themselves will exhaust the entire wealth of morbid phenomena. This results in the incommensurablity of ideas which develop from the varying ways of grasping morbid phenomena, and which give rise to the fact that a uniform understanding of morbidity is impossible [9].

The complexity of pathological phenomena, Fleck explained, induced a difference between the scientific (i.e., biological) and the medical way of thinking:

Let me use a figurative comparison: medical thinking differs in principle from scientific thinking in that it uses Gauss's coordinate system, while the latter uses the Cartesian system. Medical observation is not a point but a small circle. It is placed not in a system of coordinate straight lines inclined to one another at a constant angle, but in a system of optional, mutually intersecting curves which we do not know closely. A certain correction is induced into this picture by the fact that, strictly speaking, the multiplicity of medical phenomena can be only approximately rendered by means of Gauss's system, since its points are not univocally determinable. To all intents and purposes, scientific thinking uses, for small ranges, the Cartesian system, and for large ranges Gauss's system (as in the theory of relativity). On the contrary, medical thinking uses Gauss's system for small ranges, while in the entirety it does not find any consistent and rational way to grasp phenomena [10].

The idea of the great complexity of all pathological phenomena was not readily accepted by the medical profession. Physicians always prefer to look for simplified explanations. They developed the concept of "disease", and thereafter affirmed that diseases were entities accessible to the analytical method. But, Fleck explained, diseases are not "natural" units, which await discovery by physicians: diseases are,

> ideal fictitious pictures, known as morbid units, round which both the individual and the variable morbid phenomena are grouped, without, however, ever corresponding completely to them [11].

The idea that diseases do not exist in nature but are constructed by physicians was further developed by Fleck in the first chapter of *Genesis*, dealing with the history of syphilis, in which he showed that the view of what "syphilis" is, and which morbid phenomena had to be lumped together in a single disease, varied in different societies and in different historical periods.

Later Fleck generalized this view and argued that not only the concept of disease but all scientific knowledge is socially constructed. This generalization was made by Fleck in his second epistemological study, "On the crisis of 'reality'" [12]. In this article, Fleck radicalized his previous ideas by explaining that not only diseases, but also their etiological causes, i.e., bacteria, are constructed by scientists. It is well known, explained Fleck, that a given bacterium can be classified in more than one way. These differences in classification are not idiosyncratic, but rather depend on the purpose of the investigation, and therefore on the community to which a given scientist belongs. Thus, a fundamental scientist (interested e.g. in the biochemical activity of a given bacterium) and an epidemiologist (interested in its potential danger to public health) may classify the same bacterium differently. The scientist will prefer a more stringent definition of the bacterium because he is less worried about the exclusion from his sample of some relevant bacteria than about the inclusion of some irrelevant ones. For the epidemiologist the opposite is true: he wishes above all to avoid false-negative results, and to mistake a dangerous bacterium for a benign one. He prefers therefore a less stringent definition. Fleck gives an example of a study of scarlet fever, in which the biochemical test used by scientists revealed the presence of hemolytic streptococcus in 84% of the cases, while the test used by epidemiologists revealed the presence of the same bacterium in nearly 100% of the cases. Now such a discrepancy of classification may be interpreted by saying that the epidemiologists "adapt" the "true" definitions given by the fundamental scientists to their own practical needs. This view was rejected by Fleck, who claimed that both definitions are equally valid: the epidemiologist and the biochemist arrive at divergent and not interchangeable truths. Fleck's conclusion is that truth is relative to a given thought-style and depends on the purpose of the investigation: different views can be equally true.

affirmation was illustrated with an example taken from his own medical specialty, ophthalmology. Two well-known ophthalmologists presented two different explanations of glaucoma, each of them based on "clinical facts". One ophthalmologist explained that glaucoma is a disease of the optic nerve, because, he claimed, it is an "undisputed clinical fact" that the only way to cure this disease surgically is to section the nerve during iridectomy. The other specialist affirmed that glaucoma is a disease of the surface of the iris, because it is a "well-known clinical fact" that surgery for glaucoma is successful only if the surgeon removes the greatest possible surface of the iris. Kramsztyk concludes that,

> one of these authors must be wrong. However, iridectomy is a well-established surgical treatment, made frequently by physicians, and one might have assumed that its variants and consequences should be well-known today. One can conclude that clinical facts are very difficult to interpret and become easily deformed because one views them through the mediation of theories [18].

Not only the interpretation of isolated clinical findings, but the overall classification of pahtological states too, explained Kramsztyk, is often dependent on the theoretical presuppositions of the physician. A detailed description of a pathological case described as "trachoma" will reveal that even the clinical symptoms most commonly associated with this disease, and considered as one of its principal manifestations, can be absent in some trachoma patients and can be present in individuals who do not show clinical signs of trachoma. "Trachoma" is thus not a specific, well-defined unit, but represents a whole spectrum of pathological manifestations. The classification of the illness of a given individual as trachoma—an event which at that time had not only medical implications but social consequences too, because e.g. children with this disease were not allowed to attend school, was thus largely dependent on the views of a given physician, and might vary greatly even among renowned specialists [19].

While Fleck might have been (and in all probability was) influenced by Kramsztyk's ideas concerning the impact of theories on the construction of "scientific facts", his dynamic and holistic perception of the organism and of host-parasite relationships is very close to the ideas of another Polish philosopher of medicine, Wladyslaw Bieganski, who strongly supported a holistic approach to medicine. The anatomo-pathological school, particularly in the German-speaking countries, Bieganski explained, had a negative impact on medical practice. When we follow the recommendations of this school,

> we cannot avoid paying less attention to the sick organism, the "sick life" as Chalubinski would

say, and more attention to the sick organ, the sick tissue. To put it tersely, we lose from our sight the patient, a suffering fellow-man, and we see only the diseased organ [20].

Bieganski was convinced that a static, anatomo-physiological approach to pathology would soon be replaced by a dynamic, physiological one. This new physiological approach would radically modify the understanding of infectious diseases. Some investigators viewed diseases as a struggle between hostile armies of bacteria and phagocytes. Such a naive view, stemming from primitive metaphors, affirmed Bieganski, is in contradiction with the newest biological knowledge:

if a disease is a disturbance of function induced by the action of a pathogenic stimulus, it is senseless to speak here of any kind of struggle, just as it is senseless to express in this way our understanding of a physiological function. Nobody will claim that the digestion is the struggle of the organism against food; therefore nobody should claim that an inflammation represents such a struggle. We would not object to a metaphorical sense of the word, struggle, in the understanding of disease. This is, however, not the case. Disease is usually viewed as a regular struggle, as a regular battle waged by an army of phagocytes against bacteria. Such an idea is decidedly erroneous [21].

Encouraged by the rapid progress of physiology and physiological chemistry at the end of the 19th century, Bieganski predicted that these developments would soon bring the demise of reductionist approaches in biology and medicine, and their replacement by holistic ones. In 1899, he expressed his certainty that on the threshold of the new century:

the center of gravity of medical science is shifting from loose facts to generalizations, from organ to function, and from anatomical changes to the disturbance of functions. It is becoming increasingly probable that a disease will be viewed in the future not as a change occurring in organs, but as a sick, functionally modified life. Thus, the individual patient who almost completely disappeared in the previous, anatomical approach to disease, must regain his rights. Physicians should see not the diseased organ alone, but the diseased organism, the sick person [22].

Although Bieganski had rightly perceived the beginning of the biochemical turn in biology and medicine, his predictions that this change would lead to the development of holistic approaches to the study of life-phenomena was not realized. Thirty-five years later the bacteriologist and immunologist, Fleck, still faced the domination of reductionist approaches in biology and medicine. He firmly condemned the reductionist and static (for him, "chemical") view of the organism, and, like Bieganski earlier, he hoped that this "antiquated" perception would soon be replaced by a holistic and dynamic (for him "medical") one.

It is highly plausible that Fleck at first developed his holistic approach to the study of host-parasite relationships under the influence of his teacher, the well-known typhus specialist, Rudolf Weigl. Immediately after his graduation from the medical school at Lwow (1920), Fleck became Weigl's laboratory assistant in Przemysl, near Lwow. In 1921 Weigl was appointed professor at Lwow University, and Fleck followed him there. However, Fleck's scientific career soon came to a halt; either because of personal conflicts with other faculty members, or the anti-semitism prevailing at Lwow in this period, or both, in 1923 Fleck was obliged to leave the University and accept a position in a routine microbiological analysis laboratory. He was able to return to what he had always seen as his true vocation, scientific research, briefly in 1939 –1940 (when Lwow was annexed by the Soviet Union), then permanently in post-war Poland [23]. Thus, during the 1920's and 30's, the period during which he wrote his principal epistemological studies, Fleck's experience of scientific research was limited to his work with Weigl. One can assume therefore that he was influenced by Weigl's rather peculiar views of typhus, and of infectious diseases in general.

The majority of typhus specialists assumed that the etiological agents of this disease, the rickettsia, were a distinct class of microorganisms, different (by their size, specific morphology, and the need to grow in a living cell) from bacteria. This was not Weigl's opinion. He thought that rickettsia and bacteria were not very different, and that one might view the rickettsia as a sub-class of bacteria [24]. He further sustained this argument by arguing that *Rickettsia provazekii* (the etiological agent of the typhus fever) is not a distinct microorganism, but a stage in the complex life-cycle of a common bacterium of the digestive tract: the *Bacillus proteus X-19*. This proposition, which now may seem quite strange, was an attempt to explain a rather puzzling observation, made by Weil and Felix in 1916, that all typhus patients have in their blood high levels of antibodies against the *Bacillus proteus X-19*.

The idea that bacteria have complex life cycles and/or are able to completely change their morphology and physiology according to the conditions in which they develop was shared by a small number of bacteriologists in the 1920's and 30's, among them such famous figures as Charles Nicolle (a well known French bacteriologist and Nobel Prize laureate (1908) awarded for his discovery of the role of lice in the transmission of typhus). In his book, *Le Destin des maladies infectieuses* (1933), Nicolle explained that the pathogenic agent of typhus, the rickettsia, is an attenuated from of the *Bacilus proteus* [25], and affirmed that it is probable that not only rickettsia but all the "filterable viruses" originate from normal bacteria and represent a particularly small form of the latter.

The idea that bacteria may have complex life-cycles, and that specific forms of pathology may be attached to each stage of development, was adopted in a quite extreme version by Fleck. He believed bacteria to be highly variable and nearly infinitely malleable. Their morphology, metabolism, and the pathology they induce is, he explained, dependent on their micro-environment, and, in the case of parasitic bacteria, on their interaction with the host. Fleck summarized his views on this subject in an article "On the notion of species in bacteriology", written in 1931 [26]. There were, Fleck explained, two unorthodox approaches to this problem. One was represented by the supporters of the cyclogenic theory (Almquist, Löhnis, Enderlein), who claimed that bacteria, like certain parasites, have complex, rigid life-cycles, and thus the same bacterium can have a very different form in different stages of its life-cycle. The other approach (Kuczynski, Weigl) assumed that bacteria are endowed with a high degree of plasticity and are able to modify both their morphology and their physiology under the influence of the environment [27]. Fleck believed that bacteria may have both complex (but not rigid) life-cycles and great plasticity and capacity of adaptation to their micro-environment. He was also sympathetic to the idea (first proposed by Kuczynski) that the morphology and physiology of parasitic bacteria may be modified by the defense reactions of the host.

Fleck explained that in his opinion it would be honest to admit that up to then a widely-accepted definition of bacterial species had not existed. And, he added, not only is the notion of bacterial species not clear, but in bacteriology the definition of individual is questionable too:

> Van Longhem's promising idea of viewing the clone as an individual should allow us to distinguish between individual variations and permanent phenotypical and genotypical modifications. I believe that this idea will lead us first to a new definition of the individual and then to a new definition of species in bacteriology, a definition which would be in agreement with the definition of species in other branches of biology [28].

At that time Fleck found an echo of his opinions on the fluidity of the notion of the individual in Gradmann's article (published in 1930) "The harmonious life-units from the point of view of the exact sciences" [29]. Gradmann, a botanist, who studied plant behaviour and plant ecology, affirmed in this paper that modern science should rid itself of the idea that the living world is composed of distinct, well-defined units or "organisms". This way of seeing, explains Gradmann, is based on our subjective perception. We view ourselves as organisms, i.e., distinct units limited in space and in time: we have therefore a strong tendency to assume that the entire living world is composed of similar units. However, such an assumption has no objective ground whatsoever. For Gradmann, biologists should not study "organisms" but "harmonious life-units". The definition of what should be viewed as a "harmonious life-unit" depends on the subject of a given study. Such a unit may be composed of several entities currently defined as "organisms" (e.g., of all the components of a large ecological unit), or be a part of an "organism" (e.g., a cell). One should not forget, added Gradmann, that complex multicellular organisms evolved from associations of simple unicellular ones. Thus collaboration between parts of a given organism should not be viewed as something essentially different from collaboration between distinct organisms. For example, during embryogenesis a groups of cells develop through their mutual influence to form an organ: such cells should therefore be studied together. In a forest too, an isolated group of trees can be viewed as a unit because the development of these trees is strongly dependent on their initial position in the group. During an investigation of life phenomena, concluded Gradmann, one should determine in each case what the relevant "harmonious life-unit" is, and study the relationships of the parts to the whole in such a unit.

Fleck radicalized Gradmann's ideas. Gradmann assumed that "harmonious life-units" do exist in nature, and the scientist's task is simply to discover that unit. In contrast, Fleck affirmed that each thought-collective constructs its view of "harmonious life-units", i.e., determines its own definition of organism, which is thus strongly dependent upon a given thought style. It is not surprising, he added, that the recent modifications of the thought-style in bacteriology and the development of theories of cyclogeny and of variability of bacteria have brought about a radical change in the way organism is viewed by this specialty:

> Suddenly we do not see clearly what is species and what is individual, or how broadly the concept of life-cycle is to be taken. What just a few years ago was regarded as a natural event appears to us today as a complex of artefacts [30].

IV. FLECK'S IDEAS ON IMMUNE REACTIONS AND THEIR HISTORICAL CONTEXT

Fleck's ideas on immunity were strongly influenced by his view of bacteria as nearly infinitely plastic entities, and by his dynamic and holistic perception of the organism and of host-parasite relationships. He affirmed that when the "primitive images of war," which dominated the contemporary perception of infectious disease were replaced by a view of disease as a "complicated revolution within the complex life-unit," the "classical" concept of immunity would also be modified. Starting with the definition (which could have been borrowed from Bieganski) that "a fundamental property of all biological events is modified reaction to a repeated stimulus," he classified the immune reaction among responses to stimuli in the framework of what he called "a reciprocal action within the complex life-unit" [31]. This definition was not, however, the one accepted by the majority of specialists. Immunology, Fleck explained, often failed to situate immune reactions in a broader physiological context:

Many classical concepts of immunology were evolved during the period when, under the influence of great chemical successes in physiology, misguided attempts were made to explain the whole, or almost the whole of biology in terms of effects produced by chemically-defined substances. Toxins, amboceptors and complements were treated as chemical entities, with such adversaries as antitoxins and anticomplements. This primitive scheme based upon activating and inhibitory substances is being progressively discarded in accordance with current physico-chemical and colloidal theories in other fields. We now speak about states and structures rather than substances, to express the possibility that a complex chemico-physical-morphological state is responsible for the changed mode of reaction, instead of chemically-defined substances or their mixtures being the cause [32].

Fleck's dynamic and holistic perception of immunity brought him to the negation of the key role of chemical specificity in immune reactions. He criticized the definition of antibodies given by Wassermann's collaborator, Julius Citron: "... by antibodies we mean all the specific reaction products formed by the organism against disease germs and their products" [33], and explained that the term "specificity" is used in a distinctly mythical sense, because serologists have materialized the property of the serum of an immunized animal to decompose bacteria or blood corpuscles, "by giving the hypothetical, even 'symbolic' substance in the immune serum the name of 'bacteriolysin' or 'hemolysin'" [34].

Fleck's ideas were strongly opposed to the main trend in immunology in the 1930's, best epitomized in the title of Landsteiner's influential book *The Specificity of Serological Reactions* (1933) [35]. His judgments of earlier periods of immunology were, however, somewhat simplistic. In the years 1880–1910, immune reactions were viewed as an extension of general physiological phenomena such as the assimilation of nutrition (Ehrlich), or the evolutionary need to harmonize the functions of different parts of the organism (Metchnikoff) [36]. Scientists who then studied immune reactions were interested in cellular and humoral phenomena and by the specific and non-specific aspects of immunity.

The first ideas of self-nonself recognition were made in the framework of such physiological and evolutionary interpretations of immune phenomena, and were integrated into a holistic view of the organism. Charles Richet, the discoverer of anaphylaxis, viewed this phenomenon as aiming to prevent the accidental incorporation of foreign albumoid substances in body composition. Anaphylaxis, he explained, provides a mechanism which allows the preservation of the chemical constitution of the species:

It is in fact important that animal species are of a determined chemical entity. If, following the hazards of ingestion or injection alien proteins were found in cellular juices as a part of our humours, the chemical make-up of beings will be modified and consequently perverted. Grave danger would thus face the animal species, were they not nicely balanced in their chemical make-up. If heterogenous substances got fixed into our cells and definitively intermingled with our humours, that would be the end of the chemical constitution of each animal species, which is the fruit of slow evolution down the generations, and all the progress that had been achieved through selection and heredity would be lost [37].

But, added Richet, immune phenomena not only guarantee the stability of the chemical constitution of the species: they also shape the physiology of each individual.

Anaphylaxis, like immunization, creates humoral differences between different individuals. Encounters with antigens oblige the cell to secrete particular substances (antitoxins or, in some cases, toxic substances ["toxogénines"]) and, these toxic substances and antitoxins, which either circulate in the blood or fix themselves in the cells, create, for each organism, a specific humoral state, depending on the antigens received by this organism.

The result is that,

each of us, by our chemical make-up, above all by our blood and probably also by the protoplasm of each cell, is himself and no one else. In other words, he has a humoral personality. Every living being, though presenting the strongest resemblance to others of his species, has his own characteristics [38].

Thus, according to Richet one should distinguish the hereditary make-up of the organism—the reflection of the history of the species—from the temporary "humoral personality"—the reflection of the history of the individual. And, he added, the "humoral personality" too should become the subject of the scientist's study:

this means that henceforth study of the physiology of the species is no longer enough. Another physiology must be taken up, which is very difficult and barely broached, namely that of the individual [39].

Richet's wish was, however, long in the realization. The physiological (in Fleck's terms, "medical") thought-style in immunology was replaced in the 1910's and 20's by chemical-oriented approaches centred on the study of specific antibodies. Citron's work (published in 1910), which was criticized by Fleck for its identification of immunity with specificity, still carried a strong imprint of the physiological approach to immune phenomena. Although Citron stressed the importance of the specific humoral antibodies, he attempted to situate immunity in a global, physiological framework. Humoral and cellular immune reactions, he affirmed,

> are not confined to the field of infectious diseases proper but are to a far greater extent expressions of physiological events, whether normal or pathological. The dividing line between the physiological and the pathological cannot be biologically drawn with any precision [40].

However, in later immunological works the physiological context was increasingly replaced by *in vitro* studies of the specificity of antibodies and of serological reactions.

Fleck's understanding of the "chemical" approach to immune phenomena was, like his view of early immunology, rather simplistic and one-sided. The history of the debates between Ehrlich, Landsteiner and Bordet on the nature of the antigen-antibody bond, and the history of blood groups illustrated the long persistence of some of the ideas of the earlier period, in particular those derived from colloid chemistry, in immunology [41]. There is little doubt, however, that the vanguard of immunological research—and, at the same time, the front of applied serological research—had switched in the 1920's and 1930's to chemically-oriented studies of specific humoral antibodies. Biochemists affirmed that there was no fundamental difference between enzymatic and immunological reactions [42], while immunologists believed that progress in immunology would come from chemical studies. Hans Zinsser, the bacteriologist and immunologist, argued that the understanding of antigen chemistry and of the chemical and physical principles of the antigen-antibody reaction would undoubtedly bring more rapid advance to immunology than all the previous approaches had. His colleague, H. G. Wells, went even further and affirmed that immunology should be squarely viewed as a branch of chemistry [43].

The dominance of chemical orientations in the studies of immune phenomena did not lead to a complete disappearance of those approaches which viewed immune phenomena in their physico-pathological context. The investigations of phenomena such as auto-immunity, transplantation immunity and "resistance" to cancer were pursued in the 1920's and 30's [44]. Those domains were, however, often viewed as marginal, or even "backward" by the leaders of the profession [45].

Fleck was a typical representative of such a "backward", physiologically oriented trend in immunology. His marginality within his discipline enabled him to clearly perceive some of its problems. Immunology, Fleck affirmed in the 1930's, was a discipline in crisis. During the first period of the development, immune phenomena were viewed as directed at "resistance" to invading microorganisms. Moreover, at that time the side-chain theory of Ehrlich provided the theoretical framework for the understanding of the chemical basis of the formation of specific humoral antibodies. This theory postulated that antibodies were "lateral chains" of the cell's protoplasm. Their normal function was to assimilate nutrition, but there were in addition side-chains able to react specifically with bacterial toxins. During an infection or an immunization, the rapid loss of these side-chains would entail their hyperproduction (in agreement with the general law of "overcompensation" in pathology) and the liberation of free side-chains into the blood-stream. This view of antibody production was, however, developed in the period when it was assumed that antibodies were produced only against a limited number of pathogens. The discovery, in the early twentieth century, that the organism can produce antibodies not only against pathogenic microorganisms but against a vast array of natural substances, and, even worse, against artificial substances, undermined the original teleological perception of immunity, while no alternative perception was elaborated until the 1950's.

Today historians of immunology usually agree that the theoretical understanding of immune phenomena was shattered by the discovery of immunity against artificial antigens [46]. In the 1920's and 30's the majority of immunologists did not, however, perceive their discipline as undergoing a structural crisis, but at most as suffering from persistent minor and temporary difficulties, which would be overcome by further progress elucidating the chemical structure of antigens and antibodies [47]. In the 1930's, Fleck was probably one of the rare immunologists who perceived the inadequacy of the current theory underlying immunological theory, and of the cognitive and institutional consequences of this inadequacy:

> As a result of the breakdown of the old theory of immunity and the absence of an alternative contemporary view, the non-participation of lay persons (who mainly use an everyday-life thought-style) in the serological community and the dominant role of technique in this discipline, the fundamental terms of immunology such as "complement", "amboceptor", "serological traits", "A-B component", "serological reagent", etc., acquire the quality of hypostases, i.e., reification of states and processes, and even of laboratory operations; this reification now invariably has became characteristic of serology" [48].

Fleck strongly believed that in the modified framework of future biological theory, the "chemical" view of the immune reaction would inevitably change.

It would soon be superseded, he believed, by a more physiological concept, one that would be understood not only by a small and hermetically closed circle of serologists, but would make sense for the lay public as well. Such a renewed view of immunity would reject the "primitive" perception of immunity based on chemical complementarity between well-defined entities: "antigens" and "antibodies", and would replace it by the dynamic principles of colloid chemistry on the one hand, and by a new understanding of physical forces which shape biological reactions on the other.

Fleck's ideas on the future revolution in biology were in all probability influenced by popularized accounts of new developments in physics, such as Bohr's complementarity theory [49]. His strong rejection of chemical specificity as a basis for understanding life-phenomena in favor of the belief that the new biology would be based on new findings in physics, was not exceptional in the 1930's. Some of the pioneers of molecular biology believed too that the understanding of biological phenomena would stem from a better understanding of physical reactions. The sceptical reception of the findings on the role of DNA in the transformation of pneumococci (1944) by the members of the "phage group" led by Max Delbrück, was later attributed to their reluctance to accept "simplistic" (i.e., chemical) explanations of hereditary phenomena [50]. The rejection of chemical complementarity was not, however, the solution later adopted. The renewal of immunological theory which led to a new view of the organism and of its relationships with the external world came from a scientist who attempted to bridge the gap between recently acquired knowledge of antigen/antibody chemical structures, and the "old" physiological understanding of immune phenomena — Macfarlane Burnet.

V. THE DEVELOPMENT OF BURNET'S CONCEPT OF SELF AND NON-SELF

Sir Macfarlane Burnet liked to define himself as belonging to the last generation of "physician naturalists" [51]. His keen interest in the observation of natural phenomena, and his reluctance to utilize the latest technical innovations in his laboratory investigations should not, however, conceal the fact that— unlike the majority of scientists with similar inclinations—Burnet was always remarkably well-informed about the latest achievements of his colleagues dedicated to more reductionist studies of life phenomena [52]. For example, during his visit to the America in 1943–44, he immediately grasped the importance of Beadle's and Tatum's studies of the genetics of the bread-

[63], that the unit which preserves the "memory" of the first encounter with a given antigen must be the living cell. But for Burnet, who had "always been fascinated by the simple concepts of biology" [64], a biological understanding of immunity should not mirror the complexity of the organism but furnish a simple and elegant solution to well-defined biological problems. Consequently, in the years 1941–1956 Burnet focused his efforts on the elaboration of a theory of antibody synthesis capable of bridging the gap between physiological findings, such as the kinetics of antibody production, self-tolerance and immunological memory on the one hand, and the newest ideas on synthesis of proteins, on the other. He first attempted to do this by proposing several variants of instructive or semi-instructive theories of antibody synthesis (i.e., theories which considered that the contact with an antigen induced the synthesis of proteins with complementary chemical structure [65]).

In the first edition of his monograph (1941), Burnet favored Bergmann's and Niemann's concept of self-replicating proteins [66]. In its second edition (1949), he attempted to explain antibody formation through an analogy with inductive enzymes, then considered to be synthesized under the influence of their substrate. In his book, *Natural History of Infectious Diseases* (1953), he proposed that antigens induced the formation of stable cytoplasmic synthetic units—for him "almost a case of inheritance of acquired characteristics" [67]. Finally, in 1956, recognizing the growing importance of nucleic acids in protein synthesis, Burnet proposed a complicated model—later qualified by him as unhappy, unconvincing and over-ambitious —in which antigens associated with either cytoplasmic RNA or with DNA served as a template for synthesis of a new protein [68].

In the late 1950's, the growing evidence on uni-directional flow of information from nucleic acids to proteins had gradually made instructive and semi-instructive theories of antibody formation obsolete. The clonal selection theory, developed by Burnet (1957) on the basis of Jerne's ideas (1955), was in agreement with the new biological knowledge [69]. The proposed solution recognized the pre-existence of a vast array of antibody-producing cells able to proliferate and to produce specific antibody upon their encounter with an antigen. It was therefore in agreement with both the new dogma of molecular biology and with the physiological approach to immune phenomena. It also provided an adequate and testable explanation of self-tolerance. Finally, it was in agreement with Burnet's aspiration to simplicity viz. "to obtain the simplest interpretation of the cellular basis of antibody production which is not clearly in variance with the available facts" [70].

The reflections on the problem of self-recognition and on the evolutionary necessity of maintaining the integrity of the organism (the self), viewed as an autonomous, well-defined unit, were, Burnet explained, the starting point of his theory of immunity. He viewed the ability to distinguish the self from non-self—later recognized as the principal function of the immune system [71]—as the fundamental evolutionary mechanism enabling any organism which lives by digesting the substance of other organisms to protect its integrity and to maintain a surveillance over the orthodoxy of its chemical structure. This was done through a relatively simple mechanism of recognition of a genetically-determined set of self-markers [71]. Burnet believed that in the future, the concept of self-recognition might lead to a better understanding of embryogenesis and of differentiation phenomena. As to the present, he affirmed, this concept had already led to the birth of the "new immunology", a science that "has far greater potentialities for the understanding of living process than the classical immunochemistry which it is incorporating and superseding" [73].

The "new immunology" was even seen as conducive to a definitive solution of the principal immunological questions. In 1967 Niels Jerne explained that there were two kinds of immunologists, who hardly spoke to each other: the "trans-immunologists", who were interested in the structure of the antibody molecule and its binding to the antigen, and the "cis-immunologists", who were interested in the events following antigenic exposure. Recently, however, the consequences of acceptance of the clonal theory of antibody synthesis on the one hand, and the progress in understanding the mechanism of protein synthesis and the role of cellular receptors in the regulation of immune phenomena on the other, reduced the distance between the cis- and trans-immunologists. They were soon to meet somewhere in the middle, and their integration was to lead to the elucidation of all the mechanisms conducive to the synthesis of antibodies, and thus to the "end of immunology" [74].

VI. NETWORKS—CELLS OR IMMUNOLOGISTS?

"Sir Macfarlane Burnet", explained Niels Jerne in 1967, "must have been pleased not only to witness the vindication of his Clonal Selection Theory of Acquired Immunity, but also to see how his stimulating ideas have led to a great proliferation of immunologists" [75]. This affirmation was not exaggerated: in the 1960's and 70's there was a rapid growth of the number of scientists involved in immunological investigations. This increase was paralleled by an important increase in the number of immunological journals, the proliferation and reinforcement of immunology departments at universities and medical schools, and an large increase in resources allocated to immunological research [76].

The result of this rapid development of immunological studies was an impressive accumulation of observations concerning immune phenomena. The knowledge about immune phenomena was much increased—but the perception of their complexity grew also, and the difficulties in fitting all the observations to the framework of the elegant theory proposed by Jerne and Burnet. This was certainly not an unusual phenomenon. Fleck had already affirmed that,

> it was once thought that science would make that mysterious and intricate complex, called Nature, somehow simpler and easier to grasp for the mind. Instead science has become a structure which, as a whole, is not at all simpler than Nature, but which is much harder to attain. It is easier to find one's way in the woods than in botany. It is also easier to cure a patient than to know what his disease is [77].

And, he explained, scientific knowledge passes through successive stages of elaboration of a global theory, then its gradual rejection under the pressure of observations which cannot be fit into the theoretical framework (an idea later developed by T.S. Kuhn in his book, *The Structure of Scientific Revolutions*):

> every comprehensive theory passes first through a classical stage when only those facts are recognized which conform to it exactly, and then through a stage with complications, when the exceptions begin to come forward. The great theoretician Paul Ehrlich knew it only too well: "Unfortunately, this differs in no way from other scientific problems, since it just becomes more and more complicated" [78].

Indeed, in the 1970's and 80's immunology became more and more complicated. The immune response was seen as resulting from an interaction between several subsets of cells that communicated through a complex system of membrane receptors, which were in turn subject to various levels of genetic and somatic control. The picture was further complicated by the discovery of the role of non-specific, soluble inter-cellular mediators such as lymphokines and growth factors. The bewilderment of some of the immunologists facing this increasing complexity was aptly captured in a series of drawings by R.K. Gershon, in which the "immunological orchestra" of 1968 represented as a well-organized assembly with a single conductor, in the years 1968–1982 is gradually transformed into a rather chaotic group led by several conductors at the same time [79].

Several years after his announcement that immunologists should "wait for the end", Jerne attempted to deal with the growing complexity of immunological findings and proposed his idiotypic network theory of the immune system [80]. This theory was criticized by some immunologists [81], but nevertheless is viewed today by the majority as a crucial contribution to the understanding of immune phenomena [82]. In contrast to Burnet's self-marker theory, which stressed the importance of hereditary factors in self-nonself recognition, Jerne's theory stressed the role of the "humoral personality" (Richet), i.e., the impact of the individual's history on his immune responses. According to this theory, the immune system did not primarily evolve in order to keep the integrity of the organism (self-nonself discrimination), but its evolution was a result of "knowing of the self". Before meeting the antigen the cells of the immune system need to "talk" to each other, and the immune response regulates itself through a network of anti-idiotype responses (i.e., responses against the antigenic determinants that characterize the specific combining site of an antibody molecule). In terms of this theory, the immune response results from a complex interplay between external and internal antigens. The action of an antigen can be perfectly mimicked by its "internal image", while each idiotype acts at the same time as antigen and as antibody. In such a complex network, it is not easy any longer to decide what the "self" and what the "nonself" is.

The historian of science and feminist theoretician Donna Harraway interprets the blurring of the limits between the "self" and the "nonself" in present immunological theory, in terms of the "post-modern" view of the body:

> My thesis is that the immune system is an elaborate icon for principal systems of symbolic and material "difference" in late capitalism. That is, the immune system is a plan for meaningful action to construct and maintain the boundaries for what may count as self and other in the crucial realms of the normal and the pathological.

In modern immunological discourse these boundaries are more and more fluid and,

> the self and the other lose their rationality and become subtle plays of partially mirrored readings and responses [83].

The result is that,

> there is no part/whole relation at all, in any sense Aristotle would recognize [and] what counts as a "unit", a one, is highly problematic, not a permanent given. You or I (whatever problematic address these pronouns have) might be an individual for some purposes, but not for others [84].

Harraway's interpretation of immune reactions brings her to a perception of the organism and of the self which in many aspects is similar to the constructivist view of the organism developed by Fleck in the 1930's. But while Fleck affirmed that "organisms" are differently constructed by scientists belonging to different thought-styles, Harraway attributes these differences to political differences expressed within the framework of a single "post-modern" thought-style. Indeed systems and complex regulatory networks which are strongly present today in biological and medical discourse, are present in general discourse as well. But is the postulated "post-modern" concept of organism and self really an entirely new one? We have seen that Harraway's conclusions concerning the construction of the body are akin to the ideas developed in 1930's (Fleck, Gradmann). Moreover, the idea of an autoregulatory network of antibodies and anti-antibodies was proposed during the first "physiological" period of immunology. The idea that antibody response is regulated by anti-antibody was discussed by immunologists at the beginning of the 20th century. It was rejected by them later when it was found that the first observations of anti-antibodies were based on mistaken laboratory findings and when the research interest switched to chemical studies of antigens and antibodies, while the biological question of antibody production was neglected. Therefore, in the 1920's and 30's, the possibility of the regulation of antibody production by anti-antibodies fell into disrepute, only to be recovered sixty years later [85].

Immunology has always oscillated between biological-medical and chemical poles. One may thus view the history of this discipline as a sequential alternation between "biological" periods and "chemical" periods—in Fleck's terms between periods of predominance of a chemical and of a biological thought-style. A cis-immunologist may find speaking with a trans-immunologist difficult not because, as Jerne claimed, they study different objects, but because they have incommensurable ways of looking at the same object [86]. In some periods of immunology's history, scientists belonging to distinct thought-styles indeed did not talk with each other. In other periods, however, while the differences between thought-styles persisted, scientists belonging to different thought-collectives were able to communicate. This communication, I propose, was made possible by the existence of versatile concepts, which, by their indeterminacy, made fruitful contacts between distinct thought-styles possible. Concepts such as Ehrlich's "receptor" or Jerne's network regulation of immune phenomena are exemplars of their expansive dialogue [87]. My claim is that the network view of the immune system, based on a dynamic perception of the organism and the self, was, and still is, a framework which may facilitate the collaboration between at least partially incommensurable thought-collectives. As A.M. Moulin has put it: "the immune system was a kind of metaphor, but an ecumenical one. It solved the need for communication not only between cells, but also between the professionals of immunity" [88].

The example of AIDS may illustrate the way the network view of immune phenomena and of the self may facilitate the collaboration between chemically-oriented and medically-oriented scientists. Our present understanding of AIDS is shaped by knowledge produced by two distinct professional groups: molecular biologists who purify and analyze macromolecules, and by physicians and biologists who observe and care for AIDS patients. The first were highly successful and rapidly produced an impressive body of knowledge about the chemical structure of HIV. However, the expression, attributed to Francis Crick, "if you cannot study function, study structure" [89] has been, up to now, less successful for HIV than for the double helix. Detailed knowledge of the structure of this virus is not, in itself, sufficient to explain the pathology of HIV infection. As leading molecular biologists have recently acknowledged: "we are rapidly learning about the role of each of HIV's approximately 10,000 nucleotides, but remain largely ignorant of rudimentary aspects of the processes underlying the development of AIDS in humans" [90].

At present there is no simple explanation of the pathology of HIV infection. The hypothesis that all the pathological findings are the direct result of the interaction of the virus with target cells is not in agreement with some of the clinical and laboratory observations, such as discrepancies between the severity of infection and of disease or the existence of numerous immunological abnormalities in AIDS patients [91]. The lack of straightforward explanations of the mechanisms of HIV pathogenicity have raised doubts whether this virus is indeed the etiological cause of the disease [92]. These doubts have been firmly rejected by the great majority of the scientific community, but AIDS specialists do acknowledge the existence of serious gaps in the understanding of HIV pathogenicity and its effects on the immune mechanisms of the host [93]. Although today new information points to the possibility of a greater direct role of the virus in the destruction of immunocompetent cells, the mechanisms of HIV pathogenicity still remain largely unknown [94].

In the absence of direct explanations of the virus's pathogenicity, alternative explanations were proposed, in which a large place is given to network-based explanations, and in particular to the possibility of partially self-inflicted pathology mediated through the ability of the HIV virus to mimic self structures and to trigger auto-immune responses and self-destruction of CD4 lymphocytes. In such explanations, the simple picture of attack by an external pathogenic agent on the organism is replaced by the image of a

immunologists claim, has been superseded by new developments in the discipline since the 1960's [106]. My opinion is that such an understanding of immunology is incorrect. If one follows Einstein's advice, and examines the immunologists' deeds instead of listening to their discourse, one will not fail to notice the persistence and vitality of two thought-styles. The existence of distinct, at least partially incommensurable approaches, does not mean, however, that scientists belonging to different thought-styles are unable to work together. My analysis, although inspired by Fleck's writings, diverges from his ideas on this point. Fleck (and later Kuhn) put the emphasis on the incommensurability of thought-styles and on its consequence, the difficulty for collaboration. In contrast I have been interested in the mechanisms which facilitate contacts between communities of scientists. In immunology, I argue, one of the devices which makes the communication between different thought-collectives not only possible but fruitful is the linguistic versatility and indeterminacy of key concepts such as the organism and the self.

Hospital Necker-Enfants Malades, Paris

NOTES

1. Fleck, L. 1979 (1935), *Genesis and Development of a Scientific Fact* (transl. by F. Bradley and T. Trenn), The University of Chicago Press, Chicago and London, p. 60. Hans Gradmann's article, Die harmonische Lebenseinheit vom Standpunkt exakter Naturwissenschaft, quoted by Fleck, appeared in 1930 in *Naturwissenschaften*, 18: 641–644; 662–66.
2. Kuhn, T.S., Foreword, in Fleck, L. 1979 (1935), *Genesis and Development of a Scientific Fact, op. cit.*, vii–xi, on p. vii; Moulin, A.M., Niels Kaj Jerne, in *Nobel Laureates in Physiology or Medicine*, D. Fox, editor, Garland Press, New York. In press.
3. Kuhn, T.S., 1970, *The Structure of Scientific Revolutions*, The University of Chicago Press (second, enlarged edition), Chicago.
4. Kuhn, T.S., Foreword, in Fleck, L., *Genesis and Development of a Scientific Fact, op. cit.*, vii–xi, on p. viii.
5. Barnes, B. and Edge, D., editors, 1982, *Science in Context: Readings in the Sociology of Science*, The MIT Press, Cambridge, Mass., p. 65.
6. Fleck, L., *Genesis and Development of a Scientific Fact, op. cit.*, 59–60.
7. *Ibid.*, p. 61.
8. *Ibid.*, p. 61.
9. Fleck, L., 1927. O niektorych swoistych cechach myslenia lekarskiego. (Some specific features of the medical way of thinking), *Archiwium Historji i Filozofji Medycyny (Archives of the History and Philospohy of Medicine)*, 6: 55–64. English translation in *Cognition and Fact: Materials on Ludwik Fleck*, 1986, R.S. Cohen and T. Schnelle, editors, D. Reidel, Dordrecht. 39–46, on p. 44. [Boston Studies in Philosophy of Science, vol. 87].

10. *Ibid.*, 45–46.
11. *Ibid.*, p. 40.
12. Fleck, L., 1929, Zur Krise der "Wirklichkeit" (On the crisis of "reality"), *Naturwissenschaften*, 17: 425–430. English translation in *Cognition and Fact, op. cit.*, R.S. Cohen and T. Schnelle, editors, pp. 47–58.
13. Fleck, L., *Genesis and Development of a Scientific Fact, op. cit.*, p. 93.
14. Chalubinski, T., 1874, *Metoda wynajdywania wskazan lekarskich* (*The Method of Finding Medical Indications*). Gebethner and Wolf, Warsaw; Löwy I. *The Polish School of Philosophy of Medicine from Tytus Chalubiski* (1820–1881) *to Ludwik Fleck* (1896–1961), Kluwer Academic Publishers, Dordrecht 1990.
15. Wrzosek, A., 1924, Zadania i zamierzenia polskiego Archiwum Historji i Filozofji Medycyny (Tasks and goal of the Polish Archive of History and Philosophy of Medicine), *Archiwum Historji i Filozofji Medycyny*, 1: 1–13
16. Fritz, J., 1925, Sprawozdanie z posiedzenia organizacyjnego Towarzystwa Milosnikow Historji Medycyny w Lwowie z dnia 24.X.1925 (Report from the founding meeting of the Society of Amateurs of History of Medicine in Lwow, 24.X.1925), *Archiwum Historji i Filozofji Medycyny*, 4: 154–156.
17. Chalubinski, T., *The Method of Finding Medical Indications, op. cit.*, p. 39.
18. Kramsztyk, Z., 1899, Fakt kliniczny (A clinical fact), in Z. Kramsztyk, *Szkice krytyczne z zakresu medycyny* (*Critical Notes on Medicine*), E. Wende i Ska, Warsaw, pp. 66–71.
19. Kramsztyk, Z, 1899, O trudnosciach klinicznego badania jaglicy. (On the difficulties of a clinical study of trachoma), in Z. Kramsztyk, *Critical Notes on Medicine*, op. cit, pp. 72–88.
20. Bieganski, W., 1894, *Logika medycyny czyli Zasady ogolnej metodologii nauk lekarskich* (*The Logic of Medicine or The Principles of the General Methodology of the Medical Sciences*), E. Wende, Warsaw, p. 15.
21. Bieganski, W., 1897, *Ogolne problemy teorji nauk lekarskich* (*General Problems of the Theory of Medical Sciences*), E. Wende, Warsaw, p. 300.
22. Bieganski, W., 1957 (1899), *Mysli i aforyzmy o etyce lekarskiej* (*Thoughts and Aphorisms on Medical Ethics*), Panstwowy Zaklad Wydawnictw Lekarskich, Warsaw, pp. 32–33.
23. Schnelle, T., Microbiology and philosophy of science, Lwow and the German Holocaust: stations of a life — Ludwik Fleck 1896–1961, in *Cognition and Fact*, R.S. Cohen and T. Schnelle, editors, *op. cit.*, pp. 3–37.
24. Weigl, R., 1920, Badania nad Rickettsia prowazekii (Studies on Rickettsia provazekii), *Przeglad Epidemiologiczny*, 1: 4–17.
25. Nicolle, C., 1933, *Le Destin des maladies infectieuses*, Alcan, Paris, p. 210.
26. Fleck, L., 1931, O pojeciu gatunku w bakteriologji (On the notion of species in bacteriology), *Polska Gazeta Lekarska*, 26: 522–24; 536–39.
27. Hadley, P., 1927, Microbic dissociation, *J. Inf. Dis.*, 40: 1–312; Amsterdamska, O., 1987, Medical and biological constraints: the research on variation in bacteriology, *Soc. Stud. Sci*, 17: 657–688.
28. Fleck, L., On the notion of species in bacteriology, *op. cit.*, p. 537; p. 539.
29. Gradmann, H., Die Harmonische Lebenseinheit vom Standpunkt exacter Naturwissenschaft, *op. cit.*
30. Fleck, L., *Genesis and Development of a Scientific Fact, op. cit.*, p. 26.
31. *Ibid.*, p. 61.
32. *Ibid.*, pp. 62–63.
33. Citron, J., 1910, *The Methods of Immunodiagnostic and Immunotherapy*. Quoted by Fleck,

L., *Genesis and Development of a Scientific Fact, op. cit.*, p. 58.
34. Ibid., p. 65.
35. Landsteiner, K., 1936 (1933), *Die Spezifizität der serologischen Reaktionen.* English translation, *The Specificity of Serological Reactions*, Charles C. Thomas, Springfield, Il.
36. Chernyak, L. and Tauber, A.I. 1990, The idea of immunity: Metchnikoff's metaphysics and science, *J. Hist. Biology* 23: 187–249, Silverstein, A.M., 1989, *A History of Immunology*, Academic Press, San Diego, New York, Berkeley, etc., pp. 38–58; Moulin, A.M., *Histoire du système immunitaire*, 1986, unpublished doctoral thesis, Universtity of Lyon III, pp. 11–99.
37. Richet, C., 1967 (1913), Anaphylaxis, in *Nobel Lectures: Physiology or Medicine*, The Nobel Foundation, Elsevier Publishing Company, Amsterdam and London, pp. 469–490, on p. 489.
38. Richet, C., 1911, *L'Anaphylaxie*, Felix Alcan, Paris, pp. 250–251; Richet, C., *Anaphylaxis, op. cit.*, pp. 488–489.
39. Richet, C., *Anaphylaxis, op. cit.*, p.489.
40. Citron, J., *The Method of Immunodiagnostic and Immunotherapy, op. cit.*
41. Mazumdar, P.M.H., 1976, *Karl Landsteiner and the Problem of Species 1838–1968*, PhD thesis, Johns Hopkins University, Baltimore; Mazumdar, P.M.H., 1989. The template theory of antibody formation and the chemical synthesis of the twenties, in *Immunology 1930–1980*, P.M.H. Mazumdar, editor. Wall and Thompson, Toronto.
42. Northop, G., 1928, The mechanism of agglutinaton, in E. O. Jordan and I. S. Falk, *The Newer Knowledge of Bacteriology and Immunology*, 1972, University of Chicago Press, Chicago, p. 801.
43. Zinsser, H., 1931, *Resistance to Infectious Disease* (fourth ed.), The Macmillan Company, New York, p. ix; Gideon Wells, H., 1925, *Chemical Aspects of Immunity*, Chemical Catalog, New York, p. x.
44. Moulin, A.M., *Histoire du système immunitaire, op. cit.*, pp. 296–346; Moulin, A.M. and Silverstein, A.M., History of immunophysiology, in *Immunophysiology*, 1990, J.J. Oppenheim and E.M. Shevach, editors, Oxford University Press, New York and Oxford, pp. 3–13; Samter, M. 1979, Allergy and clinical immunology: fifty years from now, *J. Aller. Clin. Immunol.*, 64: 321–330; Currie, G.A., 1972, Eighty years of immunotherapy, *Brit. J. Canc.*, 26: 141–152; Silverstein, A.M., *A History of Immunology, op. cit.*, pp. 160–189; 214–251.
45. Medawar, P.B., 1975, Interview with J. Goodfield, in *Cancer under Siege*, Hutchinson, London, p.
46. Foster, W.D., 1979, *History of Medical Bacteriology and Immunology*, William Heineman, London, pp. 137–140; Silverstein, A.M., *A History of Immunology, op. cit.*, pp. 107–112. An alternate view has been argued that Metchnikoff's earliest formulation of immunological phenomena viewed self-definition as primary and host defense was but a secondary effect. But this aspect of his theory was either ignored or misunderstood. See Tauber, A.I. and Chernyak, L., *Metchnikoff and the Origins of Immunology: from Metaphor to Theory*, Oxford University Press, New York, 1991.
47. Jordan, E.O. and Falk, I.S., *The Newer Knowledge of Bacteriology and Immunology, op. cit.*, pp. iii–iv; Zinsser, H., Enders, J.F., and Forhergil, L.D., 1936, *Immunity: Principles and Applications in Medicine and in Public Health*, Macmillan, New York, pp. v–vi; Bordet, J., 1939, *Traité d'immunité dans les maladies infectieuses*, Masson, Paris, pp. iv–vi.

48. Fleck, L., 1939, O swoistych cechach myslenia serologicznego (Some specific features of the serological way of thinking), in *Papers of the XVth Meeting of Polish Physicians and Biologists, Lwow, July 4–7, 1937*, Nowicki W. and D. Szymkiewicz, editors, Lwow, pp. 287–289. English translation in *Science in Context*, 1988, 2: 343–344.
49. Elkana, Y., 1986, Is there a distinction between external and internal sociology of science?, in *Cognition and Fact*, R.S. Cohen and T. Schnelle, editors, *op. cit.*, pp. 309–316.
50. Stent, G., 1968, This was the molecular biology that was, *Science*, 160: 390–396; Judson, H., 1979, *The Eight Day of Creation*, Simon and Schuster, New York, pp. 50–65; Kay, L.E., 1985, Conceptual models and analytical tools: the biology of the physician Max Delbrück, *J. Hist Biol.*, 18: 207–246.
51. Burnet, F.M., 1967, The impact of ideas on immunology, *Cold Spring Harbour Symp. Quant. Biol.*, 32: 1–8.
52. Fenner, F.J., 1987, Frank Macfarlane Burnet, 3 September 1899 — 31 August 1985, *Biograph. Mem. Fell. Roy. Soc.*, 33: 101–162.
53. Burnet, F.M., 1968, *Changing Patterns: An Atypical Biography*, William Heineman, Melbourne, pp. 81–83.
54. Fenner, F.J., *Frank Macfarlane Burnet, op. cit.*, p. 112.
55. Burnet, F.M., with the collaboration of Freeman, M., Jackson, A.V. and Lush, D., 1941, *The Production of Antibodies: A Review and a Theoretical Discussion*, Macmillan, Melbourne.
56. Murphy, J.B., 1913, Transplantability of tissues to the embryo of foreign species, *J. Exp. Med.*, 17: 482–492.
57. Burnet, F.M., with the collaboration of Freeman, M., Jackson, A.V. and Lush, D., *The Production of Antibodies, op. cit.*, pp. 45–46.
58. Burnet, F.M. and Fenner, F., 1949, *The Production of Antibodies* (2nd edition), *op. cit.*, Macmillan, Melbourne.
59. Owen, R.D., 1945, Immunological consequences of vascular anastomoses between cattle twins, *Science*, 102: 400–405.
60. Burnet, F.M. and Fenner, F., *The Production of Antibodies, op. cit.*, p. 71; p. 76.
61. *Ibid.*, p. 86.
62. *Ibid.*, pp. 100–103.
63. Duclaux, E., 1986, *Pasteur, histoire d'un esprit*, Paris, p. 391.
64. Burnet, F.M., 1959, *The Clonal Selection Theory of Acquired Immunity. Cambridge University Press*, Cambridge, p. 1.
65. Mazumdar, P.M.H., 1989, The template theory of antibody formation and the chemical synthesis of the twenties., in Mazumar, P.M.H., *Immunology 1930–1980, op. cit.*, pp. 13–33; Silverstein, A.M., *A History of Immunology, op. cit.*, pp. 59–86.
66. Bergmann, M. and Niemann, C., 1937, Newer aspects of protein chemistry, *Science*, 86: 187–189.
67. Burnet, FM., 1953, *Natural History of Infectious Diseases*, Cambridge University Press. Cambridge, pp. 123–135.
68. Burnet, F.M., 1956, *Enzyme, Antigen and Virus*, Cambridge University Press, Cambridge, pp. 40–80; Burnet, F.M., 1956, Concluding remarks, in A discussion on immunological tolerance, *Proc. Roy. Soc.* (B), 146: 90–92; Burnet, F.M., *Changing Patterns, op. cit.*, pp. 204–207.
69. Jerne, N.K., 1955, The natural selection theory of antibody formation, *Proc. Natl. Acad. Sci. USA.*, 41, 849–857; Burnet, F.M., 1957, A modification of Jerne's theory of antibody

production using the concept of clonal selection, *Austr. J. Sci.*, 20: 67–68.
70. Burnet, F.M., 1959, *The Clonal Selection Theory of Acquired Immunity*, Cambridge University Press, Cambridge, p. 46.
71. Klein, J., 1982, *Immunology: The Science of Self-Nonself Discrimination*, John Wiley and Sons, New York.
72. Burnet, F.M., Immunological recognition of self, *Nobel Lectures in Physiology or Medicine, op. cit.*, vol 3, pp. 689–701, on p. 697; Burnet, F.M., *Natural History of Infectious Diseases, op. cit.*, p. 125.
73. Burnet, F.M., *Immunological recognition of self, op. cit.*, pp. 699–700; p. 689.
74. Jerne, N.K., 1967, Summary: waiting for the end, *Cold Spring Harbour Symp. Quant. Biol.*, 32: 591–603, on p. 601.
75. *Ibid.*, p. 601.
76. Mazumdar, P.M.H., 1989, Working out the theory, in: Mazumdar, P.M.H., editor, *Immunology 1930–1980, op. cit.*, pp. 1–13; National Institute of Allergy and Infectious Diseases, 1981, *New Initiatives in Immunology*, NIAID Study Group Report, NIH Publication no 81–2215, Bethesda, MD., pp. 3–9; Good, R.A., 1976, Runestones in immunology: inscriptions to journeys of discovery and analysis, *J. Immunol.*, 117: 1417–1426.
77. Fleck, L., 1939, Nauka i srodowisko (Science and environment) *Przeglad Wspolczesny*, 18: 1–9.
78. Fleck, L., *Genesis and Development of a Scientific Fact, op. cit.*, pp. 28–29.
79. Golub, E.S., 1987, *Immunology: A Synthesis*, Sinauer, Sutherland, MA.
80. Jerne, N.K., 1974, Towards a network theory of the immune system. *Ann. Immunol. Inst. Pasteur*, 125C: 373–389.
81. Cohn, M., 1981, Conversations with N.K. Jerne on immune regulations: associative versus network regulation, *Cell. Immunol.*, 61: 425–436; Cohn, M., 1985, Diversity of the immune system: "preconceived ideas" or ideas preconceived?, *Biochimie*, 67: 9–27.
82. Golub, E.S., 1980, Idiotypes and the network hypothesis, *Cell.*, 22: 641–642; Jerne, N.K., Roland, J., and Cazenave, P.A., 1982, Recurrent idiotypes and internal images, *EMBO J.*, 1: 243–246; Coutinho, A., Forni, L., Homberg, D., Ivars, F., and Vaz., N.M., 1984, From an antigen-centered, clonal perspective of immune responses to an organism-centered network perspective of autonomous reactivity of self-referential immune systems, *Immunol. Rev.*, 79: 151–168.
83. Harraway, D., 1989, The biopolitics of postmodern bodies: determination of self in immune system discourse. *Differences: A Journal of Feminist Cultural Studies*, 1: 1–43, on p. 4; p. 15; p. 23.
84. *Ibid.*, p. 25; p. 20.
85. Silverstein, A.M., *A History of Immunology, op. cit.*, pp. 252–274; Besredka, A., 1901, Les antihémolysines naturelles, *Ann. Inst. Pasteur*, 15: 785–807; Tauber, A.I. and Chernyak, L., *Metchnikoff and the Origins of Immunology, op cit.*
86. Jerne, N.K., Waiting for the end, *op. cit.*; Silverstein, A.M., *The History of Immunology, op. cit.*, pp. 82–83.
87. Moulin, A.M. and Löwy, I., 1983, La double nature de l'immunologie: histoire de la transplantation rénale, *Fundamentia Scientiae*, 4: 201–218; Moulin, A.M., 1988, Test and context in biology: in the pursuit of chimera, in *Text and Context in Biology*, P. Bieri and B. Hrushowski, editors, *Poetics Today*, 9: 145–161; Löwy, I., 1990 Variance of meaning in discovery accounts: the case of contemporary biology, *HPSP*, 21: 87–121.

88. Moulin, A.M., 1989, Immunology old and new: the beginning and the end, in *Immunology 1930–1980*, P.M.H. Mazumdar, editor, *op. cit.*, pp. 291–298, on p. 294.
89. Quoted by Jerne, N.K., Waiting for the end, *op. cit.*, p. 591.
90. Baltimore, D. and Feinberg, M. B., 1989, HIV revealed: Towards a natural history of the infection, *N. Engl. J. Med.*, 321: 1673–1675.
91. Zolla Pazner, S., Des Jarlis, D.C., and Friedman, S.R., 1987, Nonrandom development of immunological abnormalities after infection with human deficiency virus, *Proc. Natl. Acad. Sci.*, 84: 5404–5408; Levy, J.A., 1988, Mysteries of HIV: challenges for therapy and prevention. *Nature* (London) 333: 519–522.
92. Duesberg, P.H., 1988, HIV is not the cause of AIDS, *Science*, 241: 514; Duesberg, P.H., 1989, Human immunodeficiency virus and acquired immunodeficiency syndrome: correlation but not causation, *Proc. Natl. Acad. Sci. USA*, 86: 755–764.
93. Blatner, W., Gallo, R.C., and Temin, H.M., 1988, HIV causes AIDS, *Science*, 241: 115–517.
94. Ho, D.D., Moudgil, T., and Alam, M., 1989, Quantitation of Human Deficiency Virus type 1 in the blood of infected persons, *N. Engl. J. Med.*, 321: 1621–1625; Baltimore, D. and Feinberg, M.B., *HIV revealed, op. cit.*; Root-Bernstein, R.S., 1990, Do we know the cause(s) of AIDS?, *Perspectives in Biology and Medicine*, 33: 480–500.
95. Ziegler, J.H. and Stites, D.P., 1986, Hypothesis: AIDS is an autoimmune disease directed at the immune system and triggered by a lymphotropic virus, *Clin. Immunol. Immunopathol.*, 41: 305–313; Ho, D.D., Pomeranz, R.J., and Kaplan, J.C., 1987, Pathogenesis of infection with Human Immunodeficiency Virus, *N. Engl. J. Med.*, 317: 278–286.
96. Moss, A.R., 1988, Predicting who will progress to AIDS, *Brit. J. Med. Sci.*, 297: 1067–1068; de Wolf, F., Lange, J.M.A., and Houweling, J.T.M., 1989, Appearance of predictors of disease progression in relation to the development of AIDS, *AIDS*, 3: 563–569; Fahey, J.L., Taylor, J.M.G., and Detels, R., 1990, The prognostic value of cellular and serological markers in infection with Human Deficiency Virus type 1, *N. Eng. J. Med.*, 322: 166–171.
97. Duclaux, E., 1895, Louis Pasteur, *La Revue scientifique* (Paris), 56: 641.
98. Darden, L. and Maul, N., 1977, Interfield theories, *Phil. Sci.*, 44: 43–64; Maul, N., 1977, Unifying science without reduction, *Stud. Hist. Phil. Sci.*, 8: 143–160.
99. Muller, H.J., 1922, Variation due to change in the individual gene. Quoted by Kitcher, P., 1984, 1953 and all that, A tale of two sciences, *The Philosophical Review*, 93: 335–373. Italics mine I.L.
100. Fleck, L., *Genesis and Development of a Scientific Fact, op. cit.*, pp. 110–111.
101. Kuhn, T.S., Foreword, in Fleck, L., *Genesis and Development of a Scientific Fact, op.* cit., pp. ix–x.
102. Fleck, L., 1935, O obserwacji naukowej i postrzeganiu w ogole (Scientific observation and perception in general), *Przeglad Filozoficzny*, 38: 58–76. English translation in *Cognition and Fact*, R.S. Cohen and T. Schnelle, editors, *op. cit.*, pp. 59–78.
103. Fleck, L., *Genesis and Development of a Scientific Fact, op. cit.*, p. 42.
104. Ben-David, J., 1984, *The Scientist's Role in Society, A Comparative Study* (second ed.), The University of Chicago Press, Chicago.
105. Einstein, A., 1935, On the methods of theoretical physics, in Einstein, A., *The World as I See it*, London.
106. Silverstein, A.M., *A History of Immunology, op. cit.*, p. 83; Jerne, N.K. Waiting for the end, *op. cit.*

Coutinho and John Stewart from Pasteur), provides a logical link to the paper by Chernyak and Tauber, who attempt to re-formulate the question of immune Self-definition within a combined philosophical and evolutionary argument. Beginning with an extended critique of the Paris School, the conceptual infrastructure of the autopoietic argument is exposed. A re-interpretation of immunity is then offered based on the epistemological inability of either the clonal or autopoietic theories to address how mediation between Self and the Other in fact might occur. A reconsideration of the basis for individuality then follows, which relies on arguments regarding the evolution of sex as a particular property of later phyla, and is finally extended to assessing the importance of developmental identification mechanisms as the origin of immune recognition. The conclusion that protection of identity is not equivalent to the establishment of integrity emphasizes functionally different roles for the immune system and discerns the conceptual confusion inherent in merging these disparate activities. The issues central to immune/cognitive definition of the Self are most generally concerned with the decipherment of subject-object relations in a broadened metaphysical context; that immunology might contribute to this discussion bespeaks its growing importance as a model system for explicating such discussion.

FRANCISCO J. VARELA

ORGANISM: A MESHWORK OF SELFLESS SELVES

I. INTRODUCTION

Organism connotes a knotty dialectic: a living system makes itself into a entity distinct from its environment through a process that brings forth, through that very process, a world proper to the organism.

My intention in what follows is to unpack this statement, both in the sense of providing a factual, biological justification for it, and of unfolding some of its epistemological consequences. I use the term dialectic to describe properties which stand in relation so that "...one thing cannot exist without the other, that one acquires its properties from its relation to the other, that the properties of both evolve as a consequence of their interpenetration" [1, pp. 2, 3]. There is more in all this than meets the eye, as we shall presently see. In fact, my conclusion will be that the relation between organism and self turns out to be the imbrication of *two* separable dialectics: one linked to the mechanism of identity, the other linked to the mode of relationship with its world.

The issue of selfhood shows up right away: I started by saying that a living system makes it*self* into a distinct entity. Many biologists would refer to avoid this notion of 'self'. What exactly is it? The temptation is to apologize and wave one's hand: "Well some sort of, uh, not a thing, but an activity, or something like that..." Others will refer us to the self/not self distinction on immunological grounds. Few would be ready to address the self as subjectivity, the ghostly 'I' we tow around and lend our names.

A basic stance I adopt in this paper is that this issue will not go away with such vague answers and hand waving: a proper account of the constitution of an *autonomous* self and of the nature of its mode of existence, is at the very core of both biological and cognitive research. We need to avail ourselves of more powerful explanatory devices to see how such a self can, at the same time, be a *virtual* point with no localized coordinates, and yet provide a mode of identity through which an interaction can happen. In my view, recent notions of emergent properties in highly distributed, modular systems allow us to make some headway on the characterization of such virtual selves.

A second basic stance here is that — contrary to a misconstrued demand for parsimony — the prolific significance of the concept 'self' is better

Alfred I. Tauber (ed.), Organism and the Origins of Self, 79–107.

clarified by addressing it in its *entirety*, rather than in a piecemeal fashion. In other words: to discuss what self could mean it is essential to place ourselves in the larger field of all those situations where a *for*-itself (*pour-soi*) shows up. At the same time each one of the modes of self has its own characteristics as I will try to pinpoint in what follows. Thus we need to deal with a multiplicity of *regional* selves, all of them having some mode of self-constitution, and in their overall assemblage giving rise to an organism. Accordingly I want to invoke here the following "regional" selves: 1) a minimal or cellular unity, 2) a bodily self in its immunological foundations, 3) a cognitive perceptuo-motor self associated to animal behavior, 4) a socio-linguistic 'I' of subjectivity, and 5) the collective social multi-individual totality. In all these regions we are dealing with levels and processes where an identity comes about—not as substance, but as movement—and whose fabric of articulation *is* the organism. To efface the multiplicity of this meshwork is a source of confusion.

The reader need not worry: I am not about to launch into a tour of all these vast regions with some quick strokes within the brief span of my allotted space. Instead, in the following two Sections (II and III) I intend to discuss "regions" one and three in sufficient detail to procure some fundamental conclusions about the organism and self altogether, to be presented in Section IV.

II. SELF AS MINIMAL LIVING SYSTEM

II.1 *Autopoiesis as the Skeletal Bio-logic*

The bacterial cell is the simplest of living system because it possesses the capacity to produce, through a network of chemical processes, all the chemical components which lead to the constitution of a distinct, bounded unit. To avoid being trivial, the attribute 'living' in the foregoing description must address the *process* that allows such constitution, not the materialities that go into it, or an enumeration of properties. But what is this basic process? Its description must be situated at a very specific level: it must be sufficiently universal to allow us to recognize living systems as a class, without essential reference to the material components. Yet at the same time it must not be too abstract, that is, it must be explicit enough to allow us to see such dynamical patterns in action in the actual living system we know on earth, those potentially to be found in other solar systems, and eventually those created artificially by man. As stated by the organizer of a recent meeting on artificial life: "Only when we are able to view *life-as-we know-it* in the larger context of *life-as-it-could-be* will we really understand the nature of the beast" [2, p. 2].

Contemporary cell biology makes it possible to put forth the characterization of this basic living organization—a bio-logic—as that of an *autopoietic* system (from Greek: self-producing) [3, 4, 5]. An autopoietic system—the minimal living organization—is one that continuously produces the components that specify it, while at the same time realizing it (the system) as a concrete unity in space and time, which makes the network of production of components possible. More precisely defined : An autopoietic system is organized (defined as unity) as a network of processes of production (synthesis and destruction) of components such that these components:

(i) continuously regenerate and realize the network that produces them, and
(ii) constitute the system as a distinguishable unity in the domain in which they exist.

Thus, autopoiesis attempts to capture the mechanism or process that generates the *identity* of the living, and thus to serve as a categorical distinction of living from non-living [6]. This identity amounts to self-produced coherence: the autopoietic mechanism will maintain itself as a distinct unity as long as its basic concatenation of processes is kept intact in the face of perturbations, and will disappear when confronted with perturbations that go beyond a certain viable range which depends on the specific system considered. Obviously, all of the biochemical pathways and membrane formation in cells can be immediately mapped onto this definition of autopoiesis.

A different exercise—which I do not pursue here at all—is to see how this basic autopoietic organization, present at the origin of terrestrial life [6], becomes progressively complexified through reproductive mechanisms, compartmentalization, sexual dimorphism, modes of nutrition, symbiosis, and so on, giving rise to the variety of pro- and eukaryotic life on Earth today [7, 8]. In particular, I take here the view that reproduction is *not* intrinsic to the minimal logic of the living. Reproduction must be considered as an added complexity superimposed on a more basic identity, that of an autopoietic unity, a complexness which is necessary due to the constraints of the early conditions on a turbulent planet. Reproduction is essential for the viability of the living, but only when there is an identity can a unit reproduce. In this sense, identity has logical and ontological priority over reproduction, although not historical precedence.

But instead of following these historical complexities, I wish to pursue here a more pertinent question for my purpose: can a molecular structure

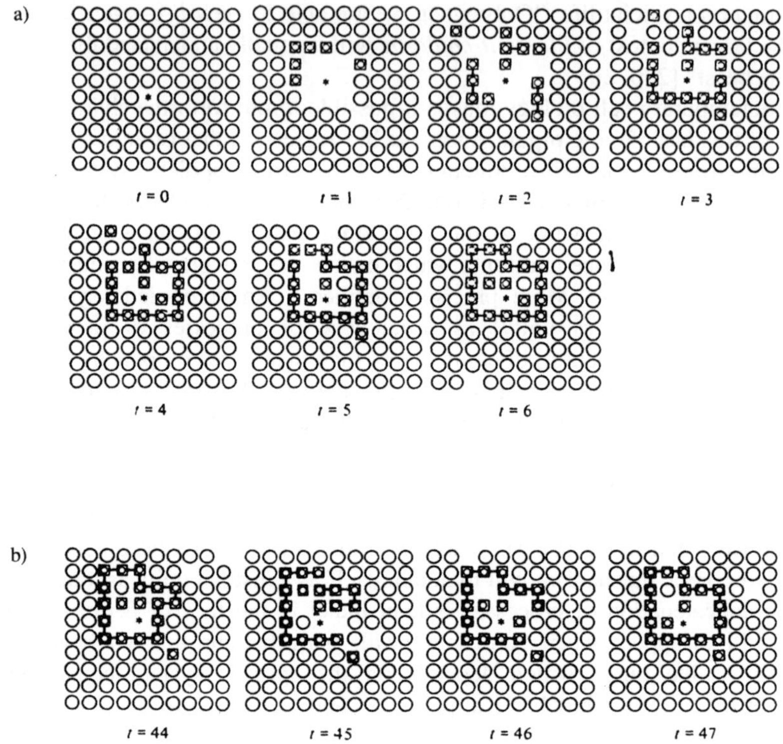

Fig. 1. Simulation of a minimal autopoietic system [5]. (a) The first seven instants of a computer simulation of the system of "reactions" described in the text, showing the spontaneous emergence of bounded unity in this two-dimensional space. Interactions between substrate 'O' and catalyst '*' produce chains of bonded links 'Ú' which eventually enclose the catalyst, thus demarcating an inside for the reactions of the substrate. (b) Four successive instants in the same runs as (a) showing the spontaneous regeneration of the boundary broken by decay of the bonded links. The ongoing production inside the unity makes it possible the reestablishment of the unity's boundary under changes of geometrical form and turnover of components.

simpler than the already intricate bacterial cell, satisfy the criteria of autopoietic organization? This question can be answered by simulation and analysis of a *minimal* autopoietic system. Minimal systems all seem to converge to the assumption of an enzyme-mediated polymerization reaction basis. Consider for example [5] a two-dimensional grid where two kinds of elements can move randomly: substrates ('O') and a few catalysts ('*'). A first transformation ('composition') produces new elements as monomers, which can link up in chains ('bonding'), until they decay, with a different kinetics than composition. Succinctly:

$$\text{composition:} \quad * + 2\,\mathrm{O} \rightarrow * + \boxed{\mathrm{O}}$$
$$\text{bonding:} \quad \boxed{\mathrm{O}}\!-\!\boxed{\mathrm{O}}\!-\!\boxed{\mathrm{O}}\!-\ldots + \boxed{\mathrm{O}} \rightarrow \boxed{\mathrm{O}}\!-\!\boxed{\mathrm{O}}\!-\!\boxed{\mathrm{O}}\!-\!\boxed{\mathrm{O}}\!-\ldots$$
$$\text{decay:} \quad \boxed{\mathrm{O}} \rightarrow 2\,\mathrm{O}$$

To visualize what these rules may generate, Fig. 1a shows a sequence of a simulation, with the spontaneous emergence of a polymer chain that loops onto itself. In this simulation we allow differential permeability through the polymer chain (i.e., null for the catalyst, very low for the monomers, high for the substrate), and an approximation to Brownian motion.

Interestingly, not only do such self-distinguishing units arise through these very simple rules, but their loopiness is endowed with a degree of self-regulation. This is seen in Fig. 1b, where a decay has occurred on a membrane segment. Given the ranges of differential diffusion and decay, the very boundary of the unity permits its own repair back into a unity, again distinct from its background.

Thus simulation is an instance of *emergent* unity, involving the parallel, distributed nature of the chemical-like processes, in common with many others examples such as Conway's well-known *Game-of-Life* [9] or more recent cellular automata classes examined in detail [10, 11]. The simulation above is an explicit attempt to produce a minimal autopoietic system, and in this sense is quite different from the other cellular automata cited above, which look for properties other than self-distinction, such as reproduction (in the case of Conway's) or spatio-temporal patterns (in the case of Wolfram's). The crucial question of emergent properties has just been touched on here; I shall return to it below.

What about such minimal autopoietic systems in actual chemical media, and hence as relevant for the origin of life? In fact, the encapsulation of macromolecules by lipid vesicles has been actively investigated as a promising candidate for an early cell [12, 13, 14]; see [15]. Luisi and Varela [16] make the case that a reverse micellar system can come close to the mark for being a minimal autopoeitic system. In particular, they discuss the case of a reverse micellar system hosting in its aqueous core a reaction which leads to the production of a surfactant, which is a boundary for the reverse micellar reaction. The interest of this case is that much is known about these chemical systems making it possible to actually put into operation a minimal autopoietic system.

The difference between environment and world is the surplus of *signification* which haunts the understanding of the living and of cognition, and which is at the root of how a self becomes one. It is quite difficult in practice to keep in view the dialectics of this mutual definition: neither rigid isolation, nor simple continuity with physical chemistry. In contrast, it is easy to conflate the unit's world with its environment since it is *so* obvious that we are studying this or that molecular interaction in the *context* of an autonomous cellular unit, and hence to miss completely the surplus *added* by the organism's perspective. There is no food significance in sucrose except when a bacterium swims upgradient and its metabolism uses the molecule in a way that allows its identity to continue. This surplus is obviously not indifferent to the regularities and texture (i.e., the "laws") that operate in the environment, that sucrose can create a gradient and traverse a cell membrane, and so on. On the contrary, the system's world is built *on* these regularities, which is what assures that it can maintain its coupling at all times.

What the autopoietic system does—due to its very mode of identity—is to constantly confront the encounters (perturbations, shocks, coupling) with its environment and treat them from a perspective which is not intrinsic to the encounters themselves. Surely rocks or crystal beads don't beckon sugar gradients out of all the infinite possibilities of physico-chemical interactions as particularly meaningful—for this to happen a perspective *from* an actively constituted identity is essential. It is tempting, at this point, to slide into some vaporous clouds about "meaning" reminiscent of the worst kind of vitalism of the past or informational jargon of the present. What I emphasize here is that what is meaningful for an organism is precisely given by its constitution as a distributed process, with an indissociable link between local processes where an interaction occurs (i.e., physico-chemical forces acting on the cell), and the coordinated entity which is the autopoietic unity, giving rise to the handling of its environment without the need to resort to a central agent that turns the handle from the outside—like an *élan vital*—or a pre-existing order at a particular localization—like a genetic program waiting to be expressed.

If we invert our perspective, this constant bringing forth of signification is what we may describe as a permanent *lack* in the living: it is constantly bringing forth a signification that is *missing*, not pre-given or pre-existent. Relevance must be provided *ex nihilo*: distinguish relevant from irrelevant molecular species, follow a gradient uphill and not downhill, increase the permeability to this ion and not to that one, and so on. There is an inevitable contretemps between an autonomous system and its environment: there is always something which the system must furnish from its perspective as a functioning whole. In fact, a molecular encounter acquires a significance in the context of the *entire* operating system and of many simultaneous interactions.

The source for this world-making is always the "breakdown" in autopoiesis, whether minor, like changes in concentration of some metabolite, or major, like disruption of the boundary. Due to the nature of autopoiesis itself—illustrated in the membrane repair of the minimal simulated example above—every breakdown can be seen as the initiation of an action on what is missing on the part of the system so that identity might be maintained. I repeat: no teleology is implied in this "so that": that's what the self-referential logic of autopoiesis entails in the first place. The action taken will be visible as an attempt to modify its world—change from place of different nutrients, increase in the flow of a metabolite for metabolic synthesis, and so on.

In brief, this permanent, relentless action on what is lacking becomes, from the observer side, the ongoing *cognitive* activity of the system, which is the basis for the incommensurable difference between the environment within which the system is observed, and the world within which the system operates. This cognitive activity is paradoxical at its very root. On the one hand the action that brings forth a world is an attempt to reestablish a coupling with an environment which defies the internal coherence through encounters and perturbations. But such actions, at the same time, demarcate and separate the system from that environment, giving rise to a distinct world.

The reader may balk at my use of the term 'cognitive' for cellular systems. As I said above, one of my main points here is that we gain by seeing the *continuity* between this fundamental level of self and the other regional selves, including the neural and linguistic where we would not hesitate to use the word 'cognitive'. I suppose others would prefer to introduce the word 'information' instead. Well, there are reasons why I believe this even more problematic. Although it is clear that we describe an **X** that perturbs from the organism's exteriority, **X** is not information. In fact, for the organism it only is a *that*, a *some*thing, a basic stuff to in-form from its own perspective. In physical terms there is stuff, but it is for nobody. Once there is body—even in this minimal form—it becomes in-formed for a self, in the reciprocal dialectics I have just explicated. Such in-formation is never a phantom signification or information bit, waiting to be harvested by a system. It is a presentation, an occasion for coupling, and it is in this *entre-deux* that signification arises [17, 18, 19].

Thus the term cognitive has two constitutive dimensions: first its *coupling* dimension, that is, a link with its environment allowing for its continuity as individual entity; second — by a slight misuse of language, I admit — its *imaginary* dimension, that is, the surplus of significance a physical interaction acquires due to the perspective provided by the global action of the organism. Many a reader will remain unconvinced, and my use of 'cognitive' and 'imaginary' may seem too much of an abuse of language. I beg those readers' indulgence, and to bear with me to the end of my argument and see if this trend of analysis may not have its merits.

III. SELF AS BASIC COGNITIVE SYSTEM

III.1 *Perception-action as Basic Neuro-logic*

In the previous Section, I have presented the fundamental interlock between identity and cognition as it appears for a minimal organism. In this Section I want to show how the more traditional level of cognitive properties, involving the brains of multicellular animals, is in some important sense the *continuation* of the very same basic process.

The shift from minimal cellularity to organism with nervous system is swift, and skips the complexity of the various manners in which multicellular organisms arise and evolve [20, 21, 22]. This is a transition in units of selection, and one that implicates the somatic balance of differentiated populations of cells in an adult organism, as well as crafty development pathways to establish a bodily structure. As Buss has stated recently: "The evolution of development is the generation of a 'somatic ecology' that mediates potential conflicts between cell and the individual, while the organism is simultaneously interacting effectively with the extrasomatic environment" [21]. Others address these questions in this volume .

For most vertebrates, this "somatic ecology" is bound together through the network of lymphocytes that constitute the core of the immune system. Again, a discussion of an immunological self is not my purpose here; it is also discussed by others in this volume. I cannot resist the temptation, nevertheless, to point out, for completeness sake, that elsewhere I have presented *in extenso* a network approach to the immune system and its role in the establishment of a flexible cellular/molecular self during the ontogeny of mammals [23]. In my view this identity is not, as traditionally stated, a demarcation of self as defense *against* the non-self of invading antigens [24, 25]. It is a self-referential, positive assertion of a coherent unity—a "somatic ecology"—mediated through free immunoglobulins and cellular markers in a dynamical exchange [26, 27]. Immune reactions against infections, although clearly important, are mediated by a "peripheral" immune system, a different sub-population of lymphocytes mobilized not through network but clonal expansion mechanisms, like a reflex reactivity acquired through evolution [28, 29]. But enough of this *excursus*. For my purposes here I will expeditiously assume the identity of a multicellular organism, distinctly different from an autopoietic minimal entity in its mode of identity, but similar in that it demarcates an autonomous entity from its environment [30].

Now, what's the specific place of the nervous system in the bodily operation of a multicellular? Whenever *motion* is an integral part of the lifestyle of a multicellular, there is a corresponding development of a nervous system linking effector (muscles, secretion) and sensory surfaces (sense organs, nerve endings). *The fundamental logic of the nervous system is that of coupling movements with a stream of sensory modulations in a circular fashion.* The net result are perception-action correlations arising from and modulated by an ensemble of intervening neurons, the *interneuron* network. Correspondingly, neurons are unique among the cells of a multicellular organism in their axonal and dendritic ramifications permitting multiple contacts and extending for large distances (relative to cellular soma sizes) providing the essential medium for this intra-organismic sensor-effector correlation.

Contrary to current habit, I wish to emphasize from the start the *situatedness* of this neuro-logic: the state of activity of sensors is brought about *most* typically by the organism's motions. To an important extent, behavior is the regulation of perception. This does not exclude, of course, independent perturbations from the environment. But what is typically described as a "stimulus" in the laboratory, a perturbation which is deliberately independent of the animal's ongoing activity, is less pertinent (outside the laboratory) for understanding the biology of cognition.

The perceptuo-motor coherencies we describe externally as behavior disguises the arising, within the interneuron net, of a large sub-set—an *ensemble* as is usually said—of transiently correlated neurons. These ensembles are both the source and the result from the activity of the sensory and effector surfaces. Consider for example *Aplysia*, a water mollusc with a "small" nervous system (a few thousand neurons) which was been intensively studied [31]. When *Aplysia* brings its siphon in touch with a surface (or the siphon is independently touched) it contracts its gill. This is the so-called gill-withdrawal reaction, one among many of the behavioral patterns normally present in these animals. In a recent study [32] voltage-sensitive dyes were

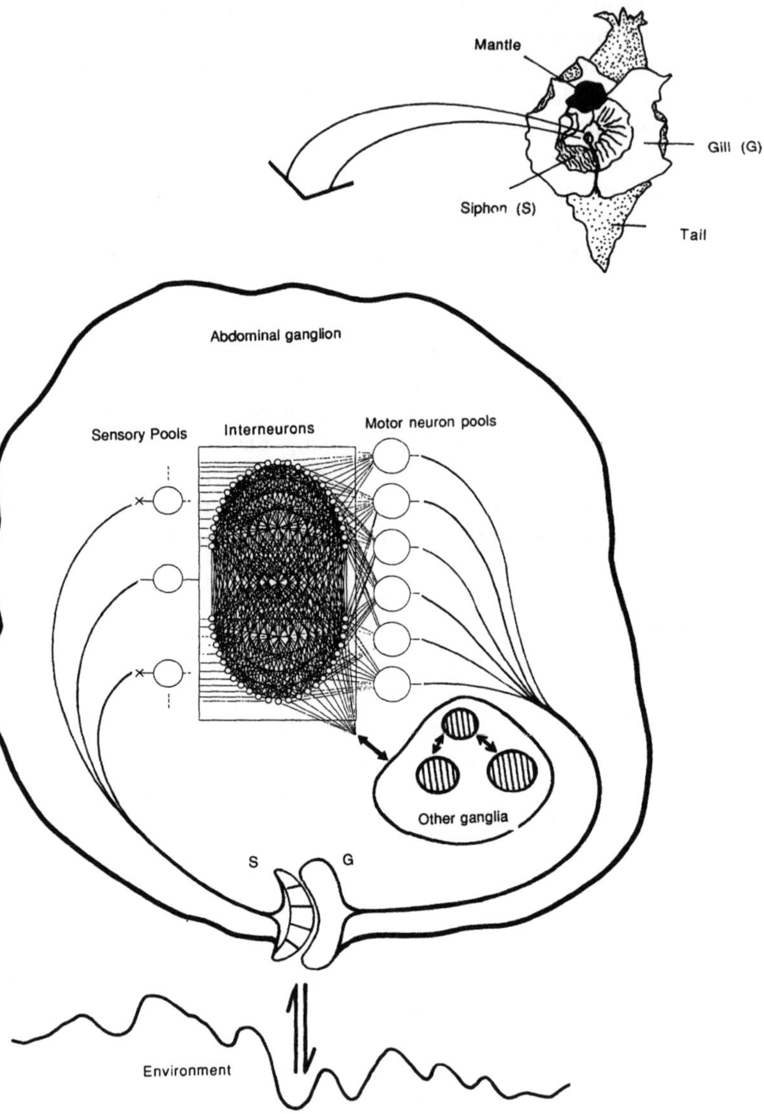
Mantle
Gill (G)
Siphon (S)
Tail
Abdominal ganglion
Sensory Pools
Interneurons
Motor neuron pools
Other ganglia
S
G
Environment

used to monitor a substantial part of the neurons in the abdominal ganglion of *Aplysia*. Contrary to the textbook image of a behavior as a reflex-arc pathway, the authors conclude than at least 300 neurons are active during this action, which represent a significant proportion of the whole interneuron network. The ensemble of neurons observed through this method arise in coordination and mutual influence, and their co-activation abates after a few seconds.

It is of course the case that those interneurons and motor neurons participating in this particular gill-withdrawal behavior, also participate in other ensembles underlying various other behaviors. Thus the interneurons of these invertebrate ganglia must be conceived as a network of overlapping ensembles which arise in various coherent configurations depending on the animal's context (Fig. 2).

The lessons from such humble molluscs carry over into the rest of animals with brains. What changes is the amount of mediating interneurons, and the specific architecture of the respective nervous system, containing various cortical regions, layers and nuclei. In humans some 10^{11} interneurons interconnect some 10^6 motorneurons which relate to 10^7 sensory neurons distributed in receptor surfaces throughout the body. This is a ratio of 10: 100,000: 1 of interneurons mediating the coupling of sensory and motor surfaces. The rise and decay of neuronal self-organization illustrated in the *Aplysia* siphon withdrawal is all the more valid in larger brains. Thus for instance John *et al* [33] find that 5–100 million neurons are active throughout the cat's brain during a simple visuo-motor task of pressing a lever. Such neural assemblies arise in a patchwork of regional areas, evincing the enormous distributed parallelism proper to vertebrate brains.

In fact, it is fair to say that a recently established fact of the brain's constitution is a *Law of Reciprocity*: if a region (say a cortical area, or a specific

←

Fig. 2. Schema of the basic organization of *Aplysia*'s abdominal ganglion, consistent with the finding that several hundred neurons are activated when the protruding siphon is brought to lightly touch a border. The sensory surface indicated correspond to mechanosensory cells with endings on the siphon skin (S). The motoneurons in the diagram are meant to indicate those innervating several muscles that regulate the gill's position (G). They receive direct (monosynaptic) connections from the sensory endings, but are fundamentally regulated by a pool of interneurons from the ganglia, whose activity patterns lead to various behaviors, such as gill-withdrawal correlative to the siphon's skin touching. Fifty interneurons are shown, each making contact with a quarter of the other interneurons; about 300 synaptic interactions are indicated. The entire sensory-motor loop is embedded in *Aplsya*'s body (upper right) in constant coupling with the envronment through its self-induced behavior. Adapted from Zecevic *et al.* (1989), p. 3688

The operational closure of the nervous system then brings forth a specific *mode* of coherence, which is embedded in the organism. This coherence is a *cognitive self*: a unit of perception/motion in space, sensory-motor invariances mediated through the interneuron network. The passage to cognition happens at the level of a behavioral entity, and not, as in the basic cellular self, as a spatially bounded entity. The key in this cognitive process is the nervous system through its neuro-logic. In other words the cognitive self is the manner in which the organism, through its own self-produced activity, becomes a distinct entity in space, but always coupled to its corresponding environment from which it remains nevertheless distinct. A distinct coherent self which, by the very same process of constituting itself, configures an external world of perception and action.

III.2 *Cognitive Self and Perceptual World*

The nature of the *identity* of the cognitive self just discussed is, like that of the basic cellular self, one of *emergence* through a distributed process. The emergent properties of an interneuron network are, however, enormously more rich, and merit further discussion at this point. What I wish to bring up here is the relatively recent (and stunning!) conclusion that lots of simple agents having simple properties may be brought together, even in a haphazard way, to *give rise* to what appears to an observer a purposeful and integrated whole, *without* the need for a central supervision. We have already touched on this theme when discussing the nature of the autopoietic process and cellular automata modelling, and later when discussing the constant arising and subsiding of neuronal ensemble underlying behavior. I wish at this point to address this issue more substantially, since it is crucial for my whole argument here. I base my conclusions on contemporary studies from various biology-inspired complex systems [2, 47, 48, 49, 50].

Let me turn to one of the best illustrations of such emergent behavior: insect colonies. They are rapidly becoming popular in recent studies of artificial life (AL) [51]. Their "superorganism" qualities remained a metaphor for a long time, until slowly retaken in the 1970's [52] with some detailed experiments which needed a global colony level for their explanation. For instance, some of the most illustrative experiments involve "sociotomy" followed by the extensive regulatory properties of the ensemble. After separating into a sub-colony, the most efficient nurses from a *Neoponera apicalis* colony radically changed their social status, becoming more prone to foraging and less involved in nursing. The contrary happened in the remaining colony: formerly low-level nurses increased their nursing activity. The whole exhibits evidence of some configurational identity with a memory, since the nurses return to their previous state when the sociotomy is undone by returning the excised element back to the colony [53]. Some behavior can be formalized with tools that are quite common for complex systems such as reaction-diffusion equations to account for the foraging behavior of ants [53]. Similar effects concern the spatial self-distinction of this colony and its ontogeny.

What is particularly striking in the insect colony case—unlike the brain—is that we readily admit that (i) its separate components are individuals and (ii) there is no center or localized self. Yet the whole behaves as a unit and for the observer it is as if there was a coordinating agent "virtually" present at the center. This is what I meant when referring to a *selfless* self—we could also speak of a virtual self: a coherent global pattern that emerges through simple local components, appearing to have a central location where none is to be found, and yet essential as a level of interaction for the behavior of the whole unity.

The import of such current models, formalisms and study cases of complex systems (i.e., emergent properties through coordinated simple elements) is, in my eyes, quite profound for our understanding of cognitive properties. It introduces an explicit alternative to the dominant computationalist tradition in the study of cognitive properties for which the central idea is that of processing an external information successively elaborated to reconstitute a centralized representation [55, 56]. This fundamental paradigm of the digital computer program will not do for biology, nor for AI.

This is, as we already discussed, evident for the neurobiologist concerned with brain. But it is also true for the researchers of artificial cognitive systems, as the current new "connectionist" schools have made clear by now [49, 57]. What we find in brains is a promiscuous tinkering of networks and sub-networks giving no evidence for a structured decomposition from top to bottom as is typical of a computer algorithm. Accordingly, one of the first messages from the study of artificial neural networks in modern connectionist terms is the absence of a principled distinction between software and hardware, or more, precisely between symbols and non-symbols. In fact, *all* we find in modern artificial neural network machines are relative activities between ensembles underlying the regularities we call their behavior or performance. We may see that some of these ensembles recur regularly enough to describe them as being program-like, but this is another matter. Although artificially built, such emerging ensembles cannot be called "computations" in the sense that their dynamics cannot be formally specifiable as the implementation of some high-level algorithm. Neural networks even in their fine detail are *not* like a machine language, since there is simply no transition between such elemental operational atoms with a semantics and the larger emergent level where behavior occurs. If there were, the classical computer wisdom would immediately apply: ignore the hardware since it adds nothing of significance to the actual computation (other than constraints of time and space). In contrast, in distributed network models these "details" are precisely what makes a global effect possible, and why they mark a sharp break with tradition in AI [58, 59]. Naturally this reinforces the similar conclusions that apply to natural neural networks in the brain, as we discussed before.

I have raised this point to caution the reader against the force of many years of dominance of computationalism, and the consequent tendency to identify the cognitive self with some computer program or high level computational description. This will not do. The cognitive self *is* its own implementation: its history and its action are of one piece. Now this demands that we clarify the second aspect of the self to be addressed: its mode of relation with the environment.

Ordinary life is necessarily one of *situated* agents, continually coming up with what to do faced with ongoing parallel activities in their various perceptuo-motor systems. This continual re-definition of what to do is not at all like a plan, stored in a repertoire of potential alternatives, but enormously dependent on contingency, improvisation, and more flexible than planning. Situatedness means that a cognitive entity has—by definition—a perspective. This means that it isn't related to its environment "objectively", that is independently of the system's location, heading, attitudes and history. Instead, it relates to it in relation to the perspective established by the constantly emerging properties of the agent itself and in terms of the role such running redefinition plays in the system's entire coherence.

Again, as we did for the minimal cellular self, we must sharply differentiate between environment and world. And again the mode of coupling is double. On the one hand, such body-in-space clearly happens through the interactions with the environment on which it depends. These interactions are of the nature of macrophysical encounters—sensory transduction, muscle force and performance, light and radiations, and so on—nothing surprising about them. However this coupling is possible only if the encounters are embraced *from the perspective* of the system itself. This amounts, quite specifically, to elaborating a *surplus signification* relative to this perspective.

Whatever is encountered must be valued one way or another—like, dislike, ignore—and acted on some way or another—attraction, rejection, neutrality. This basic assessment is inseparable from the way in which the coupling event encounters a functioning perceptuo-motor unit, and it gives rise to an *intention* (I am tempted to say "desire"), that unique quality of living cognition [60].

Phrased in other terms, the nature of the environment for a cognitive self acquires a curious status: it is that which *lends itself* (*es lehnt sich an*...) to a surplus of significance. Like jazz improvisation, environment provides the "excuse" for the neural "music" from the perspective of the cognitive system involved. At the same time, the organism cannot live without this constant coupling and the constantly emerging regularities; without the possibility of coupled activity the system would become a mere solipsistic ghost.

For instance, light and reflection (among many other macrophysical parameters such as edges and textures, but let us simplify for the argument's sake), lend themselves to a wide variety of color spaces, depending on the nervous system involved in that encounter. During their respective evolutionary paths, teleost fishes, birds, mammals, and insects have brought forth various different color spaces not only with quite distinct behavioral significance, but with different dimensionalities so that it is not a matter of more or less resolution of colors [61]. Color is demonstrably not a property that is to be "recovered" from the environmental "information" in some unique way. Color is a dimension that shows up only in the phylogenetic dialogue between an environment and the history of an active autonomous self which partly defines what counts as an environment. Light and reflections provide a mode of coupling, a perturbation which triggers, which gives an occasion for the enormous in-formative capacity of neural networks for constituting sensori-motor correlations and hence to put into action their capacity for imagining and presenting. It is only *after* all this has happened, after a mode of coupling becomes regular and repetitive, like colors in ours—and others'—worlds, that we observers, for ease of language, say colors correspond or represent an aspect of the world.

A dramatic recent example of this surplus significance and the dazzling performance of the brain as the generator of neural narratives is provided by the technology of the so-called "virtual realities" [62]. A helmet fitted with two cameras over both eyes and a glove or suit with electrical transducers for motions are linked, not through the usual coupling via the environment, as we are familiar, but through a computer. Thus each movement of hand or body correlates to images through correlation principles that can be chosen at will.

For example each time my hand, which appears as a "virtual" iconic hand in my image, points to a place; the image that follows corresponds to flying to the place pointed at. Visual perception and motions thus give rise to regularities which are proper to this new manner of perceptuo-motor coupling. What is most significant for me here is the *veracity* of the world which rapidly springs forth: we *inhabit* a body within this new world after a short time of trying this new situation (i.e., 15 minutes or so), and the experience is of truly flying through walls or of delving into fractal universes. This is so in spite of the poor quality of the image, the low sensitivity of the sensors, and the limited amount of interlinking between sensory and image surfaces through a program that runs in a personal computer. Through its closure, the nervous system is such a gifted synthesizer of regularities, that any basic material suffices as an environment to bring forth a compelling world.

This very same strategy of the situatedness of an agent which is progressively endowed with richer internal self-organizing modules is becoming a productive research program even for the very pragmatically oriented field of artificial intelligence [63, 64, 65]. To quote R. Brooks, one of the main exponents of this tendency at some length:

> I... argue for a different approach to creating Artificial Intelligence:
>
> - We must incrementally build up the capabilities of intelligent systems at each step of the way and thus automatically ensure that the pieces and their interfaces are valid.
> - At each step we should build complete intelligent systems that we let loose in the real world with real sensing and real action. Anything less provides a candidate with which we can delude ourselves.
>
> We have been following this approach and have built a series of autonomous mobile robots. We have reached an unexpected conclusion (**C**) and have a rather radical hypothesis (**H**).
>
> **C**: When we examine very simple level intelligence we find that explicit representations and models of the world simply get in the way. It turns out to be better to use the world as its own model.
>
> **H**: Representation is the wrong unit of abstraction in building the bulkiest parts of intelligent systems.
>
> Representation has been the central issue in Artificial Intelligence work over the last 15 years only because it has provided an interface between otherwise isolated modules and conference papers. [65, p. 1]

When the synthesis of intelligent behavior is approached in such an incremental manner, with strict adherence to the sensory-motor viability of an agent, the notion that the world is a source of information to be represented simply disappears. The autonomy of the cognitive self comes fully in focus. Thus in Brooks's proposal his minimal creatures join together various activities through a rule of cohabitation between them. This is homologous to an evolutionary pathway through which modular sub-networks intertwined with each other in the brain. The expected results are more truly intelligent autonomous sense-giving devices, rather than brittle informational processors which depend on a pre-given environment or an optimal plan.

It is interesting to note that in this paper Brooks also traces the origin of what he describes as the "deception of AI" to the tendency in AI (and in the rest of cognitive science as well) to abstraction, i.e., for factoring out situated perception and motor skills. As I have argued here (and as Brooks argues for his own reasons), such abstraction misses the essence of cognitive intelligence, *which resides only in its embodiment.* It is as if one could separate cognitive problems into two parts: that which can be solved through abstraction and that which cannot. The second is typically perception-action and motor skills of agents in unspecified environments. When approached from this self-situated perspective there is no place where perception could deliver a representation of the world in the traditional sense. The world shows up through the enactment of the perceptuo-motor regularities. "Just as there is no central representation there is no central system. Each activity layer connects perception to action directly. It is only the observer of the Creature who imputes a central representation or central control. The creature itself has none: it is a collection of competing behaviors. Out of the local chaos of their interactions there emerges, in the eye of the observer, a coherent pattern of behavior" [64, p. 11].

To conclude, the two main points that I have been trying to bring into full view in this Section devoted to the cognitive self are as follows: first, I have tried to spell out the nature of its *identity* as a body in motion-and-space through the operational closure of the interneuron network. This activity is observable as multiple sub-networks, acting in parallel and interwoven in complex *bricolages*, giving rise again and again to coherent patterns which manifest themselves as behaviors. Secondly, I have tried to clarify how this emergent, parallel and distributed dynamics is inseparable from the *constitution of a world*, which is none other than the surplus of meaning and intentions carried by situated behavior. If the links to the physical environment are inevitable, the uniqueness of the cognitive self is this constant genesis of meaning. Or, again to invert the description, the uniqueness of the cognitive self is this constitutive *lack* of signification which must be supplied faced with the permanent perturbations and breakdowns of the ongoing perceptuo-motor life. Cognition is action about what is *missing*, filling the fault from the perspective of a cognitive self.

the animal is "poor in its worldliness" (*Weltarm*) [70]. I have elaborated on this double-perspectival approach to human experience and cognitive mechanisms extensively elsewhere [71].

Needless to say, to do justice to these issues would take us too far afield. As was the case for the immunological self/non-self discrimination, I am providing what is bound to be a frustratingly sketchy picture open to all kinds of objections both theoretical and empirical. My purpose for bringing up this issue of the self as 'I' nevertheless is to emphasize the *continuity* of the same motif that we discussed at greater length for the cellular and basic cognitive selves. Like a fractal, this motif is repeated over and over again for the various regional selves of the organism, even if the articulation between them is far from being satisfactory as yet. But in my eyes it is very important to see how this motif is shared from our most intimate and immediate everyday experience right down to the very basic levels of life and body. Only then can we avoid splitting the selves in an organism into disjointed categories and thus avoid splitting what is a totality ranging from cells to social minds into separate pulverized realms.

IV.2 *Organism's Double Dialectics*

Organism as self, then, cannot be broached as a single process. We are forced to discover "regions" that interweave in complex manners, and, in the case of humans, that extend beyond the strict confines of the body into the socio-linguistic register.

Further, what I have argued is that behind this meshwork of the various selves we carry around, all of these selves share a common and fundamental logic while differing in their specificity. This is a case of what Wittgenstein would have called "family resemblances": rather than any characteristic being common to all instances, we deal with a *cluster* of overlapping characteristics [72]. We may also speak of this cluster of common characteristics as a shared *dialectic*, since we are dealing here with double-sided process, where co-definition is at the core of the matter. In fact, I submit that the organismic dialectic of self is a two-tiered affair: we have on the one hand the dialectics of identity of self; on the other hand the dialectics through which this identity, once established, brings forth a world from an environment. Identity and knowledge stand in relation to each other as two sides of a single process that forms the core of the dialectics of all selves.

First, a *dialectics of identity* establishes an autonomous agent, a *for* itself (*pour soi*). This identity is established through a bootstrapping of two terms:

(i) a *dynamical* term which refers to an assembly of components in network interactions and which are capable of emergent properties: metabolic nets, neural assemblies, clonal antibody networks, linguistic recursivity;
(ii) a *global* term which refers to emerging properties, a totality which conditions (downwardly) the network components: cellular membranes, sensory-motor body in space, self/non-self discrimination, personal 'I'.

These two terms are truly in a relation of co-definition. On the one hand the global level cannot exist without the network level since it comes forth through it. On the other hand the dynamical level cannot not exist and operate as such without it being contained and lodged into an encompassing unity which makes it possible.

Second, a *dialectics of knowledge* establishes a world of cognitive significance *for* this identity. This can only arise from the perspective provided by this identity, which adds a surplus of significance to the interactions of the environment proper to the constituting parts.

The key point, then, is that the organism brings forth and specifies its own domain of problems and actions to be "solved"; this cognitive domain does not exist "out there" in an environment that acts as a landing pad for organisms that somehow drops or is parachuted into the world. Instead, living beings and their worlds of meaning stand in relation to each other through *mutual specification* or *co-determination*. Thus what we describe as significant environmental regularities are not external features that have been internalized, as the dominant representationalist tradition in cognitive science — and adaptationism in evolutionary biology — assumes. Environmental regularities are the result of a conjoint history, a congruence which unfolds from a long history of co-determination. In Lewontin's [73] words, the organism is both the subject and the object of evolution.

This second tier of the organism's dialectics, then, is also established through the bootstrapping of two terms:

(i) a *significance* term which refers to the necessary emergence of a surplus meaning proper to the perspective of the constituted self: cellular semantics, behavioral perception and action, self/non-self as somatic assertion, personal identity,
(ii) a *coupling* term which refers to the necessary and permanent embeddedness and dependency of the self on its environment, since only through such coupling can its world be brought forth: physico-chemical laws for the cellular world, macroscopic physical properties for cognitive behavior, molecular interaction for immune self, socio-linguistic exchanges for our subjective selves.

35. Steriade, M. and M. Deschenes (1985), The thalamus as a neuronal oscillator, *Brain Res. Rev.* **8**: 1–63.
36. Varela, F. and W. Singer (1987), Neuronal dynamics in the visual cortico-thalamic pathway revealed through binocular rivalry, *Exp. Brain Res.* **66**: 10–20.
37. Horn, G. and R. Hill (1968), Modifications of receptive fields of cells in the visual cortex occurring spontaneously and associated with bodily tilt, *Nature* **221**: 186–188.
38. Fishman, M.C. and P. Michael (1973), Integration of auditory information in the cat's visual cortex, *Vision Research* **13**: 1415–1419.
39. Morell, F. (1972), Visual system's view of acoustic space, *Nature* **238**: 44–46.
40. Allman, J., F. Meizen, and E. McGuiness (1985), Non-classical receptive field properties, *Ann. Rev. Neuroscien.* **8**: 407–430.
41. Abeles, M. (1984), *Local Cortical Circuits*, Springer Verlag, Berlin.
42. DeYoe, E. and D.C. Van Essen (1988), Concurrent processing streams in monkey visual cortex, *Trends Neurosci.* **11**: 219–226.
43. Minsky, M. (1987), *The Society of Mind*, Simon and Schuster, New York.
44. Llinás, R. (1988), The intrinsic electrophysiological properties of mammalian neurons: insights into central nervous system function, *Science* **242**: 1654–1664.
45. Gevins, A., R. Schaffer, J. Doyle, B. Cutillo, R. Tannehill, and S. Bressler (1983), Shadows of thought: shifting lateralization of human brain electrical patterns during a brief visuo-motor task, *Science* **220**: 97–99.
46. Varela, F., A. Toro, E. John, and E. Schwartz (1981), Perceptual framing and cortical alpha rhythms, *Neuropsychologia* **19**: 675–686.
47. Goodwin, B. and P. Saunders (Eds.) (1989) *Theoretical Biology: Epigenetic and Evolutionary Order from Complex Systems*, Edinburgh University Press, Edinburgh.
48. Farmer, J., A. Lapedes, N. Packard, and B. Wendroff (Eds.) (1986), *Evolution, Games and Learning*, North-Holland, Amsterdam.
49. McClelland, J. and D. Rummelhart (1986), *Parallel Distributed Processing: Studies on the Microstructure of Cognition*, 3 vols., MIT Press, Cambridge.
50. Wolfram, S. (1984), Cellular automata as models for complexity, *Nature* **311**: 419–424.
51. Pasteels, J. and J. Deneubourg (1987), *From Individual to Collective Behavior in Social Insects*, Birkhäuser, Basel.
52. Wilson, E.O. (1971), *The Insect Societies*, Harvard Univ. Press, Cambridge.
53. Fresnau, D. and J. Lachaud (1985), La régulation sociale sociale: donnés préliminaires sur les facteurs individuels controlant l'organisation des taches chez *Neoponera apicalis, Actes Coll. Insects Sociaux* **2**: 185–193.
54. Deneubourg, J., S. Aron, S. Goss, J. Pasteels, and G. Duerinck (1986), Random behavior, amplification processes and number of participants: how they contribute to the foraging properties of ants. *In*: Farmer *et al.* (Eds.), *Evolution, Games and Learning*, *op. cit.*
55. Newell, A. (1980), Physical symbol systems, *Cognitive Scien.* **4**: 135–183.
56. Plyshyn, Z. (1984), *Computation and Cognition: Toward a Foundation for Cognitive Science*, MIT Press, Cambridge.
57. Grossberg, S. (1984), *Studies of Mind and Brain*, D. Reidel, Boston. [Boston Studies in the Philosophy of Science, vol. 70].
58. Smolensky, P. (1988), On the proper treatment of connectionism, *Beh. Brain Sci.* **11**: 1–74.
59. Dennett, D. (1990a), Mother nature versus the walking encyclopedia: A western drama, in: Ramsey, S., D. Rummelhart, and S. Stich (Eds.), *Philosophy and Connectionist Theory*, (forthcoming).
60. Dennett, D. (1987), *The Intentional Stance*, MIT Press, Cambridge.

61. Thompson, E., A. Palacios, and F. Varela (1991), Ways of coloring: Comparative color vision as a case study in cognitive science, *Beh. Brain Sci.* In press.
62. Fisher, S. (1990), *In*: G. Hattinger (Ed.), *Virtuelle Welten*, Linz.
63. Agree, Ph. (1988), *The Dynamic Structures of Everyday Life*, Report No. AI-TR 1085, MIT Artificial Intelligence Lab., Cambridge.
64. Brooks, R.A. (1986), Achieving artificial intelligence through building robots, *A.I. Memo 899*, MIT Artificial Intelligence Laboratory, May 1986 .
65. Brooks, R.A. (1987), Intelligence without representation, MIT Artificial Intelligence Report, Cambridge, MA.
66. Dennett, D. (1990b), Review of Ch. Langton (Ed.), *Artificial Life, Biology Philos.* In press.
67. Humphreys, N. and D. Dennett (1989), Speaking for ourselves: An assessment of multiple personality disorder, *Raritan* **9**: 68–98.
68. Dupuy, J.-P. and F. Varela (1990), Understandings of Origins. *In*: Varela, F. and J.P. Dupuy (Eds.), *Understanding Origins: Contemporary ideas on the genesis of life, mind and society*, Kluwer, Boston, In press.
69. Merleau-Ponty, M. (1952), *Phenomenologie de la Perception*, Gallimard, Paris.
70. Heidegger, M. (1983), *Die Grundbegriffe der Metaphysik, Gesamtausgabe* t.29/30, Klostermann, Frankfurt.
71. Varela, F., E. Thompson, and E. Rosch (1991), *The Embodied Mind: Cognitive Science and Human Experience*, MIT Press, Cambridge.
72. Wittgenstein, L. (1972), *Philosophical Investigations*, Basil Blackwell, Oxford.
73. Lewontin, R. (1983), The organism as the subject and object of evolution, *Scientia* **118**: 63-82.

of integrity. Thus the theory considering immune activity as protective implies a self-referential (self-determined) nature.

The autonomous network theory makes self-reference explicit. In this approach, immunity is considered not so much as an activity protecting organismic integrity but rather as *constituting* that integrity. From this point of view, integrity is formed by self-referential (self-determined) processes. We consider an important advantage of the autonomous network construction that it has made the self-determination concept the central intellectual issue in formulating the problem of the immunological Self. In either case however, we believe the concept of self-determination is accepted uncritically and carries connotations provided by an ancient intellectual tradition. In our analysis, we attempt to explicate this infrastructure and discuss its limitations relative to its application to the problem of the immunological Self. We then present an alternative approach to the problem of immunological self-determination. Our goal is not to judge the comparative values and virtues of each approach, nor are we concerned with assertions as to which is right or wrong. We are dealing with the possible modes of formulating theoretical questions, not with ready-formulated answers.

Our position may be summarized as follows: First, we present the controversial problem of the immunologic Self within the complementary context of the clonal selection and the autonomous network theories. Second, we accept the claim of the latter to reconsider the immune system as self-determined (self-referential), but we believe that the very concept of a self-determined system must be critically reconsidered. Third, we reformulate the problem of the immunological Self from the perspective of this reexamined concept of a self-determinated system.

There is of course a certain incongruity to our criticism, in that we are only proposing a tentative approach to formulating the problem. Although we would hope that our construction of the issue may help develop new research, at the same time, the theories subject to our critical analysis have resulted in well-functioning and highly productive experimental programs, and any serious philosophical criticism immediately pays its debt to them. But without this kind of incongruity, no new research program would appear; a serious discussion of the intellectual nucleus of a research program (to the extent that the discussion is not an external "evaluation") is in itself a sign of maturity of the research and its theory. Immunology in the late twentieth century has fully enough developed its intellectual basis to deal with reflection upon its own assumptions [1].

Our critique begins by considering the relation between the clonal selection theory and the autonomous network program, where we attempt to demonstrate that each shares the same basic assumption in dealing with the concept of organismic identity (the former, implicitly and the latter explicitly). The assumption is formulated by the proponents of the autonomous approach as opposition of *operational closure* to *interactional openness*. Although this opposition appears to express a radically different idea from that underlying the clonal selection theory, we will argue that it reflects that understanding of organismic integrity (and self-determination) which is equally shared by both approaches. In this interpretation, the concept of self-determination actually implies that dependence on the external environment, expressed in the concept of interactional openness, is compensated by operational closure. We believe that this assumption excludes any possibility to understand how a system may positively define the Other (the system's nonself) in terms of its own activity (in terms of its own functions). In these approaches, self-determination of the system would be inevitably reduced to activities to master the Other, which in turn is presented on this functional level only negatively, i.e., unknown by the system (not having been mastered yet) and inducing disturbances in the mastering (viz., immunological) activity. In this case, the discrimination between Self and nonself can never be the responsibility of such a system, because it would lack definitions of the "what-have-not-been-mastered-yet" (i.e., the Other). Such a discrimination would rather reflect a point of view of an external observer rather than offer a true discriminating capacity of the system in question. Thus in order to formulate a system as self-determined (and because of this as discriminating itself from nonself), we are forced to consider the system rather in terms of operational openness than in terms of operational closure. But is such a reinterpretation compatible with the very idea of organismic integrity? In order to consider an organic system in terms of self-determination, are we then going to destroy the very idea of organismic identity?

To answer these questions we then consider in the second portion of the essay the concept of individuality (organismic integrity) as not necessarily coinciding with the concept of organismic identity. This problem has been carefully formulated by Leo Buss showing that individuality does not coincide with genetic homogeneity. We attempt to rethink this result in order to demonstrate how positive definitions of both Self and nonself (their complementarity and discrimination) may be obtained in terms of cellular interactions and cellular differentiation. Of course, in this part of our study, we are mainly concerned with comprehension of potentials that are provided by a new approach to the problem of immunological Self-nonself discrimination in this reformulation of the concept of individuality.

In the final portion of our study, we tentatively attempt to apply this concept of nonself as positively defined in cellular interactions to specifically those of the lymphoid system. We hypothesize that the two classes of cell surface molecules encoded by the mixed histocompatibility complex (MHC) locus provide the very complementary of self-determination required for immune identification, that is those necessary for lymphoid cellular interactions corresponding to the positive definition of both Self and nonself. Our notions are still in an early phase of formulation, but the basic outline of the argument is offered as a preparatory statement of the recognition process, where the "other" is incorporated into the "self," and the Self, redefined by that encounter. This essentially dialectical encounter is defined by the metaphysical contraints of "recognition" and the problematic of integrating the Self-Other confrontation.

II. A CRITIQUE OF THE CLONAL AND AUTOPOIETIC THEORIES OF IMMUNITY

II.1 *The Problem of Self-Determination*

A powerful theoretical opposition to the classical perspective of clonal selection has been elaborated by an autopoietic more recently called an "autonomous network") approach to immunology [2]. The proponents of this latter theory attack several weak assumptions of the clonal perspective: First, current immunological theory, although carefully eluding discussion of the cognitive issues (e.g., suggesting that the term "recognition" is used exclusively as a metaphor for designating the physical affinity of certain molecular structures), actually confers cognitive properties to entities (single cells or molecules) which cannot have such features. The cognitive properties could belong only to a biological system which "may compare its previous and present experiences, and, therefore, *know* about differences" [2, 1984]. Second, the paradox of classical immunology is conditioned by its central notion of *immune responses*. The notion implies, on the one hand, knowledge of Self, because responses to *foreign* materials requires the discrimination between Self and non-Self. On the other hand, to avoid immunological self-destruction, the same notion of immune responses requires a certain ignorance of the Self [2]. Third, in the classical formulation, each immune response is thought of as an essentially independent activity of a particular clone(s); this activity is allegedly driven exclusively by external antigens. Although the response includes "regulatory" cells, but in lineal thinking, very common in occidental civilization [3], ... the idea of control is vague, and it considers exclusively influences which are "external" to the system that is being controlled.... The opposite of (external] control, however, is not necessarily lack of control; it may also be autonomy: self-determination [2, 1984].

Autonomy refers to the circular-casual mechanisms that embodies the concept of feedback; in this respect, it opposes *allonomy*, which represents rules imposed from outside the system.

According to these critics [4, 5], it was Jerne's ideotypic network theory [6, 7] that allowed applying to immunology the concept of *autopoiesis* (self-creation), or perhaps more accurately, autonomy — a design capable of revealing a basic organizational feature of living systems.

The publication of the network theory marks the beginning of the end of nearly 100 years of "horror autotoxicus." Self-nonself discrimination is no longer the central question in immunology; it is no longer even a question [5].

Generation of antibody diversity is primarily oriented not outward, but inward, and is primarily a necessary development of the lymphoid system itself [1]. With this perspective, "the lymphoid system is not an artifact to process foreign materials, but a self-reference unit, which maintains reference levels that are internally generated" [2, 1984].

What in the immune system so seductively allows for interpreting it as self-determinative? Of course, the immunological literature offers substantive data to support autopoiesis or autonomy, which may be considered as consistent with a self-determinative approach. But often, the same arguments (which we will henceforth refer to as the Paris School) could be easily adopted to more classical interpretations. For instance, a normal adult mouse conventionally bred and fed, contains $1-1.5 \times 10^6$ immunoglobulin-secreting cells; but germ-free mice, fed on chemically defined, low molecular weight diets, contain comparable numbers of secreting cells ($0.5 - 1 \times 10^6$) [8]. From the Paris School perspective, these facts are interpreted to argue that "the immune system is turned inward" and that "the larger part of this activity... is truly independent of environmental stimulation" [5]. But can we not imagine a structure whose foundation for some reason is much larger than that required for the upper part of the construction? (Could not the same logic force us—taking into consideration that each paratope constitutes only a small part of its antibody—to doubt that the functional role of paratopes is turned outward, i.e., toward antigens, and by this reason to argue against the definition of paratope as the antigen-binding site of antibodies?)

Of course the Parisian perspective has led to a well-organized and highly productive scientific program; its arguments against the classical clonal theory developed from questions and puzzles arising in experiment, theoretical meditations and mathematical modeling [9]. But reading the Paris School literature, we suspect that the main source of its inspiration is not so much facts allegedly contradicting the clonal theory, but a metaphysical [1] dissatisfaction with "lineal thinking." [3] It appears that a certain intuition insists that living systems do not behave in a simple "disturbance — responsive" way, for organic systems must have their own "inner life," their "for-itself-being". Surely, the sources of this intuition may be traced to a classical intellectual tradition (beginning with Plato and Aristotle and having reached its acme with Fichte and Hegel), which interprets different forms ("levels") of cognition or perception as mere manifestations of different forms ("levels") of self-responsibility (self-determinedness, self-sufficiency). In this tradition, the notion of recognition implies the concept of self-determination. External determination (determination by the Other) being taken as such, cannot be sufficient for constituting the phenomenon of recognition, because the latter presupposes a retention of the Other in its "otherness," and not simply some changes in the status of the recognizing system. To be changed does not necessarily mean to recognize. A recognizing system must assume responsibility for the recognition-connected changes in the system itself, because by means of those changes it retains the "whatness" of the Other. This Self-responsibility for its own changes is the true meaning of the concept of self-determination.

The most powerful temptation affecting the classical tradition throughout its history was to interpret self-responsibility (self-determination) as liberation from the Other, i.e., to interpret retention of the "whatness" of the Other as just a way to subject the Other to itself, to liberate itself from the Other by control over "physical" interactions with the Other. The logical subsequent step on the road towards this liberation was to free the Self from the tyranny of the "whatness" of the Other. Such liberation was accomplished (no later than) by Plato: images of external objects (the "whatness" of the Other) are not incorporated from the outside, but are expressions (images) of the internal activity provoked by its meetings with external reality. (The process of obtaining knowledge is only to recollect the primary images.) In this philosophic tradition, each image is a particular mode of self-determination performed in the presence of a particular Other (i.e., in "physical" interactions with the Other). By means of creating an image, the system internalizes the Other, introduces its "otherness" as an essential aspect of the system's self-determination. But in the assumption that self-determination is a liberation (on a functional, but not an interactive level) from the Other, the image changes its status of an internalized "other" and assumes a new status — it is the Self as externalized for itself: *image is that mode of Self under which Self relates to itself.* It then follows that all *metaphysical* assumptions of the Parisian approach to immunology are readily presented in this ancient philosophical tradition — the problem of cognition is reduced to the problem of self-determination, and self-determination is interpreted as "essential" liberation from the Other; the "products" of cognitive activity (the images) are interpreted not as primary reflections of the Other, but as modes of self-presentation, i.e., as modes and aspects of self-determination.

Outlining these metaphysical implications of the Paris School, we do not mean to belittle the philosophical significance of the approach, or to deny its scientific productivity for immunology. Our intention is to explicate its semantic context, in order to realize its potentials and define its restrictions. By defining the context in which both theories (i.e., clonal selection and autopoietic or autonomous network) function, allow us to realize that their alleged opposition is rather a kind of complementarity, which implies the sharing of some common, basic features [9]. First of all, both approaches (despite protestation) do accept Self-nonself discrimination as the central question of immunology. The real difference between them is *how* to formulate the question, i.e., what meaning is implied in the concept of discrimination.

Beginning with Erhlich's formulation of the side-chain hypothesis, production of antibodies was considered as adaptive responses to external antigen [10]. Thus, the object of immunological activity is not Self, but the Other. The activity is thought not as being constitutive in relation to *what* is Self, but rather as eliminating what is not self, i.e., a protective activity. From this perspective, production of antibodies is viewed as the process by which the organism defines what is the Other. The *positive* definition of the Other transforms it into the *negation* of its existence. But the same processes, being "existentially" positive in relation to Self (i.e., being protective), then define Self negatively. Otherwise, the destruction of the organism would ensue. This intellectual position forced Erhlich, as well as Burnet [11] many decades later, to seek escape from "horror autotoxicus" in the hypothesis of an incomplete antibody repertoire. Blanks in the repertoire represent the negative definition of Self. Thus, in the model, on the one hand, the Other is *essentially* (in its presentation, i.e., in the definitions provided by the immune system) retained and *existentially* eliminated; on the other hand, the Self is

activity performs its self-referential and self-determined functions. The clonal selection theory is, correspondingly, indicted for its non-Platonic nature, i.e., for its intuitive assumption of cognition as positive *representations* of the Other. But it seems that the Paris School ignores the intellectual assumptions posturing their critical relation to the clonal selection theory, a reproduction of an ancient philosophical debate. They rather present an opposition between a view of cognition newly elaborated in science and the old view of cognition.

...As soon as an immune network appears, its own self-assertion creates a clear demarcation of a cognitive domain, and hence, an entire series of discriminations between self and non-sense, proper and inadequate, friendly and foe.

This kind of enactive cognition, so clearly seen in immune networks, has to be contrasted with our usual view of cognition as being a more or less accurate representation of a world already full of signification, and where the system picks up information to solve a given problem, posed in advance (Varela, 1979, 1987). This is the understanding of immune cognition for the antigen-driven clonal selection view of the immune system, where adequate operation means optimal defense of invaders as out-there [2, 1988].

But scientific results which are considered as supporting the autonomous network (autopoietic) approach pertain rather to the "network" hypothesis, than to the concept, "autonomous." To establish that neonatal antibodies frequently react with each other, that there are correlations between connectivity and the phenomenon of tolerance (including modeling of that correlation by computer simulation of the immune network), that these is evidence of a functional idiotypic network among natural antibodies, that certain syndromes of immunodeficiency are associated with autoimmune pathology, or the therapeutic effect on autoimmune diseases of large dose infusion of normal circulating immunoglobulin, together strongly support the idea of an immune network, and, correspondingly, the concept of the immune system as self-determined [9]. But in no way do these results support that specific metaphysical interpretation of self-determination as functional closeness that underlies the autopoietic approach.

II.2 *The Self and the Other*

We doubt that the autonomous network alternative has truly accomplished explication of the metaphysical (and epistemological) potentials of the concept — "self-determination," which might provide new options for immunological theory. We agree with the autopoietic assertion that beyond the concept of recognition there is the concept of self-determination.

Although, apparently unlike the Paris School proponents, we see in their interpretation a continued fidelity to the classical tradition of thought described above. And those who wish to be solely occupied with "mechanism," must acknowledge that these very mechanisms are required to explain the phenomenon of discrimination and thus, to answer the question presented by the phenomenon — what is immunological recognition? At the same time, we believe that the identical critical arguments which the Parisian School addresses to the clonal theory can be reflected back onto their own interpretation of immunity. In the classical clonal perspective, immune activity generates definitions of the Other in immunological terms, i.e., in the terms of paratopes and lymphocyte receptors. In these terms, the Other (non-self) is the *mastered* Other, i.e., it is non-self which is drawn into Self. But because the model does not provide an interpretation of processes that positively define Self, there is no way to say what (within the sphere of this non-self as drawn into Self) is Self and what is non-self. What is, and what is not, non-self is defined not by the system, but by external judgement of the observer who *knows* (judges) that a given element of the system was generated in response to an antigen incorporated from the outside. Does the system producing, for example, antibodies to antibodies, treat Self as non-self, or is it just a "regulatory" mechanism in the immunological treatment of the Other? After all, the word "regulation" does not solve the question—what is treated by such a "regulatory" activity? Is it the Other presented in forms of Self, or the Self presented as the Other of the activity (i.e., as the Other for itself)? The former seems nonsensical, because the objects of the process are self-elements as images of non-self. But how would the latter be possible? By definition, Self is presented only by the blanks in the antibody repertoire. But the same negative dialectical result can be expected in the case of the autonomic network interpretation of immune activity. In this interpretation, immunity is that kind of activity which provides positive definitions of Self and (at best) negative definitions of non-self (non-self as such has no meaning for Self). Naturally, we can expect that the conceptual presentation of the boundaries of Self (conceptual presentation of the *what* of the Self) must be blurred in the model, in the same way as the boundaries of non-self has been obscured in the classical clonal model.

In spite of its dynamic nature, one of the basic intuitive assumptions of the Paris School approach is a "spatial" mode to discriminate the concepts of Self and non-self, i.e., not as two aspects of the same dynamic reality, but rather as two "things" having different (spatially separated) origins—one within the organism's boundaries and another originating from the environment. There is ample evidence of the immune system's activity in the absence of environmental challenge [8, 14]. *From the autonomous network point of view, the environmentally independent activity is the immunological Self* [2]. Here, non-self is referred to as "environmental," i.e., as spatially external. Correspondingly, Self is understood as the external opposition to this externally defined non-self, i.e., Self is understood as independent from the environmentally provided Other, or, more precisely, the content of the concept "self" is here actually equivalent to the intuition—"independence—from." Here we see the logical difficulty that autopoiesis cannot escape.

We agree that the only adequate definition of Self is self-definition—Self must be understood as that which constitutes (defines) itself. But it is not obvious then that this self-definition must be free of (or possible without) defining the Other, which is provided, not by an external observer, but by the very process of self-definition. In the latter case, definition of the Other is not something externally discriminated from Self (by external observation), but is a necessary component of the "internal life" of Self, and is permanently produced by Self. Otherwise, what forces us to admit that the studied structures and functions are pertaining to the activity of self-determination, self-recognition, self-definition? Only one thing—our hope to demonstrate that those structures and functions provide *independence from the Other*. Self is that which is not dependent on the Other. The model promises a positive definition of Self, and negative definition of non-self. But the autonomic network theory is pushed (by its own metaphysical implications) exactly in the opposite direction. Understanding the Self as "independence-from-non-self" is in fact dependent on the very understanding of the non-self. This ambiguity of non-self never ceases to follow autonomic reasoning—the Other (non-self) of the immune system is for the latter, on the one hand, nonsense, (e.g., cybernetic noise) and, on the other, it is a constitutive factor in establishing immunological identity.

Self, as resisting disturbances brought in by external Chaos (external is chaotic by definition), is characterized by *eigen-states* (German prefix "eigen" has the connotation of "self," "proper"). Jerne introduced the "eigen-state" term [6] for designation of the moment-to-moment equilibrium in the totality of cellular interactions. At each moment, any state of the system is defined by its *cognitive domain*, which is the *actual* antibody repertories (circulating antibody).

> Whatever may be recognize as an epitope (idiotope) may be so recognized because it has a degree of resemblance *to an already existing* [italics ours] set of epitopes (internal images] in the network. Self-recognition and the recognition of novelty are one and the same process [5].

Thus, to the extent that something does not have a degree of semblance to internal images, it does not belong to definition of Self and is not more than "nonsense" for Self. Self is always ready-made, and there is no way for the Other to enter the Self except by negation and disturbances of the Self. But each particular eigen-state is a result of a previous activity. Thus, non-self has to participate in establishing (through self-determination) this particular eigen-state. We have to conclude that there are two different sorts of non-self: non-productive and productive (i.e., that which creates just cybernetic noise, and that which participates in establishing a new eigen-state).

What is not present in the cognitive domain of the system and does not participate in the establishment of eigen states remains as nonsense (cybernetic noise) and does not participate in the immunological self [5].

Thus, there is something which is not present in the cognitive domain, i.e., is not defined as Self, and at the same time participates in self-determination of the system. Although anonymous to the system, this Other at the same time, (unlike those factors which are treated just as cybernetic noise) is "honored" and is taken into account during the process of self-reconstruction. The contradiction is apparent. The word "participation" does not solve, but rather conceals the difficulty.

How is it possible for non-self *to participate* in self-determination? What does it mean "to participate"? To the extent that participation is thought of as an external activity initiating certain changes in the system, the process (by the autopoietic definition of the concept "self-determination") cannot be considered as self-determined. But to the extent that the participation refers to the non-self finding its expression among the present repertory of internal images, the non-self is not truly an Other and cannot affect a shift in an eigen-state. Of course, it is trivial to demonstrate that the same factor can both cause disturbances and participate in establishing a new equilibrium. But the issue is not to cite examples of the factors which demonstrate something "on the one hand" and something other for another case. The question is how to describe the factors in those terms of self-determination that are proposed by the autopoietic or autonomic network approach. The true problem is how to describe an external activity (the activity of non-self) as "participating" in a process of self-determination (how to describe it as a necessary aspect of the process) if, by the model's definition, self-determination is understood as *independence from external*? Although there must be that non-self which "participates" in establishing a given state of equilibrium, the non-self to the extent that it participates in the self-determining process, cannot truly be a non-self. Where is the interface, the site of mediation?

Thus, the understanding of self-determination as independence-from-the-external forces us to model the Self by an analogy with Russian dolls. The external observer believes that Self is independent from the externally observed Other. It follows from this that the Self "sees" the Other (to the extent that the latter is essential in establishing of independence) not as non-self, but rather as Self. Thus, the external observer, in order to define a system as self-determining, must insert the system into another which is the true owner of the Self (regarding the system in question). The contradictory conclusion is inevitable—although Self is self-determination, a self-determining system determines not its own self, but the Self of a larger system in which it is inserted. Obeying the logic of this contradiction, the Paris School thesis argues [5, 15] "against current network models which build an ivory tower of the idiotypic network" [2]. The Paris School claims that their approach is "not immune system-centered, but organism-centered", and they do not recognize any contradiction in this Russian-dolls model:

Thus, considering that an autonomous immune system is largely the consequence of previous states of itself (it is self-determined), we would have to consider different self-universes of structures and interactions: the one represented within, and constituting the immune system, and the other pertaining to the rest of the organism. It is clear that in *immunosomatic* perspectives this is only an apparent problem, as all components of the soma participate, with antibodies and lymphocytes, in the same universe of interactions: the states of activity (the nature of apparent problem), as all components of the soma participate, with antibodies and lymphocytes, in the same universe of interactions: the states of activity (the nature of relationships in the immune system's elements) can still be qualified as *eigen*, as the boundaries of self have previously been (self)-determined by the immune system [2, 1984].

We believe that with such an immunosomatic perspective, the problem (of the split of self into two different universes, one of which is inserted into the other) may be considered as "only an apparent problem" due to the ambiguity provided by the verb "to participate." The ambiguity creates the illusion of a solution. If self-determination means independence from the Other, then, in terms of self-determination of the immune system, there is no essential difference between non-self, as the somatic universe, and the "true" non-self as belonging to external "environments". From the Paris School perspective, the difference is based on the requirement for a familiarity of the Other. Thus if the immune system regards these difference as real, it then acts not in its own name, but for the organism as a whole, i.e., it is but a representative of the somatic universe's self-determination.

We are simply including all self patterns that can be recognized by paratopes or mimicked by idiotopes *in the same universe* of interactions, where each structural component is important in the establishment of repertoires and their regulation. In this sense, all body components belong in the immune system, and the structure of the system is largely a consequence of the rest of the organism [5].

Therefore, having proclaimed the intention to interpret the immune system in terms of self-determination, we are asked simultaneously to regard the real enigma of the immunologic Self, the source of immune activity, as external to the intrinsic processes of the immune system. Organismic integrity is interpreted then not in terms of immunological self-determination, but rather assigns immune activity as representative of the organism's integrity or delegated to that service. This shift of "selfness" from the immune system to the organism is not due to some special data pertaining to morphological, physiological, developmental, or evolutionary aspects of our immunological knowledge; the shift has a purely metaphysical nature and is caused solely by understanding self-determination as independence from the Other.

Because the Self is not concerned with determination of the Other, it actually assumes that the Other must provide the Self with potential independence. In this interpretation of self-determination, the Other of *any* system always pays the expenses for self-determination. That is why the shift, found with the immune system-centered perspective, must also occur with an organism-centered orientation. Just as the former case splits the "other" of the immune system-organism, the autopoietic logic splits the other of organism (the "environment") into two universes, one within another. In fact autopoietic descriptions offer such attempts; consider, for example the following passage:

We should perhaps make a distinction between "foreign-familiar" materials, such as the usual components of the diet and the normal microbiota, and "foreign-foreign" materials. The former, in previous and sustained contacts, have already made their imprint, and the present eigen states consider them; they have been "assimilated." The latter, although able to connect with components of the system, thus penetrating its cognitive domain, are not assimilated by the network in its present eigen state. Thus, these foreign-foreign materials cause disturbances in the organization requiring compensatory changes, until a new equilibrium is reached permitting the assimilation of the new material into the self—turning it into foreign familiar. These disturbances correspond to what immunologists call "immune responses" [5].

In later publications, this distinction between "foreign-familiar" and "foreign-foreign" appears not in a conjectural modality but as a necessary assertion.

This foreignness-as-almost-self is dealt with by the system's metadynamics: the continuous production of novel V-regions with random diverse shapes, which, more often than not, makes little sense for the existing self as they do not interact significantly with existing molecular shapes in the network [2, 1988].

Thus this foreigners-as-almost-self is necessary for the dynamics (metadynamics) of the immune system. But accordingly to the Parisian hypothesis, to the extent that this foreigners-as-almost-self is definable in the current state of the system (i.e., identifiable in the system) it is *not* foreign, and to the extent that it *is* foreign, it is not definable (not identifiable) by the system. This "almost-self" then defines Self from the point of view of the system, which is foreign from the point of view of an external observer.

However it is not the external observer, but the system itself that is responsible for deciding which foreign materials may be adapted and which rejected.

From our vantage point, the only valid sense of *immunological self* is the one defined by the dynamics of the network itself. What does not enter into its cognitive domain is ignored (i.e., it is non-sense). This is in clear contrast to the traditional notion that IS sets a boundary between self in contradistinction to a supposed non-self. From our perspective, there is only self and its slight variations. That which is foreign is only so because it is similar to (or only slightly different from self: the *Unhemlich* of that considered as foreign can only come from this proximity [2, 1988].

If the immunological self is the one defined by the dynamics of the network itself, then the network is responsible for incorporation of new molecular patterns. But if the system accepts new material simply because it looks familiar, it means that the system is not responsible for the dynamics of its own pattern, i.e., such an immunological self is not self-responsible and, according to the very idea of autopoiesis, it is not self. Thus the central question remains as to what the system sees in new material as essential for acceptance or rejection. In recent publications of the Paris School, the question is answered by reiterating the central assumption of the theory: the system sees as essentially acceptable that which is familiar [9]. The insufficiency of this position is well-illustrated by the problem of autoimmunity: why are lymphocyte clones presenting familiar molecular configurations incorporated in a network providing "tolerance", and others are rejected to exhibit autoimmunity?

Our autopoietic view on the nature of autoimmunity opens up a whole series of new questions. Notably, we must now ask: what is it that determines whether a lymphocyte clone is adequately or inadequately connected to the network? [9, Stewart *et al.*, 1989]

Recognizing the question, does not necessarily provide an adequate answer.

Since new lymphocyte clones are continually coming out of the bone marrow, the answer to this question cannot be given "once and for all", but must be the result of an ongoing process. We shall not attempt to provide this answer here, but we do nevertheless wish to indicate the lines along which we think it may be sought.

It seems clear to us that the answer to this new question must involve a consideration of the metadynamics of the immune system: i.e., the fact that the list of lymphocyte clones comprising the immune system is not fixed (as it was in the preliminary simulations reported here), but is itself dynamically evolving—with some clones dying out, and other new clones being selectively recruited from the pool continually alimented by the bone marrow [*ibid*].

But does the Parisian approach provide a strategy by which this fundamental answer may be sought. The question does not ask about an *a priori* given list of clones which are to be incorporated in the network, but rather concerns the principles which rule the dynamics of the evolving list. It is quite possible that the same clone might be expected to be incorporated into the network, or rejected, depending on the current state of the system. So, the question is not which molecular configurations of lymphocyte receptors are acceptable, but how (according to which principles) acceptance or rejection of a clone is regulated in the system. We believe it is impossible to avoid the conclusion that to the extent the system is responsible for development of its own dynamic patterns, it must be responsible for identifying the "foreign-foreign". But to interpret self-determination as a functional closure of a system (in contradistinction to its interactive openness with the Other) will never permit the system to recognize the "foreign-foreign."

Self-determination (as independence-from) applied to *any* system will force us to consider the Self of the system (i.e., its self-determination) as a result of *participation* in (a result of its dependence on) a larger system. The contradiction is intrinsic to the Paris School perspective. But, as is usual with such contradictions, the metaphysically motivated conclusions appear to result from direct empirical observations, and from "common sense." Thus, Maturana and Varela, referring to autopoietic systems as a subset of the homeostatic machine group, formulate the same *metaphysical* contradiction as a non-contradictory "physical" generalization:

There are machines which maintain constant, or within a limited range of values, some of their variables. The way this is expressed in the organization of these machines must be such as to define the process as occurring completely within the boundaries of the machine which the very same organization specifies. Such machines are homeostatic machines and all feedback is internal to them. If one says that there is a machine M, in which there is a feedback loop through the environment so that the effects of its output affect its input, one is in fact talking about a larger machine M', which includes the environment and the feedback loop in its defining organization [16, p. 78].

We would repeat once again that the Paris School description of the immune system as belonging to a larger system, or as a kind of homeostatic machine, which includes its environment and the feedback loop, is motivated by a specific metaphysical intuition of Self. Self as "independence-from," presupposes an already provided affinity to the Other. Thus, independence becomes only that activity which retains that affinity. Beyond the self-structuring Self, there remains a constancy which is the true content of the Self. The ancient etymology of the word "identity" (Latin *idem*—the same) is still at work in this metaphysical intuition. Self is thought of as the activity which maintains the identity (in the sense—"sameness") of an entity.

> Autopoiesis attempts to capture the mechanism that generates the *identity* of the living. This identity can be loosely described as self-produced coherence: the autopoietic mechanism will maintain itself as a distinct unity as long as its basic concatenation of processes is kept intact in the face of perturbations, and will disappear when confronted with a too drastic perturbation [17].

Thus, on the one hand, self is to be defined through self-determination but, on the other hand, it always conforms to an indefinite set of processes, structures, features and properties which constitute the background "sameness" of the system in question, and which form the true content of the system's Self:

> Most immunologists would agree to extend the network to at least one class of non-immunoglobulin molecules, namely clonally distributed T cell receptors. We consider that it must be extended to all other structures in the organism which are available for interactions with the immune system; we claim that molecules which are relevant in the physiology of lymphocyte activation and differentiation to effector functions, must be treated as special cases when describing the internal activity of the system [2].

This intuition of self-determination as expressing and maintaining an identity ("sameness") leaves evolution the task of providing identity. Evolution, from this perspective, is classically the succession of identities, and is responsible for generation of patterns. Ontogeny reveals and, at most, maintains them.

> Ontogeny and evolution are completely different phenomena, both in their outlook and in their consequences. In ontogeny, as the history of transformation of a unity, the identity of the unity, in whatever space it may exist, is never interrupted. In evolution, as a process of historical change there is a succession of identities generated through sequential reproduction which constitute a historical network, and that which changes (evolves), the pattern of realization of the successively generated unities exists in a different domain than the unities that embody it [16, p. 104].

Thus, although the organism, as an autopoietic system, is thought of as self-determined, the true content of self-determination is intuited as the identity (a pattern, a "sameness") which is determined by evolution, and the latter, in its turn, is compared to ontogeny as, "a completely different phenomenon." In the same way as Self is revealed in the spatial dimension by escaping from the organism into its environment, in the time dimension of developmental processes, self escapes from a given ontogeny into its pre-developmental period of evolution. We will have occasion to further explore this issue below.

II.3 *Summary*

From the clonal selection perspective, immune activity results in the production of immunologic definition of non-self (the antibody repertoire). In terms of that activity, the immunologic definition of Self is then represented only by blanks in the repertoire, i.e., negatively. The advocates of the autopoietic approach argue (as we concur) that many difficulties, paradoxes, and *ad hoc* assumptions of the classical perspective are conditioned not only from missing information (i.e., factual knowledge), but in an essential degree, due to our neglect of the fact that any implication of the concept "self," in its turn, implies the concept of "self-determination." On the other hand, from the autopoietic perspective, the immune activity that generates the immunologic definition of Self, is, at the same time, ignorant (in any sense of a positive cognitive activity) of non-self. The latter is presented in the immunologic "cognitive domain" only as "cybernetic noise" or as rude disturbances. As we argued, the intuition which remains beyond this understanding of self-determination, equates the latter with independence-from, or, what is apparently more precise, the intuition attempts to balance the "existential" dependence on the other by an "essential" independence. In its turn, autopoiesis leads, from our critique, to certain essential contradictions. We agree to accept (at least as the most interesting hypothesis) the proposal that considers the immune system as the (or one of several) "mechanism(s)" of organismic self-determination. But to the extent that Self neglects to provide definitions of the Other, it concomitantly fails to provide self-definition. And the reverse case applies to the clonal theory. If we have serious intent to interpret the activity of the lymphoid system in terms of self-determination, we are in fact declaring our commitment to consider definitions of both Self and non-self as provided by that activity (and not reliant on an external assessment—viz., interpretation). Moreover we actually declare our commitment to consider the activity only by its own terms, i.e., as generation and interdependence of the Self and non-self of the activity. We oppose the following view:

> ...The notion of self-determination is also essential to envisage the circularity, that is, the organizational closure, of the idiotypic network. If it must (self)-determine its own actions and recognitions, the organization of the network must be closed on itself, even if open to interactions with external antigens [5].

From our vantage, exactly the opposite is the case. The notion of self-determination requires that the idiotypic network must be *organizationally open*. Our formulation not only allows for an open network for interactions with external antigens, but it demands that the network must provide presentations of the non-self as positive constituents of the network's "cognitive domain." The Other as non-self, not as another sample of self, not as essentially "the same," but rather as "otherness" must be an essential constituent of Self. So what is the Self?

III. CONCERNING INDIVIDUALITY

As we argued above, the autopoietic approach actually understands Self as the expression and maintenance of "the-sameness" of an organism, i.e., the expression and maintenance of integrity as the realization of *identity*. What counts, from this point of view, is "self-equality;" "self-non-equality" (deviations from itself) are considered temporary disturbances which must be overcome (for host preservation) to establish a (more or less) new equilibrium, which, in its turn, is just another variant of the same basic identity [18]. From our point of view, integrity of the organism (its individuality) cannot be reduced to its "identity," that is, only to establishing self-equality. This singular construction omits the other side of the equation: elimination of non-self (performed "homeostatically" to return to itself) must be mediated by (in effect) a self-alienation, where the Self is dealt with as is the Other. In this sense, it is no more appropriate to define individuality as equal to itself, than to say that it is not. We do not mean such a triviality that at any given moment individuality presents a state of deviation from a basic identity, i.e., that individuality is the forever-peculiar expression of "the same." We mean that individuality (if we are laying claim to interpret the concept in terms of self-determination) must discriminate self from Self, treating itself as non-self. This conclusion necessarily follows from the concept of self-determination. It *is* ceaselessly dialectical, for the central concept demands two *complementary* aspects of any self-determined process: (i) its ability to retain positive presentation of the Other as essential constituents of Self, and (ii) its ability to treat Self as non-self, i.e., to discriminate Self from self.

As Leo Buss demonstrated [19], the idea of individuality (as the idea is presented in the "Modern Synthesis") is affected in an essential degree by Weismann's germ-plasm theory, which provided the view of the individual as a genetically homogeneous unit. Buss argues:

The ideal of the individual as an entity that may be treated as genetically uniform is at best an approximation. It is apparent that individuality is a derived character, approximated closely only in certain taxa [19, p. 20].

The Modern Synthesis was constructed from research conducted exclusively on organisms (fruit flies, mice, and humans) with early embryonic determination of the germ line, i.e., each are "organisms in which the notion of the individual as a genetically homogeneous unit of selection is closely approximated" [19, p. 23]. But the most common mode of development among multicellular taxa is somatic embryogenesis, where a distinct germ line is lacking. In this case, the probability of genetic variations which could be inherited is very high. From this perspective, the concept of individuality does not coincide with genetic homogeneity. Thus, when various strategies are reviewed, the biological lesson concerning homogeneous individuals is that *individuality* may not be defined as genetic "sameness." Individuality as "the-sameness," being defined genetically or in any other way, is but a product of its pre-history and external circumstances. But if a pattern of individual development reflects an ongoing competition between different cell lineages, or regulatory effects over such a competition, than intrusion of the external Other is mediated in the course of development by treating itself as the Other, i.e., it is mediated by the war against itself and by the appeasement of itself. In the same way, individual development then appears not as merely a revelation of heredity, but rather as an enterprise assuming responsibility for that heredity, and its future. Thus, ontogeny is not the process negating the Other, but rather the process that actively shapes the Other—both in its spatial dimension (ecology) and in its time dimension (heritability). One of Buss' central themes is that developmental processes control evolution, and he does not forget its complement:

In Van Valen's oft-quoted terms, "evolution is the control of development by ecology." ... Evolution may also reflect the control of ecology by development [19, p. 141].

Buss attacks the traditional notion of individuality to the extent that the notion was shaped by Weismann's germ-plasm theory and later assumed in the Modern Synthesis as equivalent to genetic homogeneity. However, the metaphysical origin of the notion of individuality as self-equality and self-congruency should be sought not in Weismann's theory, but at the dawn of Western philosophy and science. In biology, the first attacks on these metaphysical implications had been made before Weismann's first publication on the germ-plasm theory in 1883 [20], and completely within the paradigm formed by the old intuition of heredity—general repetition of the parental "the-sameness." In 1881, Wilhelm Roux published *Der Kampf der Teile im Organismus (The Struggle of Parts in Organism)* [21], where he attempted do explain integrity (die Zweckmaumlssigkeit) not as something "deliberate but as becoming, as arising not teleologically but in a natural-historical, mechanical way" [21, p. 2]. He saw the theoretical basis of his undertaking in the ideas of the struggle for existence formulated by Darwin and Wallace [21, p. 3], and he believed that Virchow's idea of the sovereignty of cells provided the possibility of applying the evolutionary principle of struggle to the questions of individual development [21, p. 65]. However, as Roux himself noted, the metaphysical basis of his theory is ancient. He saw the reason for the struggle of the "parts" (the "molecules," the cells, the organs) in the trivial assumption that there could not be two absolutely equal things [21, p. 65]. These variations (understood in a pre-Mendelian fashion) generate (and, in their turn, originate in) struggle. Roux emphasized that in understanding this conflict he followed Heraclitus ("The struggle is the father of everything"), and he further believed that Darwin and Wallace simply continued in the same metaphysical tradition [21, p. 65], where integrity was freely provided [21, p. 67]. Struggle therefore is nothing but a manifestation of variations within a given, or provided, unity.

More relevant to our theme is Metchnikoff's attack (begun in the late 1860's) on the traditional concept of individuality as a revelation of a provided identity (see Tauber, this volume and 22). As one of the founders of invertebrate comparative embryology, he began with the question of how recapitulation related to the identification of developmental structures. Because a recapitulated structure carried a "biological meaning" borrowed from the context of an ancestral organism, the structure must be, on the one hand, mastered by the new developmental context (the new context must master its own history) and, on the other hand, this development is the true Other of the recapitulated structure. Thus, relation to the external environment is always mediated (in this sense, controlled) by mastering its own history (and itself) as the Other. The amoeboid mesodermic cells (according to Metchnikoff) recapitulated, in their phagocytic activity, the ancient function of intracellular digestion. Insertion of that recapitulation into developmental patterns of animals with cavitary digestion meant both a new control of developmental processes over its pre-history (over the aggressive activity of the phagocytes), and a mediation of Self by these cells (e.g., the remodeling activity of phagocytes in metamorphosis). Thus, from the Metchnikovian perspective, individuality is an incomplete entity whose primary characteristic is not an evolutionarily provided uniformity, but rather is defined by hostile relations between structures (viz., cell lineages) whose incongruence (competition) has been conditioned by their origin in different evolutionary stages. Metchnikoff's unwillingness to accept individuality as an entity with an evolutionary guaranteed uniformity pertains to the problem of heredity, but only in the sense that the question concerning criteria for identifying recapitulations became a problem of ontogenetic responsibility for phylogenically elaborated structures. The problem formed the intellectual axis of Metchnikoff's phagocytosis theory, and the basis of modern immunology. His theory of active immunity was first formulated not as an abstract speculation about the "Healing Forces of Nature," or no less abstract conjectures about the colorless blood cells as "the gendarmes of the organism" (Virchow), but as an explicitly articulated research program. The theory originated in an intellectual vision of individuality that lacked a unity provided by evolution, and which achieved unity only through control of evolution by ontogeny, i.e., control of the past and the "external" by development. Thus the first stage in the intellectual history of modern immunology was shaped by the vision of individuality as struggling to find its own integrity, and not in the ancient intuition of individuality as integrity "naturally" provided, and whose only concern was to protect itself from the foreign.

Returning to Buss, he convincingly argues that the entity which we (intuitively) consider as individuality, can not be equated with genetic homogeneity; in the majority of taxa this assumption simply does not apply. But that exercise only reveals what individuality is not, rather than what it is. The question—What is individuality?—is always implicitly present, but an explicit formulation has not been made. It appears as if in some cases, Buss defines individuality as an entity (not necessarily genetically homogeneous) characterized as a unit by its developmental pattern. But in these terms, such a definition of individuality implies a circular reference, for it considers "evolution of development as the interplay between selection on the level of the cell lineage and selection on the level of the individual" [19, pp. 30–31]. In other cases, this appears as if an entity can be considered as an individual only to the extent that genetic homogeneity is provided by its developmental pattern: "Individuality evolved as germ line sequestration acted to restrict the origin of new heritable variations in the course of ontogeny" [19, p. 30]. But surely Buss would not insist that a genetic heterogeneous "organism" is not an individual just because of its heterogeneity. Thus there is an ambiguity between his concepts of "organism" and individuality; the latter, is understood as the interplay of selective processes on the level of cell lineages and on the level of "the organism as a whole." Buss writes:

> Normal development... should reflect a complex of interactions in which a cell lineage participates in a function of *benefit to the organism as a whole* [italics ours] only if participation in that function increases its own rate of replication or if participation in that function is actively enforced by a competing cell lineage [19, p. 82].

Does the expression "the organism as a whole" mean "the totality of cells genetically homogeneous," whose homogeneity is preserved by the early sequestration of the germ-line? It is difficult however to reconcile this interpretation with another assertion: "The organism is an environment potentially populated by normal and variant cells" [19, p. 76].

From our point of view, this elusiveness in the notion of individuality reflects, first of all, the "natural" difficulties of the concept: it is something indivisible, a-tomic, simplistic and self-sufficient but, at the same time, something finite. Precisely in its self-limitation, individuality is dependent on external realities and circumstances of its genesis. In this respect, the question — What is individuality? coincides with the previously discussed issue concerning the possibility of describing a self-determining system. On the other hand, we believe the concept's elusiveness reflects, in an essential degree, the inadequacy of the concept "selection" as it is used by Buss in his interpretation of individuality.

The idea of individuality as a self-determining entity finds its parallel in Buss' interpretation of individuality as being in control of its heritability and ecology: not only the external world controls the organism, but dialectically, the organism controls ("from within") the external world. The concept of selection presents, by its very nature, the idea of control external to the origin (to the mechanism) of the variations which are supposed to be selected. Actually Buss is attempting to propose individuality as responsible for its own evolution: individuality must take part in shaping developmental patterns. Pointing at the failure to catch this elusive entity in terms of selection on the level of "organism versus external environment," he dips the relation of selection into the organism. But this forfeits our ability to comprehend this "inner life" of the organism as self-responsible. From this point of view, organism is in control over its external environment in the sense that its external relations are now mediated by competitive relations between its cell lineages. But in what sense is organism in control of its internal environment? Apparently inasmuch as the competitive processes taking place between cell lineages are mediated by selection at the level of the whole organism and its environment. But precisely in these terms, it is not the organism, but its external environment which determines control. Thus individuality escapes definition as an object of selective processes in external environments; however, it also escapes definition through internal competitive processes. Yet, despite these elusive characteristics of individuality on both levels, it is to be comprehended as their interplay. Individuality, as the "interplay" of the two selective levels (the level of relations between different cell lineages, and the level of relations between the organism and its environment) is thereby not more than the elusiveness of elusiveness. Our frustration resides not in Buss' answer to the question concerning the essence of individuality as he proposed it, but because the proper question has not been appropriately formulated. The entity of our discussion thus remains anonymous. Individuality, as hidden beyond the enigmatic word "interplay," remains but the anonymous complement to the anonymity of selection. Any combination of two anonymities cannot provide an *onyma*.

Buss writes: "No clearer example of organismal function arising through epigenetic interactions can be found than that of the immune system of mammals" [19, p. 83]. Considering immune interactions as epigenetic, Buss might view the activity of this system as primarily responsible for establishing individuality. All the more so since in his argument, competitive relations between different cell lineages, which in evolutionary past apparently played the primary role of shaping developmental patterns, characterized the relations between genetically heterogeneous lineages. But the vertebrate immune system itself is programmed to produce genetic heterogeneity. We might then expect that the immune system manifests self-determination and might play a paradigmatic role in the self-establishment of vertebrate ontogeny. But Buss explicitly deals only with the protective activity of the system, and restricts his interpretation to the classic role of immunity as defined in host-pathogen relations. This view may be too limited. The restriction resides in the fashion Buss deals with the immune system as conditioned by interpreting epigenetic relations as solely selective: if selection is the basic shaping force, then activities of the system produced by such a force are most appropriately interpreted as adaptive to the Other; specifically, they are viewed not so much as inwardly projected and constituting the very essence of individuality, but rather as external representatives of individuality. Buss directly notes certain parallels between developmental processes and immunological mechanisms: for example, "the forces which spawned the unusual life cycles of fungi are the forces which challenged metazoans to develop elaborate mechanisms of historecognition" [19, p.

158]. But again, "the forces" are anonymous, or rather simply another designation for the anonymity of the "interplay" between two levels of competitive processes—one constituting cell lineages, and the other, "the organism as a whole."

The familiar metaphysical intuition of individuality as self-equality underpins and structures Buss' vision of epigenetic relations. In this vision, each cell is striving to reproduce itself "after its kind." Thus, reproduction is an imperative of finite creatures to replicate themselves, to preserve their "sameness" despite their finiteness. Self-reproduction of a living unit is viewed as the most telling manifestation of its drive to surmount its limit, or, what is actually implied by this intuition of finitude, to surmount its dependence on the Other. All reference to cells which are destined to perform their somatic duty in order to be able to return to reproduction, is but a modern expression of the ancient metaphysical concept of an immortal identity that could realize itself only through its respective embodiment in a mortal creature, that in turn requires reproduction to immortalize. In this vision of epigenetic relations, the Other is viewed only as the factor which restricts and constrains the freedom of the Self. In its turn, freedom is understood as liberation from the Other and return to its "sameness." Mastering the Other is understood here not so much as constitutive in regard of the very essence of freedom and self-determination, but rather as the tithe paid to existential restrictions in preserving true identity. Correspondingly, to the extent that individuality is expressed ("articulated") as mastering the Other, it is presented not in its self-determination, but rather in its specific deformations externally imposed. Thus, each particular mechanism or activity, while not explicitly revealing the essence of individuality (Self-determination) demonstrates a particular kind of dependence on the Other.

As already noted, adherence to the metaphysical position hidden in Buss' interpretation of epigenetic processes is only one aspect of his approach to the problem of individuality. The other argument, which we believe is direct opposition, constructs a concept of individuality discordant with the notion of a homogeneous genetic unit. However, the idea of genetic heterogeneity becomes far more influential in Buss' argument. Noting that sex is widespread in multicellular taxa, and that "asexuality, as the sole mode of reproduction, occurs only as an evolutionary modification of the sexual process in the taxa" [19, p. 125], Buss, attempting to remain dedicated to his concept of epigenesis, in fact uses it as an external decoration of a concept with a completely opposite meaning:

Sexual reproduction allows the routine reinitiation of an individual within which all cells contribute to somatic duties, hence allowing a cellular-differentiating variant, which could not survive in a chimeric somatic environment, to succeed. [19, p. 127]

The association between sex and cellular differentiation is casual; *sex was a necessary precondition for the evolution of cellular differentiation.* The evolutionary significance of sexuality lies not only in its potential advantages to the individual and the population, but also in its impact upon selection at the level of the cell lineage. By allowing the individual to override the selective advantage of the cell lineage, sex allowed the evolution of a trait that is disadvantageous in the somatic environment. Not just any life cycle is possible (*ibid.*, pp. 128 –129).

The evolution of sex from this perspective, is a powerful example of the organism's control of its ecology: these inter-individual relations are the most significant constituent of the species' ecology. But, as Buss argues, sexual evolution was actually a means to control the *somatic* environment. Development of somatic cellular differentiation (development of the specialized industries for mastering the Other) would be *impossible* without the sexual solution. But sex (at least the function ascribed to it by Buss) would be *meaningless* without somatic differentiation. Thus, sexual reproduction and sequestration of the germ-line is not the return of the organism (or one of its sub-systems) "to itself" from the externally motivated involvement in dealing with the Other; we cannot even say that somatic differentiation is a sacrifice the organism offers in order to preserve (in a special cell lineage) its "natural" ability to reproduce itself. Sequestration of germ-line cells (which themselves are often highly specialized) is an enterprise the organism undertakes in order to exclude (at least in its progeny) the possibility of returning to the undifferentiated mode of reproduction for any cell lineage. In short, sex provides the necessary condition for somatic differentiation in metazoans. But this strategy does not confer genetic homogeneity, but rather gamete formation results (in most cases) in individual *heterogeneity*. It is especially clear in those plants whose alternating haploid and diploid generations develop within a single individual, i.e., in all higher plants. Sex is neither designed to preserve genetic identity in ontogeny, nor in the adult's offspring. In no way is sex a means to seclude "the sameness" of an individual. *Sex and sequestration of germ- line cells are evolutionarily constructed as the complement to the evolution of somatic differentiation.* If such differentiation represents the activity that masters the Other (Self as adopting, integrating, assimilating or eliminating the Other, to finally, reduce the Other to Self), the complementary appearance of sex results in the alienation of Self from itself in its self-discrimination. In this scheme, reproduction entails a self-alienation, where Self forms itself by opposing its genetic identity. And the process is possible only as complementary to mastering the Other. Thus, the very fact of sexual evolution destroys this simple idea of epigenesis. The very essence of what is reproduced and the particular way of such reproduction is the necessary complement (and in this sense—a function) of all processes of somatic differentiation.

Contrary to Buss' position, we argue that individuality is not the "interplay" between two levels of selection (cell lineages vs. "the organism as a whole"), but should be viewed along a different axis: complementarity between Self as mastering the Other, and Self as alienating itself. Mastering the Other at the cellular level (somatic differentiation) corresponds to the complementary cellular process of dealing with self-evolving specialized germ cells. But this level of relations between cell lineages is, at the same time, the level of "the organism as a whole," because the latter expression (unless it refers to the metaphysical fiction of "the-sameness") carries meaning only to the extent that a particular system in question (certain relations and processes) can be presented in terms of self-determination. Correspondingly, we do not mean that development of sex is the only form in which individuality establishes itself (establishes the Self).

Developmental processes, which lead to sequestration of germ cells and formation of gametes, reflects, as we argued, that aspect of self-determination in which Self deals with self as with the Other. But whatever importance these processes represent for a particular individual, their true *results* are addressed (in terms of specificities of cellular differentiations provided by the results) not to this individual, but to its progeny. We may search for another system whose primary function is also performed on the level of cellular interactions responsible for expression (or realization) of genetically based cellular specificities, but we wish to consider such a system under more stringent conditions. We expect that the system produces and retains such characteristics of Self that are not exclusively stocked in certain files (i.e., gametes) for future generations, but are addressed as a constitutive force to that particular individual development in which this system originates and to which it belongs. Of course, there is no *a priori* basis for such a cellular self, but were we to hypothesize its existence, the immune activity of the vertebrate lymphoid system would be the first candidate for such a role. If the immune system functions in such a capacity, it must develop and perform primarily as a physiological backbone of cellular interactions.

IV. CONCERNING RECOGNITION

We find a strong (although not direct) support for our hypothesis in recent research that has established molecular homologies between the neural cell adhesion molecule (N-CAM) and immunologlobulins and other members of the immunoglobulin superfamily, which include the T cell receptor, molecules linked to the MHC-complex, and other immune recognition molecules. Cell adhesion molecules (CAMs) of various types have been proposed as morphoregulatory [23]. Edelman assigns them the putative role of providing "a mechano-chemical link between the genetic code and the specification of three-dimensional pattern and form" [24]. In his model, binding specificities of CAMs determine the formation of borders between CAM-linked cell collectives in early development [25]. They serve as signatures for demarcating morphological boundaries, and at the same time signals for sequenced maturation. Of the various classes of molecules serving such function, only N-CAM has been shown so far to share similar structures with immunoglobulins. CAMs are not strictly histotypic.

N-CAM is central to the formation and maintenance of neural tissue but has a much wider tissue distribution and a fundamental role in very early embryogenesis as is the case with other primary CAMs [24].

And Edelman hypothesizes "the immune system did not evolve from the nervous system, but from a cell adhesion system essential to both". [24, p. 1987] Describing the picture of CAM binding, Edelman concludes about its differences and resemblance as compared to the binding provided by the members of the immunoglobulin (Ig) superfamily:

This picture of CAM binding clearly differs at the *specificity level* from Ig binding: there are vast repertoires of binding sites. Indeed, this is not surprising in view of the emphasis on regulation rather than on specific cell addressing during morphogenesis and the fact that boundaries of cell collectives depend upon the existence of different CAMs. The closer *functional* resemblance of N-CAM is therefore to those members of the Ig superfamily concerned with self recognition and cell interaction of response in the immune system (e.g., MHC molecules). Aside from antigen recognition, lymphoid cells must interact with each other for these regulatory purposes as well as with formed structures of the lymphatic system that themselves depend upon CAMs for their structure [24, 1987].

If we have been programmed to consider the immune system as a warfare machine designed for elimination of the Other, then we are doomed to see in the "self-to-self-relation" aspect of that activity only a set of subsidiary devices providing regulatory functions and embedment of the system in the internal environment of the organism. From this perspective, Edelman's comparison only refers to a hypothetical evolutionary origin of the immune system, that is, it refers to the system's "etymology" which may be quite obsolete and essentially different from its present "meaning". But another case is possible—the "meaning" does not shield its "etymology", but (as often happens with linguistic etymologies and meanings) presents it in a form which is more explicit and developed ("historically enriched"). From this point of view, N-CAM homophilic binding (N-CAM-to-N-CAM on apposed cells), as responsible for self-to-self interactions, presents the "etymology" that supports the "meaning" of the immune system as an "institutionalization" of that ontogenetic responsibility for cellular interactions. In this respect, the following facts my be particularly telling. After discussing CAM regulation, Edelman states:

Although one would expect that the principles of CAM regulation will be found to hold in a variety of animal species, both vertebrate and invertebrate, well defined CAMs have so far been isolated only in vertebrate species [24, p. 1987].

But in vertebrates, the two primary CAMs (N-CAM and L-CAM) which "appear very early in development and are seen at various times on derivatives of all three germ layers" [24] are apparently highly conserved and, do not exhibit essential differences in their respective binding within the same species or when mixed with distantly related animals (e.g., frog and mouse) [26]. Being concerned with "etymology," we might conjecture that the specificity of vertebrate morphoregulation (i.e., cellular interactions as mediated by CAMs) provided the evolutionary origin for the formation of a regulatory mechanism in the specific protective system of vertebrates. But when addressing the issue of "meaning," we hypothesize that understanding of this regulatory mechanism of vertebrate cellular morphogenesis should be sought in the function of the immune system. The fact that well-defined CAM regulation has been found so far only in vertebrates [27], and that the structures of the two primary CAMs are conserved in all vertebrate groups, appears to provide essential support for the conjecture that forces regulating cellular interactions in vertebrates are "institutionalized" in a particular (vertebrate-specific) system [28]. The fact that the "regulatory" aspects of the vertebrate immune system reflects structural similarity to the CAM-mediated cellular interactions highly suggests that immune function is (at least) an essential element of this "institutionalization."

The arguments support the idea that a recognition system must be at the same time, self-determinative: a particular (taxon-specific) mode of recognition in cellular interactions is only a reflection of that particular mode of cellular interactions which allows us to consider the interactions as self-determining. By this we do not mean (as must be clear from our analysis of the autopoietic thesis) that we will consider the immune system as exclusively oriented inward, and not antigen-oriented. As previously discussed, in order to present a system as self-determining, it must present two complementary functions—the mastery of the Other, and mastery of the Self. Specifically, we conjecture that the evolutionary transformation of the hypothetical CAM-predecessors, which evolved into modern CAMs and other recognition molecules of the immunoglobulin superfamily, reflected transformation of a system which assumed the responsibility for presenting (by means of these "cognitive" molecules) these two poles of self-determination. The system must consist in activity which discerns and relates ("in terms" of such "cognitive" molecules) these two complementary poles. The molecules of the immunoglobulin superfamily must, from this point of view, play different roles in two complementary aspects of self-determination: in reduction of the Other to definitions of Self, and in self-alienation of self-definitions.

With the present level of knowledge, it is difficult to describe another organic system which would so strikingly correspond to this notion of self-determination as is the case with the vertebrate immune system. The correspondence is clearly traced in terms of the functional roles of these "cognitive" molecules. MHC-restriction of antigen recognition is the best demonstration of the first aspect of self-determination: the primary concern of the immune system is presentation of the Other as mastered by Self. The Other (an antigen), in order to be recognized, must be inserted in the context of those cell-surface proteins which are programmed by MHC, that is, in the context of those very molecules which are responsible for tissue allorecognition and rejection of allografts. Thus, in order to be recognized, the Other is presented in that context which, actually, plays the role of "the marker of individuality". What is even more astonishing, the Other (antigen) is presented in the context of MHC-products, not as "the raw Other", but as "the boiled Other". *It is presented not in its "wild" form but as the mastered* (processed) *Other.*

The peptide fragments of those cells against which cytotoxic T-lymphocytes are activated, are of endogenous origins and must be transported to their exposed (recognizable) site (coupled to MHC-products on the cell's surface) from the cell's interior (Fig. 1) [29]. Antigen presenting cells (APC) macrophages and the dendritic cells—activate T-cytotoxic lymphocytes (T_c) and T-helper lymphocytes (T_h), by subjecting "raw" antigens to "processing" and then only fragments of the antigens containing a few amino acids are linked to the surface MHC [29]. Even capture of the "raw" antigens by the Ig receptors of B-lymphocytes does not mean (at least as a general case) that the antigens are recognized. To an essential degree, activation of B-cells is

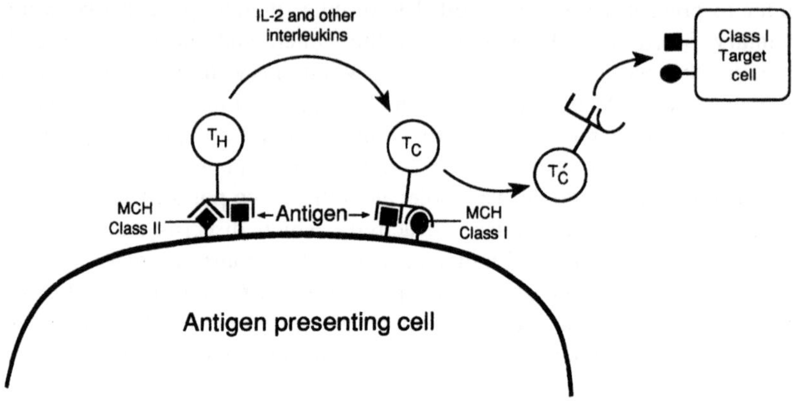

Fig. 1. *Immune stimulation.* MHC class II processing depends on receptor-mediated or fluid phase endocytosis of only exogenous antigen, which is degraded in endocytic vesicles that eventually fuse with vesicles containing class II molecules. Cytotoxic T Cell (Tc) precursors, specific for an antigen with MHC associated class I complex, are stimulated by helper T cells (T_H) that associate (and in turn are stimulated) through class II antigen presentation. This "helper" function is mediated by various humoral factors (interleukin 2 is best described of these lymphokines) and is not specific for T_c stimulation by a one to one correspondence with a given epitope. The two major classes of T cells, helper and killer, recognize two different kinds of protein antigen. Broadly, Class I-restricted cytotoxic T lymphocytes recognize cell-associated antigens such as those on virus-infected cells or cells expressing foreign alloantigens, and class II-restricted T cells recognize antigens that enter the system as soluble entities; class I peptides generally derive from endogenously synthesized proteins, while the class II universe derives from proteins taken up by the antigen-presenting cells from the external medium. Degradation and subsequent processing of both soluble and particulate antigens by macrophages, dendritic cells, or B cells is followed by T cell recognition of the associated antigen linked to the class II MHC gene product. We understand this division teleogically as in the first case (class I recognition) pathogens harbored intracellularly should be recognized and attacked by lytic cells, while in the second case (class II recognition), require positive signals from T cells to make antibody or become activated.

Class I molecules bind endogenously processed antigen, which has been committed to a pathway distinct from that used to degrade and present exogenous protein for class II recognition events [30]. As currently understood, there are three modalities which lead to antigen degradation and peptide presentation with class I molecules: 1) endogenously synthesized proteins, 2) proteins injected into the cytoplasm or perhaps fuse into the plasma membrane, or 3) phagocytosed cells (i.e., cellular debris) or microorganisms. Class II presentation requires exogenous antigens small enough to be endocytosed.

dependent on their ability to draw antigens into their cytoplasm, to process and to expose (again, in association with a surface-situated MHC-product) a peptide fragment as a potential target for T-helper cell receptors (Fig. 2).

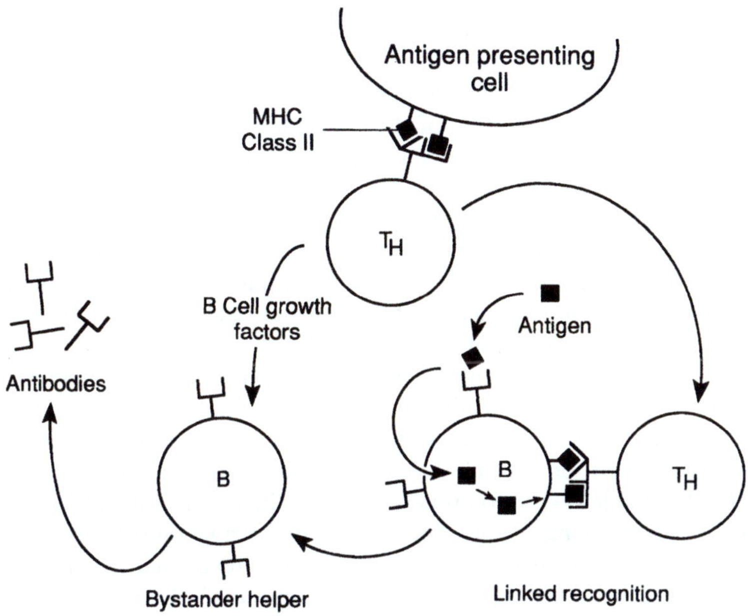

Fig. 2. *B cell activation.* B cell activation, proliferation and differentiation may be initiated by soluble, unprocessed antigen, which is first bound by surface-exposed immunoglobulin, processed (analogously to macrophage digestion) and then presented through MHC Class II receptors to helper T cells. Again interleukins are elaborated and the B cell further differentiates, divides and secretes antibody. Alternatively, B cell response is initiated just as the cytotoxic T cell is stimulated to proliferate, that is through the macrophage recognition limb, with the helper T cell directing the B cell response. In linked recognition, B cells may also serve as an APC, which is antigen specific and MHC restricted. Only those B cells with immunoglobulin receptors for epitopes on the intact antigen can bind it prior to presentation to T_H cells. By the so-called bystander mechanism, T_H cells may stimulate previously primed B lymphocytes to secrete antibodies by elaborating B cell growth and differentiation factors. B cell stimulation in this case is MHC unrestricted.

With regard to the immune response to a given antigen there are common features that are recognized by T cells and antibodies, but the mechanisms are clearly different and the molecular basis for the commonality is not clear [31]. It is possible that B cells may act as antigen presenting cells, and re-present, in association with class II, those regions of foreign protein that their antibodies recognize [32]. There are multitudinous arrays by which antigen is recognized, presented, and to which several limbs of the immune network must respond: 1) macrophages have diverse function including direct toxic effects [33], the elaboration of humoral regulatory factors [34], facilitating virus infection of lymphocytes [35], as well as the antigen-presenting function [29]; 2) lymphocyte cytotoxicity is mediated by both class I and class II associations [36]. and finally, 3) antibody is elaborated for both antibody-dependent cell-mediated cytotoxicity [37], and serum or mucosal immunoglobulin directed against various antigens generated by activated B cells [38].

If the expression "presentation of an antigen to the immune system" designates those events that mediate the physical interaction with a raw antigen on the one hand, and an immune response on the other, then antigen presentation implicitly includes mastery of the Other; the process culminates in the insertion of the digested products into a permanent context of recognition—specifically MHC-programmed proteins, (almost universal) cellular marker of "the same". Thus, the Other is presented as integrated within Self, or in other words, as the Other of the Self. But were the immunologic Self not discriminated from this integration, how would the presentation result in the immunologic Other? How can we speak about the Other as reduced to "the same", but not about "the same" as altered by the Other?

If the first side of self-determination is integration of the Other, within Self, the second aspect is discrimination of Self from self, and dealing with self as with the Other. We have alluded to this issue in relation to the evolution of sex. We considered somatic cell differentiation as such a specialization in the mastery of the Other which (normally) leads to the realization and the reproduction of genetic homogeneity. Somatic cell differentiation is the realization of the same genetic potential, or the production of functional specificities as revelation of genetic homogeneity, i.e., "the-sameness" of the organism. On the other hand, metazoan sexual production of this genetic potential is never a reproduction of the same genetic stock. In the general case, gamete production results in the production of cells that are genetically distinct from somatic cells (aside from certain genetic distinctions between gametes themselves). Furthermore, sexual reproduction, as a rule, *guarantees* genetic diversification of the progeny relative to the parents. One might argue that sexual reproduction supports a genetic consistency of a given population, but this perspective is not germane to the development of individuality per se; for this collective genetic pool defines the identity of the population, not its individuals. Thus it is difficult to see how this heterogeneity can account for the production of genetic homogeneity as allegedly the basic characteristic of individuality. The fact that the genetic identity of individuality may be produced sexually does not explain (at, least, on the level of developmental processes) why this strategy implies (or involves) this mechanism of genetic variation. On the other hand, in terms of self-determination, sexually produced heterogeneity manifests the aspect of self-alienation.

The same situation is encountered in the phenomenon of "intentional" genetic heterogeneity in lymphocytes. Ordinarily, this diversity is explained by a trivial teleological argument disguised in an "evolutionary" costume: in order to meet the enormously varied protein profile of potentially dangerous environments encountered in the course of vertebrate evolution, this method of gene shuffling was elaborated. Considering ontogenetic processes as a realization of an evolutionarily provided pattern, the argument still does not account for the "responsibility" which these processes assume for evolutionary forces. As we discussed above, illumination of such a responsibility coincides with definition of self-determination (on the level of the system) or, what is essentially the same, it coincides with definition of individuality. If antigen processing and its presentation through the MHC-restricted system represent quite clearly that side of self-determination which we described as the mastering of the Other, or reduction of the Other to Self, the generation of lymphocyte genetic diversity clearly represents self-alienation. It remains problematic whether the system is designed to eliminate somatic mutations (or intrusion of genetically foreign materials) or not. But it is evident that the system itself produces an enormous genetic diversity which exceeds (in respect to its informational loading) the universe of all the other proteins of the organism by no less than two orders of magnitude. On these merits alone, assuming that the concept of individuality coincides with the concept of genetic homogeneity, it is obscure why one would assert that the immune system, producing genetic variations that so overwhelms the informational loading of non-immune cells is that apparatus which protects individuality.

Buss argues that the relations between different lymphocyte clones is the most striking example of epigenetic interactions: each of the cell lineages is subjected to positive selection only to the extent that it has performed its somatic duty [19, p. 83–89]. We will argue (as we did against epigenetic simplification in interpreting sexual evolution) that "positive selection" means not only mediation of lymphocyte reproduction by performance of their somatic duties, but no less it denotes the complementary mediation of this somatic activity by "intentionally" produced lymphocyte genetic heterogeneity. The ordinary epigenetic interpretation discerns here the opposition of two levels: the "altruistic" on which cells perform their somatic duty, and the "egoistic" level, in which every cell lineage strives to reproduce itself. Interpreting interactions between different (including genetically identical) cell lineages in terms of self-determination, we see here not two different levels but two poles of the same relation. In these terms, the "goal" of the system is not only to use genetic self-alienation as a tool in performing certain somatic duties, but equally to use the somatic activity as the lever for establishing and sustaining this genetic self-alienation.

In the case of sex, reproduction is not only an immediate reward granted to cell lineages for the successful performance of their somatic duties, for in terms of individual genetic homogeneity, it is not *reproduction* at all (i.e., a return to a basic identity from morphological and functional fragmentation and particularization), but rather *production*, and a very specifically "institutionalized" one, resulting in a highly specialized effort to produce genetic differences. We note a similar situation in the immune system. What can the expression "somatic duties" mean in this case? Apparently, these duties designate that aspect of the system's activity that is outwardly (antigenetically) directed. We may designate in this way the work of Tc-cells after having been activated by proteins presented on cell surfaces or, following activation of mature B lymphocytes by soluble antigens, development and functions of B-cells. (This refers to reproduction of cells belonging in a particular clone(s) and their further differentiation into plasma cells and so-called "memory" cells.) But initiation of this "somatic" function in no way is guaranteed by mere contact of an antigen with (B- or Tc-cell) receptors specifically congruent with the antigen; the cells do not initiate their somatic duties with only such an isolated encounter. Correspondingly, we cannot infer that reproduction of a clone, following such contact, is immediately triggered by performance of the clone's somatic duty (or even by initiating the first stage of such a performance). There is a second circle of highly specific cellular interactions which is inserted into the first one, i.e., between, on the one hand, the primary interaction of B or Tc cell receptors with an antigen, and, on the other hand, their immunological response to the reaction: Neither B nor Tc lymphocytes, are activated simply by contact with a proper antigen, but further require specific T-helper (Th) cells. However, in order to activate, Th cells in turn must be activated. The final mechanism of stimulation involves the processing of antigens and presentation of these products to Th cells, which finally triggers reproduction of corresponding clones. Thus such a reproduction is not an immediate result of performance of a proper "somatic duty" by the clone but rather is determined by the opposite (although complementary) process—antigen presentation in a circle of interactions involving Th and antigen presenting cells (Fig. 3).

The decision concerning reproduction of a particular clone(s) is made not as an immediate result of the interactions between B or Tc cell receptors with, correspondingly, soluble antigens or antigens on a target-cell surface (the primary "presentation"), but because lymphocytes, which are putatively the final effector cells, have been exposed to preliminary processing of the antigen. The decisions *are* taken because the primary presentations to the effector cells have been mirrored in presentation to non-effector cells, i.e., Th cells. To the extent that an antigen is considered as the target of immune

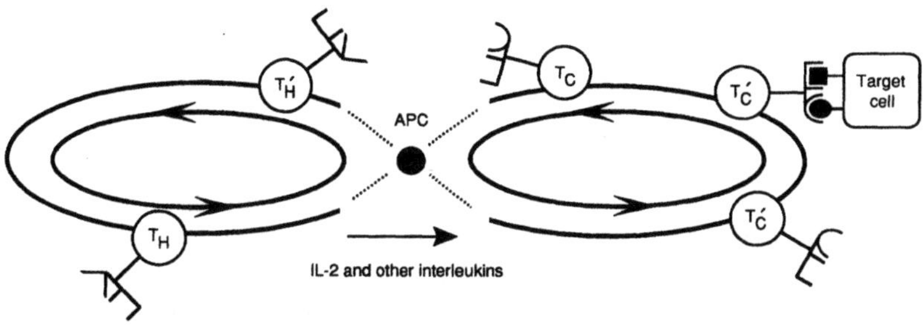

Fig. 3. *The complementarity of lymphocyte interactions.* At center of two immune cycles is the antigen presenting cell, APC (macrophage, dendritic cell, B lymphocyte) that mediates the activation of T-cytotoxic (T_c) and T-helper (T_H) lymphocytes, T_H is responsible for maintaining the immunologic image derived from APC, and T_c, activated by both T_H and APC, serves the effector function. The T_H pool serves as the library of immune images, and T_c is activated to the extent that it matches the particular stock in this T_H pool.

protective activity, presentation of the antigen on the target-cell surface, or exposure of B lymphocytes receptors to the antigen, do appear to play the role of a signal that triggers the elimination of "genetic heterogeneity". Production of such a signal in the mastery of the Other supposedly leads to the re-establishment of a somatic environment corresponding to "genetic homogeneity" of the organism. But to the extent that presentation of the antigen is mirrored as the same or another antigen in its presentation to Th lymphocytes, it leads to selection and maintenance of a specific genetic heterogeneity.

The mastering of the Other results in establishing (genetic) self-equality; the turning of the Self towards itself results in establishing and retaining (genetic) self-alienation. The "tools" to eliminate the Other in the first case, are retained, in the second, as the constitutive features of the self-alienated and self-selected Self. Through inter-cellular interactions, the "tools" (receptors of lymphocytes) take information not from "naked" antigen, but from those antigens which have been processed and inserted in the context of the "self-marker"—the MHC-programmed surface proteins. In the mirrored circle of cellular interactions producing Self-alienated self, antigen presentation must be also framed in the context of a "self-marker." But were these two contexts identical, then Th lymphocytes would remain in the difficult position of mixing their self-reflective ("intellectual") activity with the "practical" activity of Tc lymphocytes [39]. In this case, it would not be evident why the system does not work in a primitive epigenetic fashion and why Th lymphocytes are needed. But the lymphoid system in fact provides the difference between these two contexts of recognition. In the "external" immunological circle of intercellular interactions, which begins and ends with interactions of Tc cell receptors with peptides on the cell surface, these sequences are presented in the context of the proteins programmed by MHC-class I molecules. The expression of these molecules on cell surfaces is almost universal, i.e., they are, apparently, expressed on all nucleated cells including all cells of the lymphoid system [40]. Thus, they are the obvious candidates for the role of the universal marker of organismic identity (of its "sameness"). On the other hand, in the "internal" immunological circle, which begins with presentation of antigens to Th cells and ends with activation of Tc and B lymphocytes by Th, the recognized peptides are inserted in the complex programmed by MHC-class II molecules. The expression of the latter class on the surface of the cells which are responsible for antigen presentation to Th lymphocytes, are quite specific (the majority of cells in the organism are class-II-negative) [40]. Thus the MHC-complex, which, apparently, plays in the first "external" circle, the role of marking self-identity in the process of mastering the Other, in the "internal" circle plays the paradoxical role of such an "identification" of self which discriminates Self from itself.

If the sexual system, as dealing with genetic homogeneity and heterogeneity, also used MHC molecules for similar roles to mark the two aspects self-identification and self-alientation — this finding would support our scheme of immune self-determination. Such confirmation does exist. There are strong data that MHC molecules might function as individual-specific pheromones [41]. Olfactory recognition of mouse MHC (H-2) is apparently performed by detecting the soluble (partially degraded) MHC class I molecule product excreted in urine. This product putatively signals for allogeneic mate selection and the completion of pregnancy. The smell of an allogeneic male may even lead a pregnant female to abort. These data provide strong justification (from a quite unexpected source) to the broadly accepted metaphor of the MHC as the "identity card", and must be complemented by MHC class II molecules, as the "card of self-non-identity". Correspondingly, if, in functioning as a component of the sexual system, MHC class I molecules present (for the Other) a genetic identity of an individual, then we might expect that, in the same system, the MHC class II molecules present (for itself) the self-discriminated Self.

Unfortunately, we do not possess any direct information on how the formation of gametes, their maturation, and role of surrounding tissue cells in fertilization events might be affected by distribution or modulation of these two classes of the MHC molecules in vertebrate reproductive organs. There is however strong evidence testifying that the MHC class II molecules do play a central role in presentation of our hypothesized sexual self-alienated Self. As Ashida and Scofield showed [42], human sperm, being cultured with donor leukocytes or various cell lines, bind and penetrate cells expressing on their surface MHC class II (HLA-DR) programmed proteins; thus, these molecules function as sperm receptors. On the other hand, Scofield refers [43] to unpublished observations that sperm surface structures binding their targets appear to be themselves class II molecules. At the same time, although cultured human oocytes do not express MHC class II molecules either on the plasma membrane or in the zona pellucida, the molecules (HLA-DR) are expressed on many of the granulosa cells surrounding the egg [44]. The authors found these results as surprising since the expression of class II molecules had previously been considered limited to blood cells. From our point of view however, this result precisely fulfills our model.

It is interesting that the observations of Scofield and colleagues were provoked by studies of a polymorphic genetic system that controls tissue compatibility in tunicate (protochordate) colonies [45]. Tunicates permit fusion between colonies that share at least one allele at the fusibility locus, but this site also prevents the fusion of gametes sharing alleles and thus protects the hermaphroditic animals against self-fertilization [46]. Oka [47] showed that in the colonial tunicate *Botryllus primigenus*, the recognition elements are apparently expressed by sperm and the acellular chorion, or vitelline membrane that surrounds the egg. These studies were extended to show a possible mediatory role of the recognition elements in interactions between somatic cells and the gametes in tunicates. Scofield [43] noted upon incubation of sperm and blood cells from *Ciona intestinalis*, that sperm incubated in self blood did not interact with blood cells, but instead underwent striking morphological changes, but when incubated in nonself blood, bound to, and in some cases penetrated, the blood cells. Similar results were obtained with *Botryllus*, from which they concluded that these interactions in both solitary and clonal tunicates provided the first evidence for direct and specific cross-recognition between animal gametes and somatic cells [43]. Thus the finding that human MHC class II proteins may function as sperm receptors originated in the studies of a similar system (whether homological or analogical cannot as yet be determined) of tunicate allorecognition, which, at the same time, in self-sterile populations, controls this "sexual mechanism" of self discrimination.

V. CONCLUSION

We assume that the immune system of vertebrates controls intercellular interactions to the extent that those interactions are characterized by their genetic profile. We assume further that this control of cellular interactions is performed by means of specific cellular interactions leading to certain processes of cellular morphogenesis and reshaping of a genetic profile in the lymphoid system. Continuous reshaping of cellular interactions in the lymphoid system is the way by (and the limits to) which the system controls cellular interactions. It must be understood that when we are referring to "cellular interactions" these consist of a specific definition: the interactions are those to the extent that they are defined in terms of genetic identity and differences, i.e., determined by signatures characterized by genetic homogeneity/heterogeneity. Therefore such a control personified by lymphoid activity, establishes cellular interactions as self-organized or "self-institutionalized". Thus the lymphoid system becomes Self-determining. From the point of view proposed here, self-determination has two aspects. First, the immune system assumes responsibility for providing itself with its own Other; the system determines what is its Other. Definition of the Other is not the business of an external observer, contemplating what originated outside the system, but it is the business of the system itself. Thus, the definition of the Other is complementary to the definition of Self. Second, Self is the self-defining (in a sense, self-creating) self. But contrary to the autopoietic interpretation, we believe this means that the Self does not discover itself as a ready-made presupposition, but creates itself upon the "horizon" of the Other. The Other as positively defined within Self, is the self's chisel for forming the Self. From this point of view, integrity of an organism does not coincide with its identity. If we reserve the word "Self" for integrity, then Self is the complementarity of two reflections: Self as a "function" of the Other, and the Other as a "function" of the Self. Thus, self-identity (the second aspect) is only one of two complementary (self-) definitions of Self.

As we argued above, the idea of a protective system is directed by the notion of integrity as identity. It is hardly possible to doubt the reality of such systems, whose goal is to protect genetic identity. For instance, in the colonial hydroid *Hydractinia echinata*, tissue incompatibility induces activity of a defense organ (the hyperplastic stolon) whose tips are armed with nematocytes (stinging cells) [48]. As noted, competitive defense, inducible by historecognition, is a phenomenon broadly observed among sessile animal taxa. But if we wish to grant the title "immune system" only to those systems whose activity shapes organismic integrity (in terms of genetically defined cellular interactions), then there is no *a priori* reason to consider every system defending genetic identity as an immune system. Protection of identity is not equivalent to establishing integrity. According to our hypothesis, the vertebrate immune system generates complementary (and positive) definitions of Self and nonself. But there is no necessity for every system of historecognition to produce such definitions. From this point of view, if homology of the vertebrate immune and invertebrate historecognition systems was proved with sufficient certainty (which is far from being accomplished), even in this case it would not prove their analogy. Thus while such similarity exists, there is no *a priori* necessity. The point to emphasize is simply to caution against equating protective functions from those involved with establishing integrity. The distinction is crucial, but confusion dates from the earliest debates regarding modern concepts of immunity, when different metaphysical positions were opposed in forging the new discipline of immunology [22]. We are heirs to that synthesis, and conflict.

Boston University School of Medicine

REFERENCES

1. In both classical science and philosophy, the theme of self-determination was closely associated with topics of metaphysics. Thus, we may argue that to reexamine the intellectual basis of an immunological theory built upon the concept of a self-determined system, we are dealing with the *metaphysical* nucleus of the theory. In the European philosophical tradition, the term "metaphysics" was reserved for that aspect of philosophy which earlier was called by Aristotle "the first wisdom," which was to deal with "the first principles" (Aristotle) of Being. The meaning and the context of these "first principles" were provided by Plato's teaching about *eidoses*. Accordingly there are two possible approaches to the identification of an entity: The first is based on definition of that entity as characterized through its relations to others. In this perspective, we dip the entity into the Heraclidian flux: we do not define *what* is the entity but *how* it is affected by others; what was supposed to be identification of the entity was in fact countless reflexes of external effects, which were responsible for defining the entity. The second approach, which, from Plato's point of view, pursues the goal of catching the true authenticity of an entity, attempts to define the entity as it determines itself "by itself" and "for itself". The problem of an authentical definition of identity, i.e., definition of the "whatness" or the *eidos* becomes the problem of self-definition (definition of self-determination, or self-determination as de-finition of self). Thus, within this classical intellectual tradition, the relationship between a particular form of theoretical reflection and metaphysics is an intimate one.

The metaphysical nucleus of a particular theory appeared whenever *identification of its object* became a concern within the theory itself. And each time metaphysical problematics explicitly appears within a theory, it demands attention to the concept of self-determination. Vice versa, whenever self-determination is discussed, theory refers *thematically* (in full awareness or not) to traditional metaphysical topics. This observation is self-evident in classical science, where theoretical thinking was developed in explicit dialogue with traditional metaphysics; perhaps this is best illustrated in biology which was considered as an applied form of metaphysics. Discussion of self-determination in the modern (allegedly non-metaphysical) scientific context must deal with several serious questions. What is the meaning of the concept independent of its historical origin and the intellectual environment in which the concept was developed? Was the shift in the meaning of the concept provided by its insertion in a new metaphysical context or by its complete purification from metaphysical atavisms? (If the latter is true, then such a purification must be itself presented as a particular theoretical activity.) Which are the criteria that apply our concept of self-determination that we do not simply obey the inertia of this ancient intellectual tradition and are aware of reproducing its metaphysical syllogisms? The immunological system provides certain aspects of establishing organismic identity. If this commonly shared intuition is correct, then immunological activity implies self-determination. And as soon as the concept of self-determination explicitly becomes a special topic in discussions of immunity, clarification of the questions outlined above can no longer be considered as marginal to the central themes of scientific reflection. In this context, clarification of the metaphysical connotations of self-determination is nothing else but clarification of the ways in which the problem of immunological Self can be formulated and, consequently, those inner intellectual structures which determine the corresponding research programs may be more carefully examined.

2. The autopoietic position is well-presented by Francisco J. Varela in *Principles of Biological Autonomy* (New York: Elsevier-North Holland, 1979), and its immunological derivative evolved to the so-called autonomous network theory. See Coutinho, A., Forni, L., Holmberg, D., Ivars, F. and Vaz, N. From an antigen-centered, clonal perspective of immune responses to an organism-centered, network perspective of autonomous activity in a self-referential immune system". Immunol. Rev. 79: 151–168, 1984, and more recently developed Varela, F.J., Coutinho, A., Dupire, B. and Vaz, N.N. Cognitive networks: Immune, neural and otherwise, in Perelson A.S., ed., *Theoretical Immunology*, Part Two. Redwood City: Addison-Wesley Publishing Co., 1988, pp. 359–375.
3. We are puzzled by the opposition of lineal thinking, allegedly typical of Western mentality, as contrasted to the appeal to notions of autonomy and self-determination, because the latter are characteristic products of the same Western culture. In typical cases, newly born oriental alternatives to Western philosophy (i.e., Russian Platonism) or post-classical critics of classical metaphysics (i.e., Heidegger) reproach the latter for exaggeration of the problem of autonomy (correspondingly — self-determination) in elaboration of such traditional metaphysical concepts as "life", "spirit", "being", "meaning", "consciousness", "integrity". We focus upon this illusion to "occidental lineal thinking" because it reveals from our perspective a flaw in the intellectual structure of the Paris School argument. Specifically, they are apparently unaware of their metaphysical assumptions, i.e., the philosophical basis of their interpretation of Self-determination is classical and typical of an old metaphysical tradition.

4. Vaz, N.M. and Varela, F.G. Self and nonsense: an organism-centered approach to immunology. Med. Hypotheses 4, 231–267, 1978.
5. Vaz, N.M., Martinez-A. C. and Coutinho, A. The uniqueness and boundaries of the idiotypic self. In: H. Koehler Ed. *Idiotypy in Biology and Medicine*. Academic Press, New York, 1984, pp. 43–59.
6. Jerne, N.K. Towards a network theory of the immune system. Ann. Immunol. (Paris) 125C: 373–389, 1974.
7. Jerne, N.K. The immune system: a web of V domains. Harvey Lectures 70, 93–110, 1976.
8. Hooijkass, H., Benner, R., Pleasant, J.R. and Wostmann, B. Isotype and specificities of immunoglobulins produced by non-immunized germ-free mice fed chemically defined, ultrafiltered antigen-free diets. Eur. J. Immunol. 14: 1127–1130, 1984; Hooikaas, H., Boss, N., Benner, R., Pleasants, J.R., Wostman, B.S. Isotype distribution of specificity repertoire of 'spontaneously' occurring ('background') immunoglobulin-secreting cells in defined ultrafiltered 'antigen-free' diet. Prog. CLin. Biol. Res. 181: 355–358, 1985; Hooikaas, H., van der Linde-Preesman, A.A., Bitter, W.M., Benner, R., Pleasants, J.R. and Wostmann, B.S. Frequency analysis of functional immunoglobulin C- and V-gene expression by mitogen-reactive B cells in germfree mice fed chemically defined ultra-filtered "antigen-free" diet. J. Immunol. 134: 2223–2227, 1985.
9. As frequently happens in the history of science, a structural peculiarity of an approach, not explicitly formulated is realized as a natural peculiarity of an object of those studies. In recent publications, the basic ("a priori") complementarity of the autonomous network and the clonal selection hypotheses are manifest as a complimentarity of two levels in the immune system itself: 1) the central immune system (CIS) as "constituted by autonomously activated clones which display high levels of V-region connectivity characteristics which impose a complex dynamic pattern, usually referred to as 'tolerance'," and, 2) the peripheral immune system (PIS), "which needs to be stimulated by the introduction of "non-self" molecular patterns absent from the network." Varela, FJ and Coutinho, A. Times and Trends. Immune networks: Getting on to the real thing. Res. Immunol. 140: 837–845, 1989; see also Stewart, J, Varela, FJ and Coutinho A. The relationship between connectivity and tolerance as revealed by computer simulation of the immune network: Some lessons for an understanding of autoimmunity. J. Autoimmun. 2: 15–23, 1989; Stewart, J and Varela FJ. Exploring the meaning of connectivity in the immune network. Immunol. Rev. 110: 37–61, 1989; Lundkvist, I, Coutinho, A, Varela, F and Holmberg, D. Evidence for a functional idiotypic network among natural antibodies in normal mice. Proc. Natl. Acad. Sci. USA 86: 5074–5078, 1989.
10. Ehrlich, P. Die Wertbemessung des Diphtherielheilserums und deren theoretischen Grundlagen, Klin. Jahrb. 60: 299 (1897). (English translation, The assay of the activity of diphtheria-curative serum and its theoretical basis in *The Collected Papers of Paul Ehrlich, Vol.2*, compiled and edited by F. Himmelweit, M. Marquardt and H. Dale. London: Pergamon Press, 1957, pp. 107–125 and in German, pp. 86 –106). Ehrlich's side chain theory was a general extension of his concept of biological affinity of various substances for cellular elements. This mechanism was designed to meet both normal and pathogenic molecules. Specificity of this reaction was conferred by receptors that had nutritive function (in the case of competing organisms), antimicrobial function (in the case of chemotherapeutics for pathogens), and antibody elaboration (in the case of immune stimulation). In the later case, the "side-chains" (antibodies) were elaborated in response to antigen challenge.

See A.I. Tauber and L. Chernyak, *Metchnikoff and the Origins of Immunology: From Metaphor to Theory*. New York: Oxford University Press, 1991 and A. Silverstein, *A History of Immunology*. New York: Academic Press, 1989.

11. Burnet, M. *Self and Not Self. Cellular Immunology Book One*. Victoria: Melbourne University Press and Cambridge: University Press: New York, 1969; Burnet, M. *Immunological Surveillance*. Oxford: Pergamon Press, 1970; Burnet, F.M. *The Integrity of the Body. A Discussion of Modern Immunological Ideas*. Cambridge: Harvard University Press, 1962.
12. Smith, H.R. and A.D. Steinberg. Autoimmunity—a perspective. Ann. Rev. Immunol. 1: 175–210, 1983; Marshak-Rothstein, A. Origins of autoantibodies Cur. Opin. Immunol. 1: 697–700, 1989; Lohse, A.W. Autoreactive T cells. Cur. Opin. Immunol. 1: 718–762, 1989.
13. Jerne, N.K. The natural-selection theory of antibody formation. Proc. Natl. Acad. Sci.(U.S.A.) 41, 849–857, 1955.
14. Forni, L., Coutinho, A., Koehler, G., Jerne, N.K. IgM antibodies induce the production of antibodies of the same specificity. Proc. Natl. Acad. Sci. (U.S.A.) 77: 1125–1128, 1980; Benner, R., Rijinbeek, A.M., Bernabe, R.R., Martinez-A., C., Coutinho, A. Frequencies of background Ig-secreting cells in mice, as a function of organ, age and immune status. Immunobiology. 158: 225–238, 1981; Benner, R., van Oudenaren, A., Bjoerklund, M., Ivars, F., Holmberg, D. Background immunoglobulin production: measurement, biological significance and regulation". Immunol. Today. 3, 243–249, 1982; Coutinho, A., Pobor, G., Pettersson, S., Leanderson, T., Forsgren, S., Pereira, P., Bandeira, A., Martinez-A., C., T cell-dependent B cell activation. Immunol. Rev. 78: 211–224, 1984; Pereira, P., Larsson, E.L., Forni, L., Bandeira, A., Coutinho, A. Natural effector T lymphocytes in normal mice. Proc. Natl. Acad. Sci. (U.S.A.) 82, 7691–7695, 1985; Pereira, P., Forni, L., Larsson, E.L., Cooper, M., Heusser, Chr., Coutinho, A. Autonomous activation of B and T cells in antigen-free mice. Eur. J. Immunol. 16, 685–688, 1986.
15. Coutinho, A. The self-nonself discrimination and the nature and acquisition of the antibody repertoire. Ann. Immunol. (Paris) 131D: 235–253, 1980.
16. Maturana, H.R. and Varela, F.J. *Autopoiesis and Cognition. The Realization of the Living*. Boston Studies in the Philosophy of Science. ed. by R.S. Cohen and M.W. Wartofsky. Vol. 42. D. Reidel Publishing Company. Dordrecht, Boston, London. 1980.
17. Luisi, P.L. and Varela, F.J. Self-replicating micelle—a chemical version of a minimal autopoietic system. Origins of Life and Evolution of the Biosphere 19: 633–643, 1989.
18. Vaz *et al.* write that because of

> compensatory changes in its clonal structure, the system incorporates, more often than not, new materials in the activity of the self. The ontogeny of immune system is, therefore, never completed [italics ours]; it actually began generations earlier with the individual's ancestors. (5)

Does the incompleteness of the immune system in ontogeny mean that the immunologic self is (despite all that was previously argued) open to the Other, or, what is essentially the same, that the organizational principle of immunological activity implies the establishment of non-equality with the Self? Absolutely not. The context of the autopoietic approach does not leave any doubt that the incompleteness of the immunologic self refers to nothing else but the ontogenetic incompleteness of variations of the basic "sameness":

> In fact, to the potential and available repertoires defined by Jerne, we could add another, constituted by the set of specificities that are being used (the actual repertoire) to describe three sets of specificities: 1)

what can be made; 2) what has been made and can be used (precursor B-cell repertoire); and 3) what is actually being used (circulating antibody repertoire). The first is genetically determined, reflects the history of the species, and draws the limits of the others; the second is derived ontogenically in a process which, through maternal influences, implicates many generations and reflects the history of the recent ancestors. The third and the last set of specificities is ephemeral, determined at each moment by the present experiences which define the history of the individual, but on the basis of the "recursive" properties of the immune system, it will influence all future actual (and available) repertoires. Thus, ephemeral as it may be, this last system of contacts constitutes, at any moment, the immunological (idiotypic) self (5).

And in recent publications, the theme of Self as identity still remains unchanged: "... It is the immune system itself which defines the molecular identity, or 'Self', as the totality of participating molecules." (9, Stewart *et al*, 1989).

19. Buss, L.W. *The Evolution of Individuality*. Princeton: Princeton University Press, 1987.
20. Weismann, A. Úber die Vererbung, reprinted in Aufsaumltze ber Vererbung und verwandte biologische Fragen (Jena: Gustav Fischer, 1893). Discussed by F.B. Churchill, Weissman, hydromedusae, and the biogenetic imperative: a reconsideration, in T.J. Hordes, J.A. Witkowsky and C.C. Wylie (eds.) *A History of Embryology*. Cambridge: Cambridge University Press, 1986.
21. Roux, W. *Der Kampf der Teile im Organismus*. Leipzig, Verlag von Wilhelm Engelmann, 1881.
22. Chernyak, L. and Tauber AI. The idea of immunity: Metchnikoff's metaphysics and science. J. Hist. Biol. 23: 189–246, 1990 and Tauber and Chernyak *Metchnikoff and the Origins of Immunology*. op. cit.
23. Edelman, G.M. Specific cell adhesion in histogenesis and morphogenesis, in Edelman, G.M. and Thiery, J.P. eds. *The Cell in Contact: Adhesions and Junctions as Morphogenetic Determinants*. John Wiley: New York, 1985, pp. 138–168.
24. Edelman, G.M. CAMs and Igs: Cell Adhesion and the Evolutionary Origins of Immunity. Immunol. Rev. (Copenhagen, Denmark), pp. 11–45, 1987; *Topobiology. An Introduction to Molecular Embryology*. New York: Basic Books, 1988.
25. Edelman argues (24, 1987) that "two major means that exist for altering binding behavior or forming borders. The first is to express a new CAM at cell surface having a different binding specificity than that on neighboring cells (e.g., L-CAM instead of N-CAM or vice-versa). The second is to change the mode by which a particular CAM is expressed at the cell surface by mechanisms of cell surface modulation."
26. Hoffman, S., Chuong, C.M. and Edelman, G.M. Evolutionary conservation of key structures and binding functions of neural cell adhesion molecules. Proc. Natl. Acad. Sci. USA 81, 6881–6885, 1984.
27. Although CAMs have only been identified in vertebrates does not preclude the possibility that in invertebrate taxa there may be other kinds of molecules responsible for cell recognition possibly homologous to vertebrate CAMs. In that case, the immunoglobulin superfamily would have originated from molecules which were common predecessors of cell recognition molecules for all matazoan species. For the discussion of such a possibility see: Matsunaga, T. and Mori, N. The origin of the immune system. The possibility that immunoglobulin superfamily molecules and cell adhesion molecules of chicken and slime mould are all related. Scand. J. Immunol. 25: 485–485, 1987; Williams, A.F. and Barclay, A.N. The immunoglobulin superfamily—Domains for cell superface recognition. Ann. Rev. Immunol. 6: 381–405, 1988.

28. "There is great diversity of sequence in the immunoglobulin superfamily molecules, and the question arises—why is a conserved pattern seen at all? The biological functions... require unique recognition specificities, and these cannot directly be responsible for conserved sequence patterns." Williams, A.F. and Barclay, A.N. The immunoglobulin superfamily, ibid.
29. Klein, J. *Natural History of the Major Histocompatibility Complex*. New York: John Wiley and Sons, 1986. Adams, D.O. and T.A. Hamilton. The cell biology of macrophage activation. Ann. Rev. Immunol. 2: 284–318, 1984. E.R. Unanue. Antigen-presenting function of the macrophage. Ann. Rev. Immunol. 2: 395–428, 1984.
30. The major breakthrough in understanding the form of class I-associated foreign antigen was obtained in cytotoxic T lymphocytes studies of influenza A virus. Townsend, A.R.M. and Skehel, J.J. Influenza A specific cytotoxic T cell clones that do not recognize viral glycoproteins. Nature 300: 655–657, 1982.
31. Skehel, J.J., Barnet, B.C., Burt, D.S., Daniels, R.S., Douglas, A.R., Graham, C.M., Hodgson, J., Knossow, M., Mills, G.H.G., Riska, P.F., Thomas, D.B., Weis, W., Wiley, D.C. and Wrigley, N.G. Immune recognition of influenza virus haemagglutinin. Phil. Trans. R. Soc. Lond. B323: 479–485, 1989.
32. Ceppellini, R., Frumento, G., Ferrara, G.B., Tosi, R., Chersi, A. and Pernis, B. Binding of labelled influenza matrix peptide to HLA DR in living B lymphoid cells. Nature 339: 392–394, 1989.
33. Mogensen, S.C. Role of macrophages in natural resistance to virus infections. Microbiol. Rev. 43: 1–26, 1979.
34. Nathan, C.F. Secretory products of macrophages. J. Clin. Invest. 79: 319–326, 1987.
35. Mock D.J., Domurat, F., Roberts Jr, N.J., Walsh E.E., Licht, M.R. and Keng, P. Macrophages are required for influenza virus infection of human lymphocytes. J. Clin. Invest. 79: 620–624, 1987.
36. Kane, K.P., Vitiello, A., Sherman, L.A. and Mescher, M.F. Cytolytic T-lymphocyte response to isolated class I H-2 proteins and influenza peptides. Nature 340: 157–159, 1989; Chen, B.P. and Parham, P. Direct binding of influenza peptides to class I HLA molecules. Nature 337: 743–745, 1989. Eisenlohr, L.C. and Hackett, C.J. Class II major histocompatibility complex-restricted T cells specific for a virion structural protein that do not recognize exogenous influenza virus. J. Exp. Med. 169: 921–931, 1989; French, R.A., Tang, X-Lin, Anders, E.M., Jackson, D.C., White, D.O., Drummer, H., Wade, J.D., Tregear, G.W. and Brown, L.E. Class II-restricted T-cell clones to a synthetic peptide of influenza virus hemagglutinin differ in their fine specificities and in the ability to respond to virus. J. Virol. 63: 3087–3094, 1989.
37. Hashimoto, G., Wright, P.F. and Karzon, D.T. Antibody-dependent cell-mediated cytotoxicity against influenza virus-infected cells. J. Inf. Dis. 148: 785–794, 1983.
38. Kilbourne, E.D. (1987) *Influenza*, New York: Plenum Medical Book Co., pp. 184–188.
39. These two aspects of immunological activity echo the opposition found in speculative philosophy—the opposition of "practical reason" and "theoretical reason." For the former, the external are mere objects of human desires and needs. Specificities of the objects' appearance signal that they have not been mastered nor subjected to negation and assimilation in the self-identity of the subject. On the other hand, theoretical reason strives to retain what an object is "in itself," i.e., as independent of human intentions. In this case, the characteristics of an object become the proper characteristics of reason itself. With practical

reason, the Self is only manifest as the negation of the Other, while with theoretical reason, the Self shapes itself with the patterns of the Other. But the complimentarity of these two aspects of mind is only apparent in speculative philosophy: Both are ultimately reducible to their underlying unity, which are but one-sided manifestations made explicit in the philosophical construction of "the speculative mind." The only (but from our point of view, the most important) difference between this speculative constitution of mind and the two aspects of immune activity described above by our model of self-determination, is that we assume no such implicit or already provided unity.

40. About distribution of class I and class II MHC molecules among cells of different tissues see Klein (ref. 29), pp. 152–175.
41. Bruce, H.M. An exteroceptive block to pregnancy in the mouse. Nature 184: 105, 1959; Yamazaki, K., Boyse, E.A., Milke, V., Thaler, H.T., Mathieson, B.J., Abott, J., Boyse, J., Zayas, Z.A. and Thomas, L. Control of mating preferences in mice by genes in the major histocompatibility complex. J. Exp. Med. 144: 1324–1335, 1976; Yamazaki, K., Yamaguchi, M., Andrews, P.W., Peake, B. and Boyse, E.A. Mating preferences of F_2 segregants of crosses between MHC-congenic mouse strains. Immunogenetics 6: 253–259, 1978; Yamazaki, K., Yamaguchi, M., Baranoski, L., Bard, J., Boyse, E.A. and Thomas, L. Recognition among mice: Evidence from the use of a Y-maze differentially scented by congenic mice of different major histocompatibility types. J. Exp. Med. 150: 755–760, 1979; Yamazaki, K., Beauchamp, G.K., Wysocki, C.J., Bard, J., Thomas, L., and Boyse, E.A. Recognition of H-2 types in relation to the blocking of pregnancy in mice. Science 221: 186–188, 1983; Yamazaki, K., Beauchamp, G.K., Matsuzaki, O., Kupniewski, D., Bard, J., Thomas, L. and Boyse, E.A. Influence of a genetic difference confined to mutation of H-2K on the incidence of pregnancy block in mice. Proc. Natl. Acad. Sci. USA 83: 740–741, 1986; Weissman, I.L., Scofield, V., Saito, Y., Boyd, H., Rinkevich, B. Speculations on the relationships of two Botryllus allorecognition reactions—colony specificity and resorption — to vertebrate histocompatibility. In *Invertebrate Historecognition*. Ed. R.K. Grosberg, D. Hedgecock, K. Nelson. New York: Plenum Press, 1988, pp. 67–78.
42. Ashida, E.R. and V.L. Scofield. Human major histocompatibility complex encoded class II structures may act as sperm receptors. Proc. Natl. Acad. Sci. USA, 84: 3395–3399, 1987.
43. Scofield, V.L. Evolution of MHC genes and functions: the disease connection, In *Invertebrate Historecognition*. Ed. by R.K. Grosberg, D. Hedgecock, Keith Nelson. New York: Plenum Press, 1988, pp. 81–91.
44. Dohr, G.A., W. Motter, S. Leitinger, G. Desoye, W. Urdl, R. Winter, M.M. Wilders-Truschnid, B. Uchanska-Ziegler, A. Ziegler. Lack of expression of histocompatibility leukocyte antigen class I and class II molecules on the human oocyte. J. Immunol. 138: 3766–3770, 1987.
45. Taneda, Y. and H. Watanabe. Studies on colony specificity in the compound ascidian, Botryllus primigenus Oka. Dev. Comp. Immunol. 6: 243–252, 1982; Scofield, V.L., Schlumpberger, JM, West, L.A. and Weissman, I.L. Protochordate allorecognition is controlled by a MHC-like gene system. Nature 295: 499–502, 1982.
46. Müller, W. Experimentelle Untersuchungen ber Stockentwicklung, Polypendifferenzierung und Sexualchimaren bei *Hydractinia echinata*. Wilhelm Roux' Arch. Entwicklungsmech. 155: 181–268, 1964.
47. Oka, H. Colony specificity in compound ascidians: the genetic control of fusibility. In,

Profiles of Japanese Science and Scientists. H. Yukawa (ed.). Tokyo: Kodansha, 1970, pp. 196–206.

48. Buss, L.W., McFadden, C.S., Keene, D.R. Biology of hydractiniid hydroids. 2. Histocompatibility effector system/competitive mechanism mediated by nematocyst discharge. Biol. Bull. 167: 139 –158, 1984.

EDITOR'S COMMENTS TO ROOT-BERNSTEIN

Abnormal immune regulation was noted in the nascent years of immunology. Anaphylaxis (1902), the Arthus reaction (1903), and serum sickness (1906) were regarded as "allergy", or altered reactivity, and auto-antibodies, e.g., antibodies directed at autologous tissue, was first shown as clinically relevant in 1904 by Donath and Landsteiner. They demonstrated that a rare hemolytic anemia, paroxsymal cold hemoglobinuria, was caused by an auto-antibody that binds to erythrocytes at low body temperatures and initiates complement-mediated lysis upon warming. These early successes of immunopathology were followed by what has been called the "Dark Ages of Autoimmunity", where, following the first decade of this century, little progress (or attention) was made in this area of investigation for 40 years. The pendulum swing from a biological to a chemical orientation was reversed in the 1960's, when observations in the diverse areas of tissue transplantation, tolerance, immunodeficiency and immunopathology demanded a more biological orientation. Root-Bernstein offers a lucid up-date of this complex area and suggests a novel theory to coordinate the diverse characteristics of the autoimmune problem that strikes at the very notion of the Self in complex animals. The conceptual issues however date to the intellectual infrastructure of immunology as first formulated by the early immunologists.

The concept Paul Ehrlich expounded of "horror autotoxicus" as dysteleologic reflects the alarm most immunologists felt of the immune reaction acting against the organism's own elements. The general community misinterpreted "horror autotoxicus" as prohibiting auto-antibodies altogether, a position Ehrlich never held. Why was autoimmunity viewed as dysteleological? From the alternate Metchnikovian point of view, the immune reaction was based on warring disharmonious centers: the immune response arose from the disharmonic state. For Metchnikoff, phagocyte function in infectious diseases, scavenging of damaged, cancerous, or effete cells, or participation in developmental metamorphosis were various aspects of "physiological inflammation"; immune mediation was but a special case. If antibodies were in fact only facilitators of these processes, then such "auto-antibodies" were expected. In fact, search for such factors in Metchnikoff's laboratory were of great interest. Research directed at defining how autoantibodies might be controlled, or even prevented from being formed, was an active focus of early immunologic research. Despite the common conclusion from both the Ehrlich and Metchnikoff camps that anti-auto-antibodies (i.e., antibodies that bound immunoglobulin directed against host targets) were normally formed, Ehrlich projected this finding primarily as the means of thwarting a "dysteleological" condition, whereas the Metchnikovians viewed phagocyte (immune) attack against host tissue as normal, consistently teleologic in the broader context of their true evolutionary function. Auto-antibody in this latter case served as opposing (in this case, harmonizing) the natural phagocyte activity.

Thus while the lesson of immune reactivity to pathogens and foreign insult was generally accepted by the immunological community by 1900, the metaphysical infrastructure of that understanding was certainly diverse. Beyond the different scientific traditions of descriptive biology vesus chemico-reductionism, there resided a different orientation towards the basic teleological vector of the organism. Here we see Ehrlich invoking biological purpose in its most arcane form—how (or why) should an organism self-destruct? Metchnikoff had turned the question on its head—how (or why) are the disharmonious centers constituting the organism harmonized? We are left with the same dilemma.

ROBERT S. ROOT-BERNSTEIN

SELF, NONSELF AND THE PARADOXES OF AUTOIMMUNITY

> Paradoxes play a key role in the advancement of science. They are associated with excitement, and with the knowledge that we must be looking at something the wrong way. The clear formulation of a paradox can herald important advances, since the resolution of the paradox is generally a conceptual step forward. It is therefore of paramount importance to identify paradoxes and focus attention on them.
>
> G.W. Hoffmann, J.G. Levy, and G.T. Nepom, *Paradoxes in Immunology*, 1986.

I.

It is difficult to know where to begin an essay on the problem of immunological discrimination between "self" and "nonself" and the paradoxes created for this discrimination by the phenomenon of autoimmunity. As Sherlock Holmes once commented, "These are much deeper waters than I had thought." [1] Much of my subject is mysterious not only to me but apparently to everyone who has investigated it. Therein lies a clue. The subject itself is a mystery, and therefore calls for the approach of the mystery writer. But this "who done it" differs from most in drawing its characters not only from the human society, but from the animal, microbial, cellular, and even chemical realms as well. I will trace, as novelists are wont to do, their lineage, their personal histories and, of course, their interactions. As a result, this is not the usual story of autoimmunity that is laid out in the few histories of the field that exist [2] or in textbook accounts, especially since I have focused upon the contradictions and paradoxes inherent in our current knowledge, and the gaps in it, rather than upon the myriad facts and inadequate explanations that are the usual fare. But my approach has its inadequacies, too. Most importantly, it provides no conclusion; only clues. I can only hope that, perhaps, if

Alfred I. Tauber (ed.), Organism and the Origins of Self, 159–209.

the plot is devised properly, some budding Sherlock Holmes will perceive the importance of the clues that even the writer overlooks.

The crime of autoimmunity, if we may call it that (for there is some doubt as to whether all manifestations of autoimmunity are undesirable) has probably been committed for as long as higher organisms possessing specialized lymphocytes have existed. But the crime was a silent one, unrecognized and unreported until this century. The first clue that something a little sinister might be going on in the dark recesses of the lymph system followed from findings in 1899 by the Frenchman Jules Bordet. Following hard upon the discovery of antitoxins (antibodies) by Emil Behring, Bordet performed a series of now-classic experiments demonstrating that if red blood cells (erythrocytes) from one species of animal were injected into the blood stream of another species, these cells would be destroyed by antitoxins (antibodies) produced by the host's immune system. These results stimulated the German scientists Paul Ehrlich and Julius Morgenroth to consider the possibility that the same sort of immunological mechanism might provide a normal means for individual animals to eliminate their own old or defective erythrocytes [3]. The answer is still not in on that question. Some immunologists maintain that autoantibodies—that is, antibodies created in response to self antigens—play precisely this role (for example, Lewis Thomas's concept of "immune surveillance," in which the immune system functions to eliminate abnormal "self" cells such as those in cancerous tissues); others that autoantibodies are often formed but are non-functional in normal animals; and others that such true autoantibodies never exist except in disease states [4]. But I digress.

Ehrlich and Morgenroth devised a series of experiments to test their hypothesis. These experiments were technically simple. In one series, blood was removed from an individual animal and then reinoculated into the same animal some time later. In another set of experiments, blood was removed from one individual and inoculated into a different member of the same species. Although Ehrlich and Morgenroth often observed the formation of isoantibodies—that is to say, antibodies directed at the antigenic variations that distinguish one member of a species from another (e.g., the blood group factors, such as those later identified as ABO, MN, and Rh)—they found no evidence of the formation of autoantibodies. Autotransfusion does not lead to autoimmunity. They therefore concluded in 1901 that "the organism possesses certain contrivances by means of which the immunity reaction, so easily produced by all kinds of cells, is prevented from acting against the organism's own elements and so giving rise to autotoxins . . . so that we might be justified in speaking of a 'horror autotoxicus' of the organism.

These contrivances are naturally of the highest importance for the individual." [5]

There has been ongoing debate as to what, exactly, Ehrlich meant by the term "autotoxins" [6]. Did he mean what we now mean by the term autoantibodies, or more generally the entire toxic immunologic reaction which begins with the recognition of a self antigen by an autoantibody and is followed by the series of steps in which a fraction of blood serum called complement binds to the antibody initiating a series of chemical reactions that eventually result in destruction of the antigen? For, as we shall see, it is possible for antibodies to bind to antigens without setting off the complement reaction, and therefore without causing harm to the antigen or its host. The production of autoantibody is not, therefore, sufficient to induce an active autoimmune reaction. But Ehrlich did not know this, and his experiments were not designed to distinguish between the presence of non-immunotoxic autoantibody and the absence of a complement reaction. Despite this important distinction, most immunologists have interpreted the enunciation of "horror autotoxicus" as representing the origin of the modern concepts of immunological "self" and "nonself". The organism, according to most immunology textbooks written during the past ninety years, will not normally react immunologically against its own chemical components.

Yet the immune system apparently does not behave exactly as Ehrlich described, or at least not always. A year before Ehrlich announced his concept of "horror autotoxicus" Serge Metalnikoff had performed a seemingly contradictory experiment in the Pasteur Institute laboratories of Elie Metchnikoff. He and Elie Metchnikoff and their colleagues found that not only erythrocytes, but almost any tissue from any animal could be used to induce an immune response in an animal of another species (heteroimmunization) or in most individuals of the same species (isoimmunization). Like Ehrlich and Morgenroth, they were also interested in what happens as a result of autoimmunization. So Metalnikoff tried inoculating a guinea pig with its own sperm. He found that the guinea pig's serum reacted to the sperm. There was an interesting twist to his observations, however. These autoantibodies only destroyed the sperm when the two were combined with complement in a test tube. Inside the body of the guinea pig, the autoantibodies did not appear to react immunologically to the sperm, nor (unlike isoantibodies to sperm) did they appear to produce any pathological symptoms. Although at first sight these results contradicted the concept of "horror autotoxicus," Ehrlich interpreted Metalnikoff's results in a very different way. He noted in a footnote that the failure to produce an *active* immune response was consistent with his notion of "horror autotoxicus" since his principle stipulated nothing about the production of autotoxins, only that such autotoxins might not "act against the organism's own elements" [7].

Just three years later, even this refuge disappeared when a self-reactive immune reaction was finally demonstrated. In 1904, Julius Donath and Karl Landsteiner, whose very different philosophical leanings led them to be constant opponents of Erhlich's theories, produced evidence for the presence of self-reactive antibodies in human patients suffering from paroxysmal cold hemoglobinuria (PKH). PKH is a disease associated with a small fraction of syphilitic patients. When patients suffering from PKH experienced a drop in body temperature, or sometimes if they simply drank a cold glass of liquid, they would subsequently develop severe pains in their limbs, back, and head; they would vomit; and they would pass dark brown urine as their body temperature returned to normal. These symptoms were correlated with destruction of red blood cells. Donath, who saw a number of these patients in his clinic, collaborated with Landsteiner, to demonstrate that antibodies directed at erythrocytes were present in the blood of PKH patients. For unknown reasons, these antibodies would only interact with the erythrocytes when the patient's body temperature dropped below a certain temperature. Then, when the patient's temperature rose back to normal, a complement-activated immune reaction resulted in the destruction of the erythrocytes to which antibody had previously bound. The resulting destruction of red blood cells caused localized pain and the hemoglobin released into the serum then passed into the urine causing brownish hemoglobinuria. Despite the fact that the cause or causes that initiate PKH were (and are) totally unknown, the 1904 report by Landsteiner and Donath on PKH is often cited as the first verified instance of the formation of "autoantibodies" in human beings, and thus of the phenomenon of human "autoimmunity". This phenomenon led Donath and Landsteiner to conclude that the concept of "horror autotoxicus" could not be correct, or at least not under all circumstances [8].

Ah! But what were these circumstances? When did this immunological Dr. Jekyll, who protected the body from infection the vast majority of the time, suddenly transform into the pathological Mr. Hyde, with his self-destructive tendencies? Does autoimmunity result from true autoimmunization (perhaps in the case of PKH to tissues damaged during syphilitic infection, for example), or is autoimmunity an unusual complication of iso- or heteroimmunization in which the resulting antibodies "cross-react" with self determinants as well? Further research conducted before World War II, both clinical and laboratory, provided clues but no firm answers. Autoimmune reactions to brain tissue, the lens of the eye, the thyroid, and to various tissues following syphilis infections were all reported [9]. Oddly, the data did not (and for many more recently discovered autoimmune conditions, still does not) clearly support one or the other of the positions elaborated during the first four years of autoimmunity research. Our ignorance is stunning. Some of the experiments seem to confirm the most rigorous interpretations of Ehrlich's principle of "horror autotoxicus"; others Metalnikoff's finding of inactive autoantibodies; and still others, the Donath-Landsteiner observation of active autoimmune reactions, but almost always in "naturally" occurring disease in which there was no identifiable initiator. One additional twist has also been added to complicate the already complicated picture: During the middle of this century, it became clear that immunity is not regulated solely by antibody producing cells (B-cells), but also by cells that react directly with antigens and are activated in the thymus (T-cells). In some cases in which autoantibodies were absent, auto-reactive T-cells were eventually identified.

Perhaps the only thing that can be said safely about the results of the past ninety years of research on autoimmunity is that today, just as at the turn of the century, the phenomena of autoimmunity are still at odds with immunological concepts of self and nonself. How can the body both distinguish immunologically between self and nonself and yet, under certain circumstances, produce an immune reaction against itself? Most textbooks "solve" this paradox by speaking of autoimmunity as an "unnatural" or "pathological" state that results from "chance" or "accidental" triggering of immunological reactions against "foreign" antigens that unfortunately "cross-react" with "self" proteins and "override" or "bypass" the "normal" functioning of the "suppressor" system that should prevent the activation of "forbidden clones" that normally cannot react immunologically against "self". Why such self-destructive immune responses should survive the deletion or inactivation steps that supposedly prevent self-reactive clones from proliferating under "normal" circumstances is never adequately explained and often not addressed at all. Thus, autoimmunity is often portrayed as a mistake that only occurs when the immune system fails to distinguish self from nonself as it ought [10].

But who or what determines how the immune system "ought" to work? As appealing as this teleological approach to immune function may be, it is fraught with difficulties, not the least of which is that autoimmunity is so common. It is now thought that as many as one in ten human beings suffer from one or more diseases that have autoimmune complications, and in certain syndromes, such as acquired immunodeficiency syndrome (AIDS), autoimmunity (sometimes directed against multiple tissues) is the rule rather than the exception. A phenomenon that is so common can hardly be attributed to abnormal causes or to accident. On the contrary, it is becoming clear that something is amiss with both our understanding of the self-nonself distinction and our desire to attribute autoimmunity to unnatural or accidental factors. Such terms represent nothing more than a failure to grapple successfully with the phenomena. The very language with which we discuss the problem reveals the need for a new approach.

A new approach is needed not only because the outstanding problems presented by autoimmunity demand a new approach, but also because current immunological theory and practice has become too uniform. The history of immunological research demonstrates repeatedly that controversies such as that represented by the debates over "horror autotoxicus" between the teams of Ehrlich-Morgenroth and Donath-Landsteiner drive research in the most fruitful directions by pinpointing the most problematic areas of ignorance and yielding the most interesting and paradoxical experimental results.

In fact, it is interesting to note that what has been called the "Dark Age" of autoimmunity research—a generation-long period of virtual stagnation that occurred between the World Wars—coincided with the deaths or emigration from the field of the major figures who had defined the issues and phenomenological details of autoimmunity: Ehrlich died in 1915; Metchnikoff in 1916; Behring in 1917; and Morgenroth in 1924. Bordet's career effectively ended with his acceptance of the Nobel Prize in 1919. Landsteiner was forced to leave Europe after World War I to take up a new position at the Rockefeller Institute in New York where he performed no further autoimmunity research. And Metalnikoff, who had returned to Russia, was forced by the Revolution to flee back to Paris, where his new focus of research was on the neurological control of the immune response. In consequence, all of the major controversies that had driven autoimmunity research for the prior two decades came to an abrupt and unresolved end. It is probable that most students of immunology sighed with relief at this unexpected hiatus, and viewed the clearly difficult problems of autoimmunity as barren soil no longer worth the effort of tilling [11].

Research turned to the new fields of immunochemistry and genetics, both of which were to play essential roles in later developments in autoimmunity research, but not until the totally new theories of antibody formation and self recognition proposed by Felix Haurowitz and Linus Pauling came into conflict with the emerging concepts of clonal selection and a natural selection theory of antibody formation devised by Macfarlane Burnet and Niels Jerne.

Out of that renewed controversy came the modern theories of immunology that dominate the field today; but these theories, too, have failed to explain the still puzzling problem of autoimmunity. It is therefore time for new theories and new controversies to restimulate the field.

II. AUTOIMMUNITY IN LIGHT OF EXPERIMENTAL ALLERGIC ENCEPHALOMYELITIS (EAE)

I would therefore like to propose a new theory of autoimmunity—at present half-baked and certainly incomplete, but perhaps detailed enough to foster the necessary theoretical and experimental rethinking. It has developed out of detailed studies of one particular autoimmune disease that was discovered, as chance would have it, during the "Dark Age" of autoimmunity just mentioned, and which has attracted a great deal of research since World War II: experimental allergic encephalomyelitis, or EAE. EAE is just one of a group of very well-studied animal models of autoimmunity, which include experimental autoimmune thyroiditis, experimental peripheral neuritis, experimental autoimmune orchitis, and several others. All of these diseases have very similar pathological, histological, and experimental characteristics, with the one major exception that each disease affects a different tissue and is initiated by a different antigen [12]. The many similarities in disease progression and response to various treatments suggests that autoimmunity—or experimental autoimmunity, at least—has a common mechanism underlying it, if no single cause.

EAE developed from studies of the antibody response to brain tissue antigens. It had been observed since the beginning of the century that single inoculations of homologous brain tissue would cause the development of antibodies against that tissue, but not the development of autoimmunity. Thus, when T.M. Rivers created EAE in 1933 by repeated subcutaneous injections of emulsified homologous brain tissue, it was a major breakthrough. EAE developed several months after the initial inoculations, resulting in a slowly progressive paralysis, which histological studies revealed to be caused by demyelinization of nerves in the brain, spinal cord, and peripheral nerves. The pathology was also characterized by early disruption of the blood brain barrier, microglial abnormalities, and perivascular cuffing caused by quite conspicuous cellular infiltration by lymphocytes of central nervous system vasculature and surrounding tissue (Fig. 1). Oddly, and importantly as we shall see, lymphocytes were almost never observed to directly attack the myelin sheaths of nerves, but somehow damaged them as a

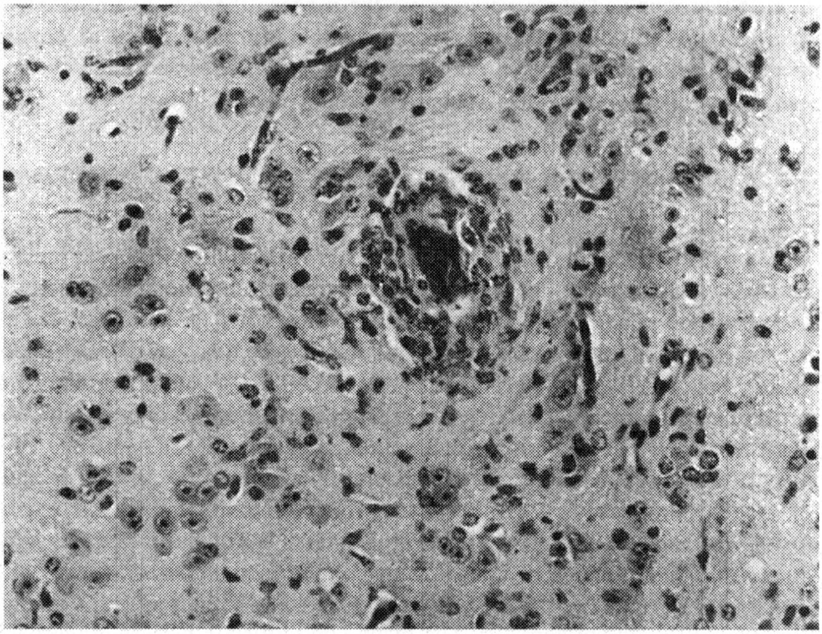

Fig. 1. Perivascular cuffing within the brain of a guinea pig that is typical of the histology of advanced experimental allergic encephalomyelitis (EAE). The stained cells are mostly lymphocytes that have congregated around a vein shown at the center of the photograph. These lympho cyte complexes are responsible for neighboring demylinization.

result of the neighboring perivascular cuffing. Nonetheless, EAE became known as a disease of myelin, and almost all descriptions of the disease state that the immune response is directed against myelin proteins [13].

The actual cause of the immune response took some time to elucidate. Although anti-brain antibodies were detected early in history of EAE research, no correlation could be found between antibody levels and pathogenesis, and passive transfer of antibodies to healthy animals did not result in disease. EAE was eventually demonstrated to be caused by T-cells, following the Patterson experiments in 1960 showing that the disease could be transferred passively from one animal to another by lymphocytes but not serum. Because similar paralysis, demyelinization, and disease progression characterizes multiple sclerosis (MS), studies of EAE are often justified as possibly illuminating the etiology of MS. It has been suggested, however, that EAE may be more similar to post-rabies encephalomyelitis, which haunted the use of Pasteur's rabies vaccine from its first widespread use in 1885 period. Some batches of the Pasteur vaccine may have contained trace quantities of rabbit brain, just sufficient to induce EAE after repeated administration to humans and animals [14].

III. MICROBIAL 'ADJUVANTS' AND THE EVOLUTION OF MOLECULAR MIMICKRY

From a purely practical standpoint, EAE was a difficult disease to study during the initial phase of its history. Experimental diseases that take months to develop and years to be fatal are not the most useful ones to a scientist (and in the modern age do not produce sufficient papers to assure grant renewals) and so EAE research might not have developed very quickly except for another exciting, but rather confusing discovery by Jules Freund and his colleagues in 1942. They found that sensitization to horse serum could be facilitated by mixing the serum in an oil-water emulsion with some bacteria. The oil-water emulsion apparently served to prevent the antigens from being swept into the circulation or lymph and cleared by non-immunological means such as enzymatic breakdown. The role of the bacterium was somewhat puzzling, since a single species of bacterium such as *Mycobacterium tuberculosis* was eventually found to have an immunopotentiating effect on several apparently different antigens. On the other hand, not every bacterial species would act as an adjuvant. By trying enough different bacterial combinations, however, Freund (and also Voisin) was able to demonstrate that mixing adjuvants with sperm altered the results of Metalnikoff's experiment to produce cytotoxic antibodies. And Elvin Kabat demonstrated in 1947 that the time course of EAE induction could be decreased from Rivers' months to a matter of days when homologous or heterologous brain tissue was homogenized in a complete Freund's adjuvant prior to inoculation. This new, acute form of EAE has been the focus of a great deal of study in the decades since, and the techniques of adjuvant use pioneered by Freund and Kabat have been applied to the induction of almost all experimental autoimmune diseases [15].

Adjuvants are interesting for more than just their acceleration of the disease course in autoimmunity. Their primary component, microorganisms, play other interesting roles in the story of autoimmunity research and theory. The use of bacterial adjuvants was not new with Freund (although he was the first to standardize their use) but is, as far as I have been able to determine, hidden in the mists of unwritten history. The story I have heard, but cannot vouch for from documentary sources, is that the use of adjuvants began in Pasteur's laboratory during the 1880s as a result of vaccine manufacture. At about that time, producers of antisera noticed that some animals produced much higher quantities of immunoglobulins than did others, and that the high producers invariably developed an infection at the site of the inoculation. The infectious bacterial material was therefore cultured and purposely added to subsequent inoculations to boost antiserum production. As a result of this use, adjuvants became defined as non-specific immune potentiators that boosted the response to particular antigens. The question of how adjuvants do this immunological boosting [16] is one that will be taken up in various contexts below. It holds a number of unexpected surprises for anyone trying to explain autoimmunity.

Microorganisms are relevant to autoimmunity in other ways also. At about the same time that Freund and Kabat were revolutionizing the production of animals models of autoimmunity by using adjuvants, other investigators were beginning to ponder the possibilities that human autoimmune diseases such as multiple sclerosis, post-pertussis and post-measles encephalopathies, and amyotrophic lateral sclerosis (ALS), were triggered by viral or bacterial antigens. There was certainly evidence that Guillain-Barré syndrome, a progressive but usually non-fatal paralysis, was an autoimmune syndrome that followed a variety of viral infections [17]. But how could species as different from a human being as are viruses and bacteria cause autoimmunity?

The answer comes oddly enough from considerations of the theory of evolution by natural selection. Imagine for a moment, a time hundreds of thousands of years ago when there existed (at least presumably) a pathogenic family of viruses or bacteria (either one will do for the moment) whose genotypes differ from one another just as human genotypes differ from individual to individual. Some members of this microbial family produced proteins that are highly antigenic to the genus Homo; some members of the family were, by chance, less antigenic. The less antigenic members had a better chance of surviving and being transmitted to another human being, and so preserving the family name. At the same time, a similar selection process occurred among the individual members of the genus Homo. Those most susceptible to infection died without reproducing; those less susceptible, or with more active immune systems, survived to reproduce. Over the eons, a balance was struck between antigenicity and pathogenicity. In some cases, this balance no doubt resulted in immune tolerance to the microorganism and, if the pathogenicity was low enough (as is the case with *E. coli* and various other common gram negative bacteria) to at least partial symbiosis.

Now, add to this story the concept of "horror autotoxicus." If we accept, at least provisionally, that the immune system will not, under normal circumstances, mount an active response to self antigens, then a clever individual might reason that those pathogenic members of the virus or bacteria family with the best chance of surviving over the millennia are those whose proteins mimic most closely the proteins of their host and are therefore the least antigenic. The more alike the hominid and microbial proteins become, the less likely the host immune system will be to recognize them as nonself, and the less likely it will be to mount an effective immune response. Therefore, one would expect that one of the consequences of Darwinian natural selection as applied to infectious disease would be the development of molecular mimickry between pathogen and host. So argued the Australian parasitologist J.F.A. Sprent in a seminal paper in 1962 [18].

In fact, Sprent appears to have been right. Subsequent studies of many human pathogens have demonstrated significant molecular mimickry to human proteins and cell-surface markers. For example, more than a hundred of the gram negative bacteria (including various *E. coli* strains) produce immune responses cross-reactive with human blood group antigens, and it is thought that infections with these bacteria (which are virtually ubiquitous) may account for the presence of blood group immunity very early in life when no isologous blood contact has occurred. Cross reactions have also been observed between Type XIV pneumococcal polysaccharide and human blood groups. Substantial homologies have been identified between Staphylococcal antigens and a wide range of human proteins, and a number of strains of *Streptococci* induce antibodies that cross-react with heart muscle during rheumatic fever. *Mycobacterium leprae* has a human T cell epitope cross-reactive with HLA-DR2 peptides, and it has been suggested that such cross reactivity may help to explain observed genetic susceptibility in the epidemiological patterns of leprosy. And a large number of viruses, including influenzas A and B, adenovirus, hepatitis B virus, Epstein-Barr virus, and measles virus also contain proteins with significant and often multiple homologies to human proteins. Altogether, one study reported that about three percent of anti-virus antibodies have been found to cross-react with human tissues [19].

The details of this molecular mimickry are still being investigated, but it is already clear that microorganisms are incredibly clever at hiding from the immune system. Viruses often make up their outer membrane directly from the cell they infect. Bacteria and protozoa seem to be able to vary their surface proteins quite rapidly, and there is speculation and some evidence that some even take up host proteins into their outer membranes or cell walls as immunological camouflage.

Mammals have also co-evolved to utilize common bacterial antigens, such as the adjuvant peptide (N-acetyl-muramyl-L-Ala-D-isoGln), a cell-wall component commonly produced by gram positive bacteria, in clever ways that respond to this molecular mimickry. Several investigators have recently demonstrated that this adjuvant peptide has many of the properties of the neurotransmitter and immune stimulating hormone serotonin (5-hydroxytryptamine). Both cause macrophages to release interleukins; both stimulate slow wave sleep; both induce fever; and the adjuvant peptide has even been shown to bind to serotonin binding sites on lymphocytes and certain types of brain cells. These findings suggest that higher animals have evolved to respond to the presence of bacterial breakdown products by reacting not only in the specific ways defined by the production of particular antibodies or T-cell lines, but also in a general physiological manner consistent with increasing the immune response and conserving energy by increasing sleep [20]. It is possible that the general mode of bacterial adjuvant activities may therefore be directly related to their mimickry of the immune-stimulating hormones and neurotransmitters of the host organism. This would explain the wide-ranging and non-specific nature of adjuvants in many of the cases in which they are used. Obviously, this is an area that merits a great deal more research than has so far been expended on it, and one that will be of great interest to evolutionary biologists, immunologists, microbiologists, and vaccinologists alike. The immunostimulating effects of adjuvant peptide will also turn out to be important for the understanding of autoimmunity, as we shall see below, but the role of the adjuvant in autoimmunity appears to be much more specific than suggested by the observations summarized above.

IV. LIMITATIONS OF MOLECULAR MIMICKRY AS AN EXPLANATION FOR AUTOIMMUNITY

But to return to our main theme, how does this molecular mimickry explain the initiation of autoimmunity? In itself, it does not. But it does get us a step forward toward such an explanation, for one might expect that the more a pathogen mimics its host, the harder it will be for the immune system to react successfully to it; and that the more successful an immune response is to a molecular mimic, the more likely it is to be self-reactive. Thus, autoimmunity itself may be a natural consequence of Darwinian selection in response to microbial antigenic mimickry. Moreover, the existence of microbial-host molecular mimickry provides a quite plausible mechanism for creating heteroimmunization that may produce antibodies capable of causing autoimmunity.

But is heteroimmunization with self-mimicking antigens *sufficient* to produce autoimmunity? Although several investigators have postulated such an idea [21], that would be too simple, and our mystery would end before it had really become interesting. No, the evidence indicates that something besides molecular mimickry is necessary for an infectious agent to induce autoimmunity. Recall, for example, the failure of isoimmunization by erythrocytes to cause autoimmunity in Ehrlich and Morgenroth's experiments, or the autoimmunization experiments of Metalnikoff. Antigenic identity or similarity is clearly not sufficient. Whatever the mechanism that prevents autoantibodies stimulated by autoimmunization or isoimmunization from producing pathological consequences (and I emphasize that this mechanism remains a mystery in its own right), also seems to work to prevent autoimmunity from resulting from heteroimmunization against antigens that mimic self proteins.

The evidence on the inadequacy of heteroimmunization with self-mimicking antigens to induce autoimmunity is quite clear. Detailed studies of the incidence of antibodies to streptococci strains that cross-react with heart tissue have demonstrated that the presence of such antibodies are unrelated to whether patients develop rheumatic fever or heart damage. Similarly, anti-myelin antibodies have been observed in many cancer patients, although these patients appear to have no nerve damage. And other evidence leads to the same conclusion. For example, although measles virus has significant homologies with human myelin basic protein, the incidence of post-infectious measles encephalitis is on the order of 1 in 10,000 cases. The incidence of post-vaccinal measles encephalitis is on the order of 1 in 100,000 vaccinations. The same approximate figures apply to almost all infectious diseases, whether viral or bacterial. Clearly, then, the majority of people—the *vast* majority—escape the vast majority of their infections without autoimmune complications and in spite of developing what appear to be autoantibodies [22].

Part of the protection (and therefore, conversely, the susceptibility) against autoimmunity is undoubtedly due to genetic diversity. Thus, the studies of Jaworski *et al.* on a highly inbred population of Old Colony Mennonites in Canada have demonstrated unusually high incidences of multiple sclerosis, insulin-dependent diabetes, rheumatoid arthritis, and red cell aplasia. The most notable correlation between genetics and autoimmunity is undoubtedly the case of ankylosing spondylitis, which is associated with HLA B27 to the extent that the relative risk of developing the disease is nearly ninety percent among individuals having this HLA type. On the other hand, ankylosing spondylitis is a notable exception among autoimmune diseases, and most other HLA associations are much less predictive. HLA Cw6 provides a marker associated with about thirteen percent risk of developing psoriasis vulgaris, for example, HLA DR2 about a four percent risk for developing multiple sclerosis, and HLA DR4 about a six percent risk for developing insulin-dependent diabetes. These are much more typical of the sorts of genetic predisposing factors that have been elucidated for autoimmune diseases thus far. Genetic factors, in short, do not cause autoimmunity, but most probably determine susceptibility to other, as yet undetermined, triggering factors [23].

The interpretation of genetic susceptibility data just summarized is consistent with the studies of species susceptibility to EAE. Numerous studies of the requirements for EAE induction have been carried out. These demonstrate that auto- and iso-immunization are not capable of inducing EAE, but that heteroimmunization with myelin basic protein (or a peptide fragment of MBP) in combination with one of several Freund's complete adjuvants is usually sufficient. These data suggest both that some degree of *difference* between self and nonself antigens is necessary to induce autoimmunity (an interesting observation in and of itself), and that differences in the antigen often demand differences in the adjuvant. Thus, whereas one peptide fragment of MBP is effective with a Mycobacterial adjuvant in some species of animals, another peptide fragment combined with *Bacillus pertussis* may be necessary to induce the disease in another species (Table I). For example, the minimal requirements for inducing EAE in guinea pigs was a different peptide derived from myelin basic protein than that for rats, and rather than employing a Mycobacterial adjuvant, required an unidentified component of *Bacillus pertussis.* And EAE in monkeys could be induced with yet a different peptide subunit of MBP in combination with *Mycobacterium tuberculosis.* While some investigators continue to maintain that certain species, or inbred strains of certain species, are resistant to EAE induction, time and the ingenuity of researchers have demonstrated that most animals can be provoked into developing EAE if the correct encephalitogen-adjuvant combination is given. The reasons for these differences in species susceptibility to varied molecular determinants of EAE induction is as yet a mystery.

Two further lines of evidence suggest that just as molecular mimickry is insufficient to explain the incidence of autoimmune complications following infection, so is genetic susceptibility. One reason is that within a species, such as guinea pigs or rats, susceptibility to experimental autoimmunity induction,

TABLE I
Species specificity of known major encephalitogenic regions within myelin basic protein. Reproduced from Root-Bernstein and Westall, 1986 [38].

Sequence	Residues	Species in which EAE is produced			
		Monkey	Rabbit	Guinea pig	Lewis rat
PHE-LYS-LEU-GLY-GLY-ARG-ASP-SER-ARG-SER-GLY-SER-PRO-MET	154–167	+	NT	NT	NT
PHE-LYS-LEU-GLY-GLY-ARG-ASP-SER-ARG	154–162	NT	+	–	+
PHE-SER-TRP-GLY-ALA-GLY-GLY-GLN-ARG	114–122	—	+	+	–
THR-THR-HIS-TYR-GLY-SER-LEU-PRO-GLN-LYS	65–74	–	+	–	–
PRO-GLN-LYS-SER-GLN-ARG-SER-GLN-ASP-GLY-ASN-PRO-VAL-VAL	71–89	NT	–	–	+*
Mid-peptide	45–89	+	+	–	+*

NT = Not tested.
* = Requires *B. pertussis* instead of *M. Tuberculosis as* 'Adjuvant' for production of Hyperacute form of EAE.

while variable in absolute clinical manifestations from individual to individual, is nearly universal. For example, various inbred strains of guinea pigs are more susceptible to the induction of EAE than others, but an experimenter can expect that a protocol that induces EAE in a susceptible strain of guinea pigs will have reasonable success on outbred (that is to say, genetically diverse) strains as well.

The more impressive evidence against genetic susceptibility as the most important controlling factor for autoimmunity is the case of AIDS patients. The majority of AIDS patients have autoimmune complications. The vast majority (probably 70 to 80 percent) of AIDS patients develop dementias of one type or another, and nearly half of these dementias involve a demyelinating process affecting either the central nervous system, the peripheral nervous system, or both. The cause of this demyelinization has not been found, but will be discussed in detail below. Virtually all gay men who develop AIDS (over 90 percent) have antibody to sperm present in their sera, and these antibodies have been demonstrated to cross-react with T-cells, which may result in at least some portion of the immunosuppression that typifies AIDS. In addition, biopsies and autopsies on gay men who have developed AIDS have demonstrated that aspermatogenesis, testicular atrophy, and damaged seminiferous tubes are the rule. The histological details are very similar to what is seen in experimental autoimmune orchitis induced by repeated inoculations of heterologous semen or by single inoculations of semen in Freund's complete adjuvant (FCA). And it has been found that other autoimmune diseases, such as idiopathic thrombocytopenic purpura (which results in the destruction of platelets thereby interfering with blood clotting) occur at much higher rates in AIDS patients (as well as lymphoma patients) than in the general public. All of these data strongly suggest that, just as in animal models, the correct combination of factors will induce widespread autoimmunity in human beings within a fairly heterogeneous genetic population [24]. The crucial question is what this correct combination of factors may be. Or, stated another way, what characterizes AIDS patients that is unusual in other human populations?

V. PARADOXES POSED BY THE PREVENTION AND SUPPRESSION OF EAE

Posing the question this way gets us somewhere. But before the clues provided by AIDS cases can be understood, many aspects of animals models of autoimmunity need to be examined first. While all these discoveries of adjuvants, genetics, molecular mimickry, and human autoimmunity were being followed up, various odd observations began to be made about the characteristics of our main character, EAE. It was beginning to behave quite suspiciously. For example, many investigators noted that if an animal was inoculated with brain material or its purified encephalitogenic components some days or weeks prior to inoculation of the autoimmunity-inducing dose (that is to say, the encephalitogen-adjuvant combination), the animal did not develop EAE. Similarly, if the animal was inoculated with the adjuvant material several days or weeks prior to receiving the autoimmunity-inducing dose, the animal did not develop EAE. In short, prior administration of either component of the antigen-adjuvant mixture actually *protected* the animal from developing autoimmunity. Ben-Nun and his colleagues were even able to demonstrate that this protective capacity could be passively transferred from myelin basic protein-inoculated animals to those receiving an encephalitogen-adjuvant combination by means of T-cells activated against myelin basic protein, an encephalitogenic component of brain tissue [25].

Now, these observations concerning the protective value of pre-administering the encephalitogen or the adjuvant are particularly odd in light of several previous observations. One was Richet's discovery of anaphylaxis. He had observed in 1903 that prior sensitization of an animal with an antigen sometimes resulted in such an overwhelming immune response on reinoculation with that antigen that the animal died within minutes [26]. In most cases, however, the result of reinoculating any antigen was to boost the immune response to it. In this instance, however, prior sensitization actually resulted in the opposite phenomenon—loss of the expected (auto)immune response or, in other words, suppression. Here is one of the most outstanding paradoxes of immunology: how can a repeated dose of antigen result in three, mutually exclusive immunological responses?

Then there were Rivers' previous protocols to consider [27]. Rivers had used multiple inoculations of brain material to *overcome* "horror autotoxicus" and so produce active autoimmunity. Why, then, should not inoculations of brain material followed by brain material in FCA also result in autoimmunity in the same way?

Finally, there was the matter of the adjuvant itself. Adjuvants had come to be defined as non-specific immune potentiators. Why, then, did not multiple inoculations of the encephalitogen, especially potentiated by the presence of the adjuvant in the second inoculation, simply speed up the rate at which autoimmunity developed? One often voiced "explanation" provided for these prevention experiments is that T-suppressor cells are induced by the initial inoculation. Logically, however, T-suppressor cells do not help to clarify the situation, since one has to suppose that while a single inoculation of encephalitogen in the absence of adjuvant causes immunosuppression by activating suppressor T-cells, in the presence of adjuvant, T-killer cells were activated instead. Was the role of the adjuvant, then, to override some natural suppression system? Perhaps, but if so, that result would again be paradoxical, because adjuvants such as Mycobacteria are well known to induce their own, quite active immune response. And how can they override the suppressor system when in the presence of the encephalitogen, and yet cause immunosuppression when inoculated by themselves several days or weeks prior to the encephalitogen-adjuvant dose? How, in short, do the same antigens and the same adjuvants inoculated in the same doses (but at different times and in different combinations) result one time in active immunity (but not autoimmunity), another time in immunosuppression, and yet in other instances in autoimmunity? Clearly timing has something to do with it, but how does the immune system tell time?

The most difficult problem posed by the observation that pre-administration of adjuvant protected against induction of EAE is the implication for understanding the difference between suppression (or tolerance) and active immunity. A single dose of Mycobacteria provokes active immunity against Mycobacteria. A single dose of heterologous myelin basic protein (MBP) provokes active immunity against this MBP. And yet, the implication of prevention studies is that both of these active immunizations lead to suppression (or tolerance) if followed by an encephalitogenic inoculation. There are only three possible interpretations of these data: 1) that there is no difference between active immunization and tolerance induction, which implies that T-killer cells and T-suppressor cells are actually of the same origin; 2) that T-killer cells can transform into T-suppressor cells in response to some subsequent signal; or 3) that the immune system can foretell the future, and therefore knows in advance whether to make killer or suppressor cells. None of these possibilities are currently acceptable according to modern theory, and thus these data are tremendously important and exciting.

And that is only the beginning of the oddities. In 1960, Shaw *et al.* found that when guinea pigs inoculated with an encephalitogenic dose were subsequently subjected to skin tests with brain extract, the clinical and histological symptoms of the animals were greatly reduced or even eliminated. In other words, *post-induction* suppression of EAE was possible using the encephalitogen. This observation was followed up by many other investigators, who demonstrated that either repeated doses or single large doses of any protein or peptide cross-reactive with the encephalitogen would cause such immunosuppression of autoimmunity. But they also observed, most unexpectedly, that cell-mediated immunity to myelin basic protein remained at essentially the same levels after successful suppression as they had been before [28]. If the autoimmunity that characterizes EAE is caused by a T-cell response to myelin, as most descriptions of the disease maintain, then how is the continued presence of autoreactive T-cells to myelin in EAE-free animals to be explained? Once again, the notion of T-suppressor cells are often introduced, but once again, the problem remains of why myelin basic protein produces a T-suppressor response if inoculated prior to encephalitogen-adjuvant challenge; a T-killer response if inoculated alone without subsequent challenge; a T-killer response in combination with adjuvant; and a T-suppressor response as a result of reinoculation following an encephalitogen-adjuvant challenge? One must either assume that the immune system is incredibly "smart" and makes "decisions" concerning its response based upon "criteria" that are currently mysterious to immunologists; or one can assume that our scientific analysis of the various situations described above is inaccurate and needlessly complex. Since nature generally employs very simple principles in a diversity of fascinating ways, my own prejudice is to believe that it is our understanding that is inadequate, and that the immune system is neither "smart" nor does it make "decisions."

Another important and confusing aspect of EAE concerns the role of the adjuvant. By what mechanism do adjuvants exert their effects? This is another of those unsolved problems that have lain hidden under the fabric of autoimmunity research for decades. So far, attempts to address it have all failed and as with many immunological theories, mainly because the very language we use conflates theory and experience and replaces explanations with names. In this instance, to label something an adjuvant effect is no better than for Moliere's doctor to assert that sleeping potions cause sleep because they are somnogenic. For example, the most obvious theory that could explain both the induction of EAE by an antigen-adjuvant combination, and the fact that antigen and adjuvant were each, independently, capable of suppressing the disease was that the antigen and adjuvant were immunologically cross-reactive. An immune response to one would therefore result in an increased immune response to the other by means of cross-reactivity. Suppression could be explained similarly on the well-established observation that large or repeated doses of an antigen often results in development of tolerance. Unfortunately, this very plausible explanation was quickly demonstrated to be nonviable. [29]

Mycobacterium adjuvants do indeed elicit cell-mediate immunity, but a large number of studies (none of which, interestingly, cite the others, suggesting that negative results are all too often ignored even when they are published) demonstrated that adjuvant-induced immunity is not cross-reactive with myelin basic protein [30]. Moreover, by 1980, the structures of the antigen and adjuvant, and the minimal component of these necessary to cause EAE, had been defined and demonstrated to be very different molecules. For example, in guinea pigs and rabbits, the minimal requirements for EAE induction consisted of the so-called tryptophan peptide of MBP (residues 111 –121, Phe-Ser-Trp-Gly-Ala-Glu-Gly-Gln-Arg) and the cell-wall breakdown product, N-acetyl-muramyl-L-Ala-D-iso-Gln, or "adjuvant peptide," that mimics serotonin. Similar research on autoimmune orchitis demonstrated that the minimal molecular determinants for induction of this disease were luteinizing hormone-releasing hormone (LHRH) complexed with adjuvant peptide [31]. In short, the adjuvants neither looked structurally similar to the antigens, nor was there any evidence that they promoted the immune response to the antigen by inducing cross-reactive immunity. Interestingly, however, the tryptophan peptide of MBP and LHRH are themselves very

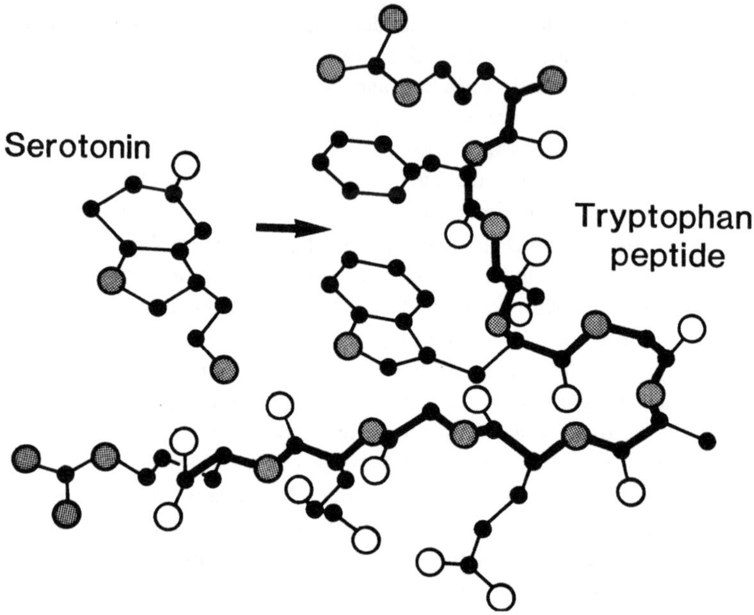

Fig. 2A. Illustration of the intercalation of serotonin (5-hydroxytryptamine) into a "molecular sandwich" composed of the phenylalanine-serine-tryptophan sequence of the tryptophan peptide of MBP (Fig. 1). Reproduced from Root-Bernstein and Westall, 1984 [20].

similar in structure, both containing a sequence called a "molecular sandwich" which contains two aromatic side chains flanked by common hydrophilic side chains, and both of these "molecular sandwiches" act as binding sites for serotonin (Figs 2A and 2B). In fact, adjuvant peptide also binds to this serotonin binding site [32], Thus, structural similarity in the antigen is correlated with sharing a common adjuvant, and the antigen and adjuvant are chemically *complementary*, rather than being chemically similar (Fig. 3).

Yet other sets of experiments also argued against adjuvants being simply non-specific immune potentiators in the induction of autoimmunity. Kabat and Freund had demonstrated during the late 1940s that mycobacteria could be replaced as adjuvants for EAE with only a very limited number of other microorganisms [33]. This suggested not non-specific immune potentiation, but some sort of specific collaboration between antigen and adjuvant. Other

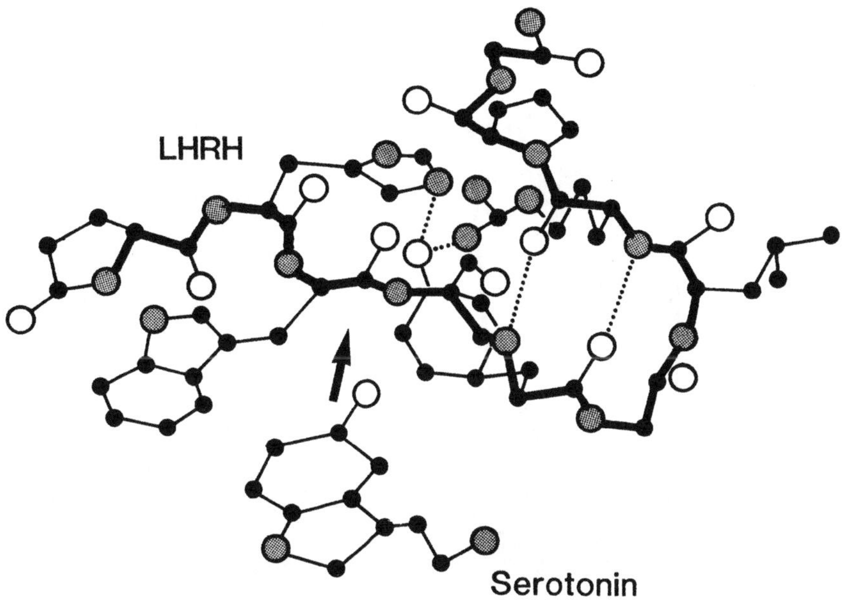

Fig. 2B. Illustration of the intercalation of serotonin (5HT) into a "molecular sandwich" composed of the tryptophan-serine-tyrosine sequence of luteinizing hormone-releasing hormone (LHRH). Note the similarity of the structure shown here with the binding site shown in Fig. 2A. Reproduced from Root-Bernstein and Westall, 1984 [20].

experiments could be interpreted along similar lines. For example, although it is possible to inject the antigen and adjuvant separately and induce EAE, they must be injected at the same site and within a very limited time of one another [34]. There is no logical reason that a general immune potentiator should have to be inoculated either at a particular time or place in the animal. Moreover, if the role of the adjuvant is merely to stimulate a more intense immunological reaction to the antigen, then one would expect that simply inoculating animals with more antigen would result in the same effect. In fact, Salk and his colleagues tested this logical inference and found that not only did increased doses of the encephalitogen (MBP) not induce EAE, they actually caused profound tolerance. These results led Salk to postulate that it might be possible to cure multiple sclerosis by very large inoculations of purified porcine myelin basic protein, but these trials were unsuccessful, either because porcine myelin basic protein is not sufficiently cross-reactive

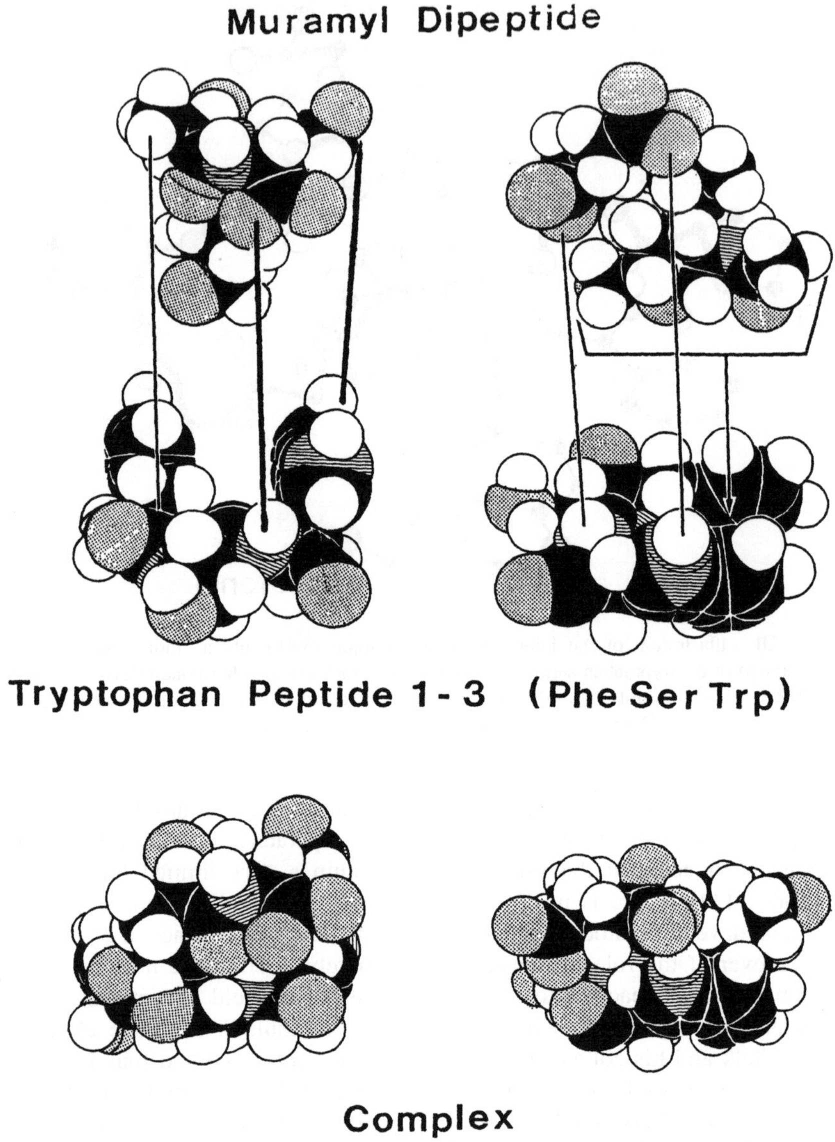

Fig. 3. Illustration of the intercalation of muramyl dipeptide (the "adjuvant peptide") into the "molecular sandwich" composed of phenylalanine-serine-tryptophan. This is the same site into which serotonin binds (Fig. 2A), which helps to explain the physiological mimicry that exists between muramyl dipeptide and serotonin (see text). Reproduced from Root-Bernstein and Westall, 1990 [45].

to induce tolerance to human MBP, or perhaps because multiple sclerosis is not, as is usually stated, as close an analogue to EAE as some investigators may wish to believe [35].

Now, if we look back carefully at what has been said so far, it is possible to recognize a number of inadvertent confusions that have dogged immunology for decades. The most salient concerns the nature of adjuvants. As noted above, adjuvants were initially discovered as a result of their immunopotentiating effect of increasing antiserum production. Note, however, that in these cases the presence of adjuvant material, while boosting the immune response, did not lead to autoimmunity. In other words, not every combination of antigen and adjuvant is autoimmunity inducing. That is a very important point indicating that the role of the adjuvant in the induction of autoimmunity is somewhat different from its role as a nonspecific immune potentiator. Historically, there is some confusion on this point. When Freund and Kabat demonstrated that adding adjuvants to tissues such as brain or testes speeded up the induction of autoimmunity from months to days, they could argue that the role of the adjuvant in autoimmunity was the same as in antiserum production; that is to say, one of potentiating an existing immune response. After all, the protocol of Rivers resulted in autoimmunity (eventually) whether an adjuvant was employed or not. But more recent research clearly demonstrates that the use of adjuvants in experimental models of autoimmunity is not mere potentiation of an existing immune response (that is, their effect is not merely quantitative), but rather that a combination of antigen and an appropriate adjuvant actually creates a qualitatively different sort of immune response than is created by either antigen or adjuvant alone.

The evidence for this qualitative difference is now quite strong. Myelin basic protein (MBP) or the tryptophan peptide of MBP alone will not induce EAE in guinea pigs in any known dose or series of doses, nor will FCA or the adjuvant peptide alone in any known dose. The same is true of experimental autoimmune orchitis: LHRH will not induce the disease alone, nor will the adjuvant peptide. Likewise, experimental autoimmune thyroiditis, which can be induced by a thyroglobin-adjuvant combination, cannot be induced by either component alone. Experimental uveitis, which can be induced by retinal S-antigen in FCA cannot be induced by either component alone. Experimental autoimmune neuritis, which can be induced by a combination of the P2 protein of peripheral nerves with CFA, also cannot be induced by either component alone. Also like EAE, each of these experimental autoimmune diseases can be prevented by either component inoculated several days or weeks in advance of disease induction [36]. In other words, so-called adjuvants have very specific roles in autoimmunity induction that cannot be accounted for in terms of general immunopotentiation. They are *necessary* components of autoimmunity induction for purified antigens. Why? What mechanism explains these activities?

VI. A DUAL-ANTIGEN THEORY OF AUTOIMMUNITY INDUCTION

Fred Westall, one of the people who had worked out the molecular identity of the encephalitogenic peptides of myelin basic protein, and I have proposed a novel solution to many of these outstanding paradoxes of experimental autoimmunity. The essence of our idea is very simple (as all useful scientific ideas must be), and originates in Paul Ehrlich's invaluable idea of antigen-antibody complementarity. Ehrlich stated, many years before evidence existed to support his idea, that each antigen would elicit the production of a different antibody that possessed a chemically complementary recognition site. The individual recognition site on each antibody is now called its idiotype—meaning the chemical determinant that sets it apart from all other antibodies. These idiotypes can, themselves, act as antigens to induce complementary antibodies known as antiidiotype antibodies. In other words, it is possible for antibodies to be chemically complementary to one another. Moreover, it has been demonstrated that an anti-idiotypic antibody can mimic the chemical properties of the original antigen to which the first antibody was created. Thus, molecular complementarity is conserved during the immune response.

Now this idea of chemical complementarity in the immune system intrigued me, since I had been working for some time on chemical complementarity between peptides and other small molecules. I find the notion of complementarity in all of its aspects, chemical and philosophical, fascinating [37]. Thus, while the vast majority of my colleagues are busy tilling the rows of a reductionistic philosophy that maintains that knowledge of the individual parts will provide deterministic insight into whole systems, I have bee investigating the unsuspected properties of combinations of these parts. All too often, I have found the old saw, "a system is more than the sum of its parts" to be true. Thus, one day, pondering the unsolved questions of EAE with Dr. Westall, I raised the question of what would happen if the body were faced with two, mutually complementary antigens simultaneously. From that thought experiment, Westall and I drew the following surprising conclusions (Fig. 6). Each antigen would elicit a clone that would produce antibody (or T-cells) complementary to it. Since the antigens were complementary, the

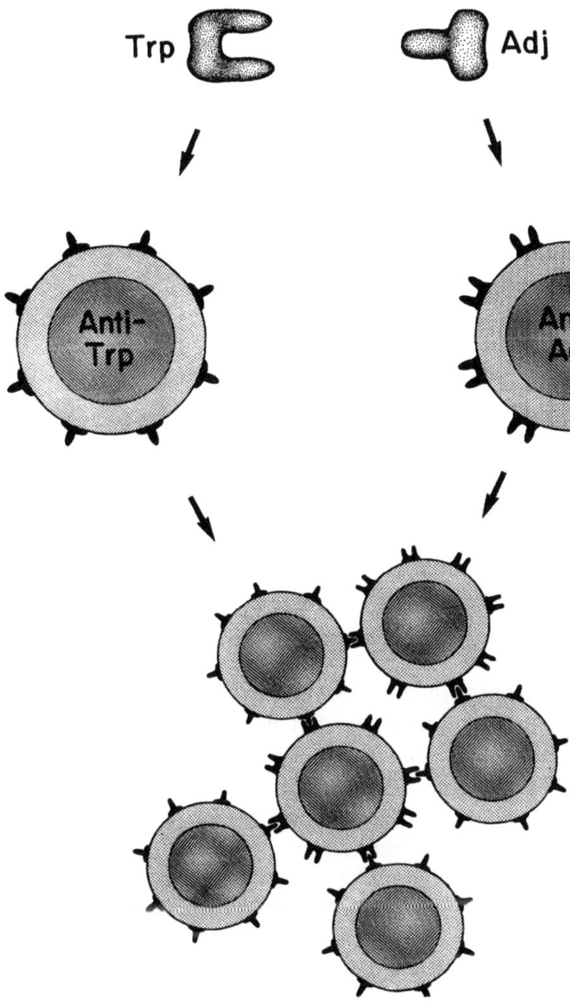

Fig. 4. An model of the consequences of simultaneous activation of two lymphocytes lines to chemically complementary antigens, such as the tryptophan peptide and adjuvant peptide (see Fig. 3). Reproduced from Westall and Root-Bernstein, 1983 [25].

idiotypes of the resulting antibodies or T-cells would also be complementary. That is to say, each would be the anti-idiotype of the other. Each anti-idiotype would, in turn, mimic one of the original antigens. The body would then be faced with an unsolvable paradox: in order to react immunologically against either of the new antigens, it would also have to react immunologically against the newly formed antibodies or T-cell clones. The result would be autoimmunity of a kind very different than is usually discussed: autoimmunity involving the immune system attacking itself [38].

Westall and I therefore proposed that autoimmunity results only when three basic criteria are satisfied: 1) that two or more chemically complementary immunogens are present simultaneously in the host; 2) that at least one of these immunogens be homologous to some self-protein so that immunological cross-reactivity results from response to the immunogen; and 3) that the other immunogen(s) be capable of acting as adjuvant(s) for the immunogen(s) described in the second criterium. The location of the resulting autoimmune reaction will be determined by the location(s) of the self-protein(s) to which the antigen(s) are homologous. We summarize these principles under the name "dual antigen" theory of autoimmunity induction, and I have recently suggested that this theory is a subset of all possible "multiple antigen mediated autoimmune" (MAMA) states [39].

This novel immunological theory redefines autoimmunity. In the first place, it transforms autoimmunity from an inappropriate and pathological response that overrides the self-recognition system, into a correct, normal, and necessary response no different from any other immune response. The only thing unusual is that this response is elicited only in the presence of two or more chemically complementary antigens. But even this condition of mutually complementary antigens may not be as rare as one might at first think. Consider, in the first place, that the integrity of any multicellular organism depends upon cell-cell recognition, which in turn depends upon molecular complementarity. Many molecules in our bodies must be chemically complementary to others. Consider, in the second place, that all human pathogens have evolved in tandem with the genus Homo, and that part of this evolution has involved selection for molecular mimickry. It would hardly be surprising to find that different microorganisms have evolved to mimic different molecular constituents of human tissues. And finally, consider that even something as simple as a virus or a bacterium depends upon internal molecular complementarity for self-assembly. In light of all of these factors, it is no wonder that autoimmune reactions, taken as a group, are far from rare. Conversely, in light of the specific complementarity between antigens that is necessary to induce any particular autoimmune reaction, it is no wonder that each individual autoimmune disease is rare.

Another important alteration demanded by the dual antigen theory of autoimmunity is that it redefines the nature of the role of adjuvants in autoimmunity. Some chemicals and microorganisms may indeed act as "non-specific immune potentiators" under a variety of immunological conditions, but this is not an accurate description of the role of bacterial antigens or "adjuvant" peptides in experimental autoimmune diseases. The fact that these so-called "adjuvants" are absolutely necessary to the induction of experimental forms of autoimmunity involving purified antigenic components, and that they have all of the immunological characteristics of the so-called antigen (or, in EAE, the "encephalitogen"), including the ability to prevent disease induction, clearly argues for their equality in the combination. Westall and I therefore believe that the use of the terms "antigen" and "adjuvant" are inappropriate in the context of autoimmunity, and that the necessity for both requires the use of the term "antigen" for both. In this case, language clearly betrays the preconceptions that have prevented previous investigators from perceiving the mechanism that we see at work in autoimmunity. I will continue to use the terms "encephalitogen" and "adjuvant" in the following discussion for the sake of clarity, and despite the inaccurate denotations these terms convey in light of the dual antigen theory.

The most important aspect of the dual antigen theory is that it explains many of the paradoxes elaborated above, including the failure of any dose of the "encephalitogen" alone to induce EAE, the necessity for concomitant administration of the dual antigens, and the otherwise puzzling details of prevention and suppression experiments. If autoimmunity in fact results from complementary sets of lymphocytes or antibodies attacking each other, then an antigenic response to any one antigenic determinant cannot result in autoimmunity. The preventative effect of pre-administering either of the two antigens alone can, in turn, be explained in either of two ways [40].

Assume, for the moment, that the so-called "encephalitogen" (myelin basic protein) is preadministered. A T-cell clone complementary to this antigen will be activated. This T-cell clone will exist in large numbers prior to the second inoculation consisting of the "encephalitogen"-"adjuvant" combination. The "encephalitogen" component will further induce the existing T-cell clone at a time when the "adjuvant" component is first activating its complementary T-cell clone. The previously existing anti-encephalitogenic T-cells might then totally inactivate or destroy the anti-adjuvant T-cell clone before it proliferates significantly. According to this scenario, prevention is due to direct inactivation of one clone by the previously induced clone. Clearly, the same mechanism would work, regardless of which of the antigens was used as the preventative agent.

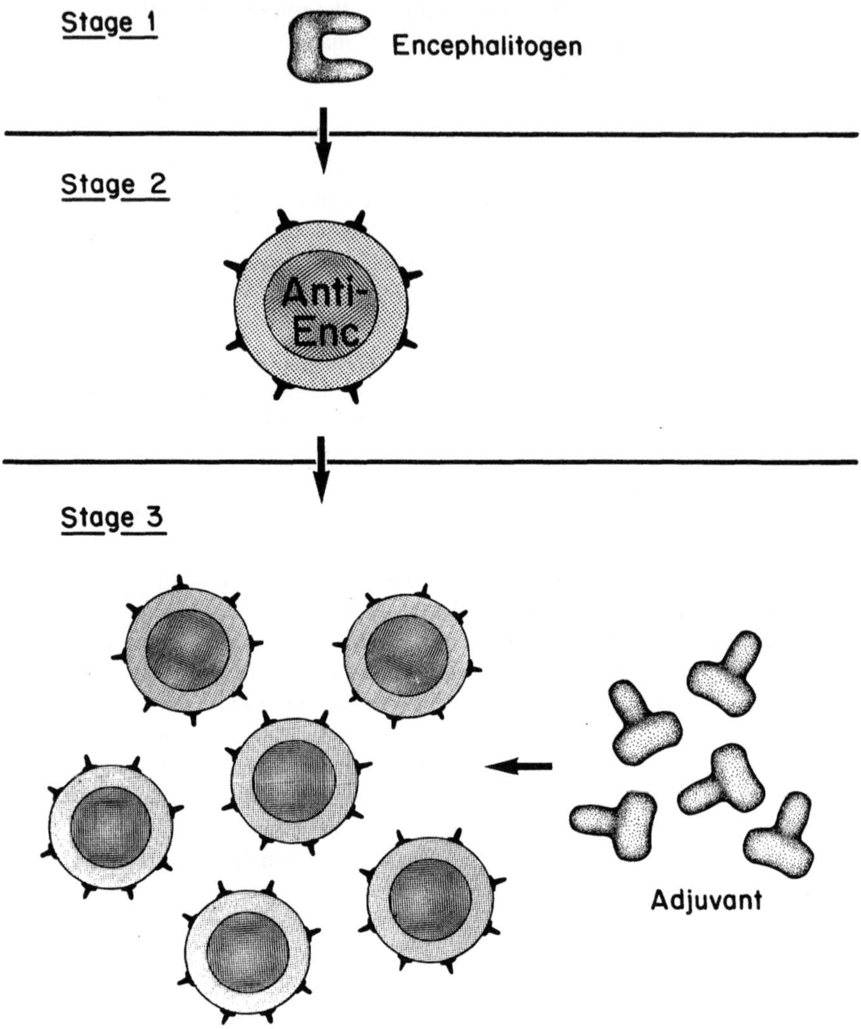

Fig. 5. One possible model for explaining the protection against EAE afforded by prior inoculation of the encephalitogen (stage 1). This antigen causes the production of anti-encephalitogen antibodies or T-cell lines (stage 2). The idiotype of these antibodies or T-cell molecules will be chemically complementary to the encephalitogen. Because the encephalitogen is also chemically complementary to the "adjuvant" molecule in EAE induction (see Figs 5 and 6), the adjuvant will be a structural mimic of the antibody or T-cell molecule idiotypes. Thus, by the time the "adjuvant" is inoculated (in combination with the encephalitogen — not shown) in stage 3, it will be recognized as "self". The same mechanism, *mutatis mutandis*, might explain prevention of EAE when the "adjuvant" is preinoculated. Reproduced from Westall and Root-Bernstein, 1983 [25].

On the other hand, a quite different mechanism might explain the same observations. If we consider the idiotypic structure of the T-cell receptors for the "encephalitogen", it is clear that, due to the chemical complementarity of the antigens, the anti-encephalitogen idiotype will mimic the "adjuvant" (Fig. 5). If we assume (as I think is necessary) that as a T-cell clone proliferates, it is actively recognized as self (by whatever mechanism), then clearly the anti-encephalitogen idiotype will be recognized as self. Thus, when the "encephalitogen"-"adjuvant" combination is subsequently administered, the immune system will be tolerant to the "adjuvant," recognizing it as a mimic of the "self" anti-encephalitogen idiotype. The same argument can be made, *mutatis mutandis* for the preventative action of the prior administration of the "adjuvant."

The same two mechanisms can be invoked to explain the suppressive effects of further doses of the "encephalitogen" (or its analogues) after induction of EAE. According to one mechanism, the subsequent doses cause preferential proliferation of the anti-encephalitogenic clone allowing it to overwhelm and possibly eliminate the anti-adjuvant clone. This is a direct competition model. According to a different mechanism, subsequent doses would cause the development of tolerance to the encephalitogen, leaving the anti-adjuvant clone intact. In either case, the odd observation that anti-encephalitogenic T-cells are present in animals following successful suppression of EAE with encephalitogen is now no longer anomalous, but predictable [41].

It is important to point out that whichever mechanism of prevention and suppression turns out to be correct, both have important implications for our understanding of how autoimmunity apparently avoids the strictures of self recognition. Both mechanisms suggest that one consequence of the activation of any particular lymphocyte clone is the concomitant restriction of complementary (anti-idiotypic) clones. The fact that autoimmunity can occur when complementary antigens are inoculated simultaneously suggests that this restriction is not made until after the clone proliferates. Thus, the window in time in which autoimmunity to any particular antigenic combination may be induced is quite short (probably only a few days or perhaps a week) following immunologic response to the first antigen. Another implication of these mechanisms is that every antigenic stimulus will result in restriction of more anti-idiotype clones, thus resulting in ever-decreasing immunological responsiveness with age. Not only does this prediction match observation, but it has a corollary that is also born out by epidemiological data, and that is that although the incidence of auto-antibodies increases with age (a direct consequence of encountering ever-more self-mimicking antigens) the induction of autoimmunity becomes less likely with increasing age (since the more antigens a person has encountered, the more autoimmune diseases he or she is protected against). Thus, most autoimmune diseases should have their onset during the first two or three decades of life rather than in the last decades of life. This observation is certainly born out by the data concerning the incidence of diseases such as multiple sclerosis and Guillain-Barré syndrome [42].

Various tests of the dual antigen theory have been made. One prediction that has been successfully tested is that the "adjuvant" should be just as effective as a post-induction suppressor of autoimmunity as is the encephalitogen, which it is. Moreover, this prediction follows from nothing that was previously thought about adjuvants, since one would hardly expect a "non-specific immune potentiator" to also act as a potent immunosuppressant [43].

Another test of the dual antigen theory is to determine whether the disease can be caused in genetically identical animals by inoculating one animal with "encephalitogen", a second animal with "adjuvant" and then passively transferring the disease to a third by means of a combination of the T-cells from the first two. The elegance of this experiment is that, neither of the first two animals will have autoimmunity, but if it works, the third animal will. This prediction certainly follows from nothing that is currently known about autoimmunity. So far, this experiment has not been attempted, but a simpler analogue has. Genetically identical strain 13 guinea pigs were inoculated in three groups: one group with the tryptophan peptide of myelin basic protein; one with muramyl dipeptide (adjuvant peptide); and a third group with saline solution. The T-cells were then recovered from the lymph nodes of the animals and cultured. All possible permutations of their combinations were made at the same time. Daily cell counts were performed, and observations made microscopically. The experiment demonstrated that significant cell-aggregation occurred in the encephalitogen-adjuvant lymphocyte combinations but in no others (Figs 6A-D) and that the encephalitogen-adjuvant lymphocytes died off at a rate about 30 percent faster than the other mixtures (Table II). These data suggest that the sort of T-cell complementarity proposed by the dual antigen theory does occur. Moreover, this experimentally produced lymphocyte aggregation may explain the puzzling observation that most autoimmune diseases are characterized by circulating immune complexes and perivascular cuffing unassociated with the supposed target of the immune reaction (Fig. 1). It is important to note that other investigators (e.g., Suckling *et al.*) have also isolated and reproduced photographs of such T-cell aggregates from peripheral blood in EAE animals. One confounding

a)

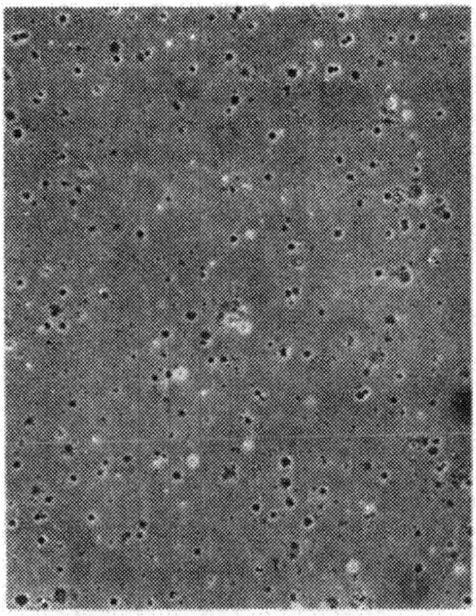

b)

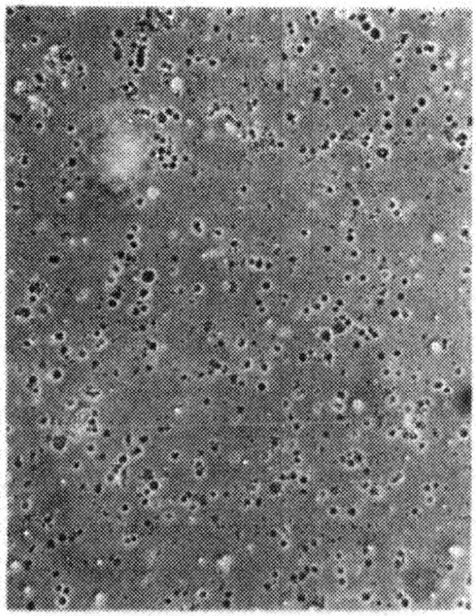

c)

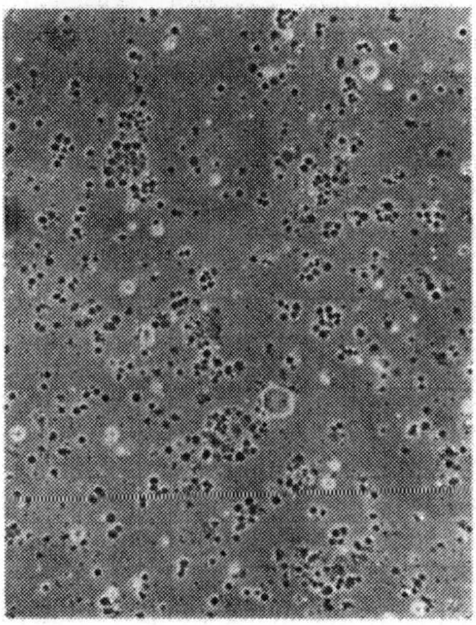

d)

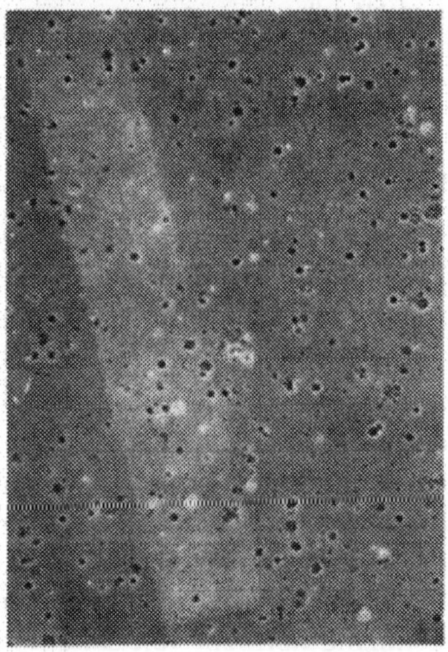

Fig. 6. Lymphocytes isolated from the lymph nodes of genetically identical strain 13 guinea pigs and stained with trypan blue (see the caption to Table II). (A) Inoculated with tryptophan peptide. (B) Inoculated with muramyl dipeptide (adjuvant peptide). (C) Combination of lymphocytes from the two animals in A and B. Note the pronounced lymphocyte aggregation that mimics that seen in perivascular cuffing in EAE (Fig. 1). (D) Combination of lymphocytes from the animal in A and an untreated animal. Note the complete absence of clumping. The same lack of clumping was observed when lymphocytes from animal B were mixed with those of an untreated animal.

TABLE II

Lymphocytes stimulated by tryptophan peptide and muramyl dipeptide demonstrate increased mortality in combination. Genetically identical strain 13 guinea pigs were treated either with tryptophan peptide, muramyl dipeptide, or were untreated. Lymphocytes were then harvested from lymph nodes immediately prior to sacrifice and placed in Eagles complete medium. Cell counts were done daily, and viability determined by trypan blue exclusion. Combined lymphocytes from tryptophan peptide animals combined with those from muramyl dipeptide treated animals had about thirty percent lower viability and lower cell counts than controls or combinations with lymphocytes from untreated animals. Reproduced from Root-Bernstein and Westall, 1986 [38].

	Day 0 (1 hour post-excision)	Day 1 (pre-isolymph)	Day 1 (post-isolymph)	Day 2
Tryptophan peptide (4650)	1.24×10^8	4.25×10^7	2.31×10^7	1.89×10^7
Viability	92%	90%	100%	92%
No. viable cells	1.14×10^8	3.83×10^7	2.31×10^7	1.74×10^7
Adjuvant peptide	1.39×10^8	5.20×10^7	4.73×10^7	3.21×10^7
Viability	90%	90%	99%	91%
No. viable cells	1.25×10^8	4.68×10^7	4.68×107\	2.92×10^7
Tryp. + Adj. peptide (4650 + 4660)	1.39×10^8	3.61×10^7	2.21×10^7	1.78×10^7
Viability	88%	87%	100%	88%
No. viable cells	1.22×10^8	3.14×10^7	3.21×10^7	1.57×10^7
Expected	1.20×10^8	4.26×10^7	3.50×10^7	2.33×10^7
Difference	2%	27%	37%	33%

factor, however, is that perivascular microglial cells of the central nervous system are bone-marrow derived and can present antigen themselves in *vivo*, suggesting that some CNS cells may be an active part of the autoimmune reaction and may help to localize the antigens that centralize the reaction [44].

The best tested aspect of the dual antigen theory is the prediction that the two antigens are chemically complementary (Fig. 2). Westall and I have used nuclear magnetic resonance spectroscopy to demonstrate that muramyl dipeptide, which acts as the "adjuvant" for both myelin basic protein and for luteinizing hormone-releasing hormone (experimental autoimmune orchitis), binds directly to both of these peptides [45].

VII. AIDS-RELATED NEUROPATHIES AS A TEST OF THE DUAL ANTIGEN THEORY

Despite the theoretical simplicity and the observational tests of the dual-antigen theory of autoimmunity, it did not, however, resolve the outstanding problem of what might initiate such autoimmunity in "naturally" occurring diseases such as multiple sclerosis, amyotrophic lateral sclerosis, or post-infectious and post-vaccinal neuropathies. EAE, for all that it has been studied, is not a reasonable model of these diseases. Clearly, human beings do not usually (or even rarely!) run around inoculating themselves with brain tissue (or purified extracts from it) mixed with bacteria, as Rivers and all subsequent investigators have done to create EAE. (One possible exception are the Fore tribesmen of New Guinea whose contact with human brain matter during their cannibalistic rituals led many to develop the neurologically destructive disease kuru [46]). In other words, there was no direct connection between the mode of induction of any of the well-studied animal models of autoimmunity and the presumed mode of induction of any human counterpart. Thus, although various reports did suggest that certain post-vaccinal neuropathies were often associated with combined viral and bacterial infections, these reports were insufficient to pinpoint possible combinations of microorganisms that might induce autoimmunity [47].

This is where AIDS reenters the picture. For some years now I have been studying animal models of autoimmunity and thinking about AIDS and, like the vast majority of my colleagues, keeping the two subjects in separate, non-intersecting compartments of my mind. There is no excuse for this sort of compartmentalization of information, and yet our textbooks, journals, and even departmental structures foster it. Such compartmentalization of knowledge quite possibly represents the greatest obstacle to discovery we face as scientists. Answers to many of our outstanding questions doubtless exist and may even be known to us if only we were able to perceive them in the proper context, or juxtapose the necessary information properly [48]. If so, then we probably need to spend more time rethinking what we think we know, rather than rushing about creating new data as quickly as possible. In any event, it finally occurred to me very recently that AIDS patients, who are both unusually prone to autoimmune complications and to multiple, concurrent infections, were exactly the sort of patients most likely to contract autoimmunity if the dual antigen theory of autoimmunity was accurate. Therefore, AIDS patients might provide the keys to unlocking some of the "natural" causes of human autoimmunity. This insight may or may not be correct, but the data are certainly intriguing.

Every AIDS patient has multiple infections. Most AIDS patients develop an AIDS dementia of one type or another. A large subset of these dementias have as their main histological characteristic demyelinization of central nervous system or peripheral nervous system nerves, and sometimes of both. An analysis of autopsy reports accompanied by complete neurological evaluations and details of infectious histories demonstrates that AIDS patients who develop such demyelinating and microglial neuropathies almost always have combined cytomegalovirus infection accompanied by a mycobacterial or other bacterial infection. Thus, of twenty-eight patients with autopsy-demonstrated demyelinization, all had CMV infections, whereas only eighteen of forty-four patients without demyelinization had such infections. Twenty-three of the twenty-eight patients with demyelinization had concurrent mycobacterial or other bacterial infections, whereas only five of the forty-four patients without demyelinization had such a concurrent infection. No significant correlation was observed between the incidence of pneumocystis pneumonia, candidiasis, or toxoplasmosis, and the development of demyelinating or microglial neuropathies; no significant correlation was found between fungal or protozoal infections and either CMV or bacterial infections and the neuropathies; nor was there a significant correlation with the incidence of Kaposi's sarcoma either alone or in combination with any of the other diseases. HIV data were not available for most of these patients, but if we assume that most or all were HIV seropositive, then HIV is as prevalent in those developing neuropathies as in those who do not. AIDS patients who develop dementias due to other causes, such as cryptococcus infections of the brain or lymphomas, also lack the CMV-Mycobacteria combination [49].

This finding is obviously striking, since mycobacteria are primary components of the Freund's complete adjuvants used to induce EAE and other experimental autoimmune disease. The crucial question is whether CMV and other herpes viruses have significant homology with myelin basic protein or the P2 protein of peripheral nerves, and so act as the infectious molecular mimics for inducing a sort of EAE in human beings. That, in fact, appears to be the case, and not only for CMV but for Epstein-Barr virus and hepatitis B virus as well (Table III). In short, those AIDS patients who develop demyelinating neuropathies mimic (at least by analogy) exactly the conditions previously worked out in animal models for the induction of EAE and EAN. They have multiple, concurrent antigenic stimuli; one of the infections (the viral one) results in an immune response cross-reactive with self-proteins; the other infection (the Mycobacterium) is, by analogy to EAE, very likely to be an active "adjuvant" for the first. Moreover, like animals who develop EAE and EAN, many AIDS patients also develop optic neuritis as well.

Male homosexuals who develop AIDS have additional autoimmune complications that may also mimic aspects of experimental models of orchitis [50]. As I mentioned above, people do not usually inoculate themselves with foreign tissue, but it has been suggested that homosexual men who engage in

TABLE III

Homologies between human myelin basic protein, P2 protein from peripheral nerves, HLA-DR beta protein, and some viral antigens commonly found in AIDS-related infections: hepatitis B virus (HBV), human cytomegalovirus (HCMV), and Epstein-Barr virus (EBV) Reproduced from Root-Bernstein, 1990 [39].

Protein	Peptide sequence
MBP 66–75	Thr-Thr-His-Tyr-Gly-Ser-Leu-Pro-Gln-Lys
MBV Poly 589–598	Ile-Gly-Cys-Tyr-Gly-Ser-Leu-Pro-Gln-Glu
HCMV HFRFO 961–70	Arg-Thr-His-Tyr-Arg-Asn-Leu-Gln-Gln-Gln
MBP 107–116	Arg-Gly-Leu-Ser-Leu-Ser-Arg-Phe-Ser-Trp
EBV EC-RF2 432–441	Arg-His-Gly-Glu-Leu-Phe-Arg-Phe-Ile-Trp
MBP 113–121	Phe-Ser-Trp-Gly-Ala-Glu-Gly-Gln-Arg
HCMV HRFO 1070–77	Phe-Ser-Tyr-Gly-Leu-Glu-Phe-Arg
MBP 122–131	Arg-Pro-Gly-Phe-Gly-Tyr-Gly-Gly-Arg
HCMV HFLF2 (pol) 1048–57	Arg-Thr-Gly-Tyr-Gly-Tyr-Ile-Ser-Thr-Lys
MBP 126–135	Gly-Tyr-Gly-Gly-Arg-Ala-Ser-Asp-Tyr-Lys
EBV EC-LF2 194–203	Gly-Tyr-Gly-Asn-His-Ala-Gly-Asp-Tyr-His
P2 29 – 38	Thr-Arg-Lys-Leu-Gly-Asn-Leu-Ala-Lys-Pro
EBV EC-LF1 473–482	Asn-Arg-Met-Leu-Gly-Asp-Leu-Ala-Arg-Ala
P2 91–101	Leu-Asn-Gln-Val-Gln-Arg-Trp-Asn-Gly-Lys
HCMV HFLF2 (pol) 204–13	Leu-Arg-Arg-Leu-Ile-Arg-Trp-Phe-Gly-Arg
HCMV HaLF1 1198–1206	Asp-Glu-Val-Asp-Arg-Trp-Ile-Arg-His
P2 95–105	Gln-Arg-Trp-Asn-Gly-Lys-Glu-Thr-Thr-Ile-Lys
EBV EC-LF2 649–658	Ser-Trp-Leu-Ala-Lys-Arg-Lys-Ala-Ile-Lys
HCMV HFLF2 (pol) 209–18	Arg-Trp-Phe-Gly-Arg-Pro-Val-Gly-Tyr-Lys
HCMV HFLF2 (pol) 977–85	Gln-Arg-Trp-Asp-Tyr-Arg-Gln-Val-Ser
HLA-DR BETA 50–59	Pro-Asp-Pro-Leu-Gly-Arg-Pro-Asp-Ala-Glu
HCMV IE-2 79–88	Val-Thr-Glu-Leu-Gly-Arg-Pro-Asp-Glu-Asp

unprotected anal intercourse may, unintentionally, be doing exactly that. The typical gay man who develops AIDS is very promiscuous, often having hundreds or thousands of partners. Thus, they may be exposed to just the sorts of recurrent inoculations of sperm that investigators used to create experimental orchitis in the early decades of the present century. Rectal trauma may allow semen to penetrate into the lymph or blood stream to be immunologically processed. Recent animal studies indicate that semen inoculated into the rectum does result in such an immune response. Almost all homosexual men who engage in receptive anal intercourse demonstrate the presence of anti-semen antibodies. Autopsies show that nearly every gay man who dies of AIDS suffers from aspermatogenesis, testicular atrophy, and destruction of the seminiferous tubes. Since most of these men, almost without exception, also have concurrent infections with cytomegalovirus, sexually transmitted diseases, recurrent bouts of "gay bowel syndrome," and various bacterial infections of the urogenital and bowel regions, it is highly likely that semen inoculation is accompanied by "adjuvants" [51].

Oddly, there may be an unexpected connection between the development of AIDS dementias and AIDS orchitis. Lewis demonstrated in 1934 that cross-reactivity exists between testicular antigens and brain antigens, but between these antigens and no other tissues tested [52]. It is therefore possible that the development of autoimmunity against one organ or the other may also be responsible, indirectly, for the immunopathogenic response observed to the other. As far as I can determine, no one has investigated this possibility with regard to AIDS patients or in other diseases of the central nervous system such as multiple sclerosis.

Perhaps the most important possibility raised by the application of the dual antigen theory of autoimmunity to AIDS is that, once the causal infections have been identified, the extensive studies on prevention and suppression of autoimmunity described above provide guidelines for possible prevention and suppression of autoimmunity in AIDS. Prophylactic vaccinations with Mycobacteria or killed or attenuated CMV might protect those at risk for AIDS from developing dementias. And for those patients who have already developed autoimmune neuropathies, the same vaccines might be effective as suppressive agents. Thus, the theory is not only useful for understanding how such autoimmunity may be initiated in AIDS patients, but may also provide sufficient understanding of the causes to allow specific medical intervention [53].

It is also possible that the same type of in depth pathological and histological studies that are very frequent for AIDS patients but extremely rare in other diseases may someday allow us to determine the constellation of infections that induce the autoimmune reactions we call multiple sclerosis, Guillain-Barre syndrome, amyotrophic lateral sclerosis, and so forth, in genetically susceptible individuals. Equally possible is that genetic studies will provide a molecular basis for understanding why certain genotypes are more or less susceptible to autoimmunity when the relevant HLA protein sequences are compared with the newly identified viral and bacterial triggers. Perhaps, as in EAE, a certain degree of antigenic difference is necessary to induce an immune response, but a certain degree of homology is necessary to provoke an autoimmune reaction. In any event, as the inducing agents are identified, specific preventative and suppressive treatments will follow, if the dual antigen theory is correct.

VIII. SOME REMAINING PARADOXES

Now, as far as these observations get us toward understanding the possible initiation of autoimmune conditions, and even perhaps towards their prevention and treatment, they still leave huge gaps in our understanding. One of the most interesting can only be recognized by comparing experimental autoimmune diseases such as EAE and another well-characterized cousin, adjuvant arthritis. Adjuvant arthritis can be induced in susceptible animals such as Lewis rats with an single inoculation of *Mycobacterium tuberculosis* in oil-water emulsion. It cannot, however, be induced with purified fractions of mycobacteria, such as the adjuvant peptide. This indicates that multiple, chemically complementary antigens exist within some microorganisms, and that at least one of these is cross-reactive with collagen in animals and perhaps in human beings as well [54]. So far there is nothing surprising here. But what is surprising is that when Freund's complete adjuvant made with *Mycobacteria* in the same doses that induce adjuvant arthritis are mixed with myelin basic protein and inoculated as a combination, the animals develop only EAE and no clinical or histological signs of arthritis! Oddly, no one seems ever to have pointed out this paradoxical observation in print although several groups, such as that working with Irun Cohen, work with both adjuvant arthritis and EAE. Unfortunately, the minimal chemical requirements for adjuvant arthritis induction have not yet been elucidated, but when they are, it can only be hoped that their structures will resolve this paradox.

Other gaps in our knowledge are also illuminated by the data and theory presented above. One of these can best be illustrated by returning to the example with which we began: PKH. By 1949, syphilis was generally recognized to be the primary cause of the disease, although it was equally evident that the vast majority of syphilitic patients did not develop PKH. Thus, syphilis may be necessary, but it is not sufficient to account for PKH. It had also been determined that intensive antisyphilitic treatment usually resulted in complete disappearance of the autoimmune symptoms, suggesting that the presence of the offending bacterium was in some way responsible for the chronic nature of the autoimmune reaction. The success of antibacterial treatment of an autoimmune condition appears to be unique to the class of autoimmune conditions; but perhaps that is only because many autoimmune diseases have viral etiologies against which we have no useful antibiotics. Most strange was the observation that even when the symptoms of PKH had disappeared, the Donath-Landsteiner reaction would still occur if the blood of the cured patient was placed in a cold test tube and then warmed in the presence of complement. Thus, the cause of the autoimmunity was still present within the patient but, like Metalnikoff's antisperm antibodies, something prevented the autoimmunity from being expressed [55].

These observations are difficult to explain by any modern theory of immunology, but several lines of research may shed some light on them. In the first place, it is evident that the presence of self-recognizing antibodies is not sufficient to induce autoimmunity. Why not? To answer this question, I believe that we must seriously consider enlarging the scope of what controls autoimmunity from merely antibody or T-cell activation to include the complement system. Recent research has demonstrated the existence of molecules that suppress the complement reaction. These molecules are known collectively as the "regulator of complement activation" or RCA family of proteins [56]. The existence of the RCA proteins suggests that the recognition of self may reside not only at the genetic level and be determined not only by B-cells and T-cells, but may also depend on the presence or absence of RCA proteins in the tissue or on the cell surface of potential autoantigens. If these speculative conjectures have any merit, one would expect to observe that defects in the RCA system would be manifested in increased incidence of autoimmune condition. If so, then RCA proteins may provide exogenous treatments for some autoimmune diseases. Or, perhaps the RCA system is directly or indirectly regulated by (some) HLA types, which would explain a great deal about why autoimmunity appears to be regulated partially by HLA determinants. One might also predict that those tissues against which autoimmunity occurs most frequently (the so-called "hidden antigens") lack RCA proteins (and these would include neurons, testes, the lens of the eye, and the thyroid, as well as the immune systems cells themselves, which could hardly be immunologically active if they produced their own inhibitors!).

A related clue needs to be investigated as well, and that is the temperature dependence of the PKH reaction (which is only one of a family of such temperature-dependent autoimmune reactions). Since autoantibodies are always present in PKH patients, but the complement reaction is only activated after cooling and rewarming of the local blood temperature, it follows that the regulatory molecule is one that is involved in temperature regulation as well as immune stimulation. One can imagine that under normal temperature conditions, this regulatory molecule is not released in significant amounts into the blood stream, but that as the local tissue become cool, it is released to increase local circulation (and thereby, temperature). The presence of significant concentrations of this regulatory molecule as tissue warming occurs allows the immune reaction to occur, but only as warming of the tissue is accomplished. Thereafter, the regulatory molecule is no longer released in the blood stream, and no further immune reaction occurs. These criteria should be sufficient to make identification of the candidate molecule possible. Serotonin has many of the requisite properties, since it is released during tissue injury, regulates temperature and local blood flow, and is immunologically stimulatory. Anti-serotoninergic compounds might therefore prevent temperature-dependent autoimmune reactions from occurring.

Another intriguing line of research that may shed some light on the current paradoxes of autoimmunity concerns the presence of what appear to be autoantibodies in all syphilitic patients, with or without PKH. In 1906, Wassermann and his colleagues developed their famous blood test for syphilis, but with the oddest of results. The Wassermann reaction, as it is now known, was developed to recognize the presence of antibody to the cause of syphilis, *Treponema pallidum*, but the reaction is not antigen specific. Extracts of tissue from *any* organ of a syphilitic patient, even organs demonstrably devoid of the *Treponema pallidum*, will yield a positive reaction. "Now, eighty years later," Silverstein writes in his history of immunology,

> an acceptable explanation for the presence of these serodiagnostic antibodies in syphilis (which have been shown to differ from those that react specifically with known treponemal antigens) is still awaited. The antigen active in complement fixation tests for syphilis was quickly shown to be a lipid and later was purified and named cardiolipin, but the origin of the antibodies from syphilitic patients with which it reacts is still a mystery [57].

These data both tell us that so-called autoantibodies are a normal response to certain antigenic stimuli and warn us that the observation of such so-called autoantibodies or T-cell lines, even in well-established autoimmune diseases such as multiple sclerosis, may be artifacts of the disease process rather than their direct cause. In other words, we must be extremely careful in our interpretations of the presence of so-called autoantibodies and auto-T-cells and not attribute to them activity that they may not have. Indeed, the presence of such so-called autoantibodies may have a very interesting and previously unsuspected beneficial effect for the individual who develops them. Perhaps—this is pure speculation—perhaps secondary immunity persists best for those antigens that induce a weakly self-cross-reactive immune response constantly fuelled by an almost subliminal antigenic stimulation. In this case, molecular mimickry would have distinctly beneficial effects for the host responding to the camouflage techniques of his pathogens. The primary point is that interpreting immunological data in autoimmune states is fraught with peril, and consequently with intriguing possibilities as well.

IX. KOCH'S AND WITEBSKY'S POSTULATES REVISED

One very important issue raised by the data and theories summarized above is the problem of establishing etiologies for human autoimmune diseases. How do we know when we have discovered the cause of a human disease, or have developed a reasonable animal model for it? This is as deep an epistemological question as exists in medicine, and it is a particularly difficult question for those studying autoimmunity. In 1957, Witebsky proposed a series of postulates, based on Koch's postulates, to fulfil this function [58]. He suggested that an autoimmune response be considered the cause of human disease if 1) it is usually associated with the disease; 2) if the affected tissue, or some isolate of that tissue, can induce autoimmunity in experimental animals when inoculated into them; 3) if the animal develops a pathology corresponding to the human disease as a result of the induced autoimmunity; and 4) if the autoimmune pathology can be transmitted to another experimental animal by means of serum or lymphocytes. Since these postulates are based upon a single-factor theory of autoimmunity induction, it should be evident that these criteria are no longer satisfactory in light of the dual antigen theory of autoimmunity [59]. Theories, in other words, redefine the philosophical criteria of evaluation, just as philosophical assumptions set boundaries on what theories can be imagined. Witebsky's postulates also fail in terms of his assumption that heterologous tissue is a reasonable substitute for the unknown agents that induce human autoimmunity.

I propose a different set of criteria for establishing the etiology of human autoimmune diseases: 1) that an active autoimmune response be demonstrable in the disease, accompanied by antibody or lymphocyte aggregations; 2) that some complex of transmissible antigens (these could be some combination of viruses, bacteria, other infectious agents or, under unusual circumstances, human cells such as sperm or, among tribes such as the Fore, brain tissue) be demonstrable prior to or concomitant with the onset of the autoimmune response; 3) that this complex of transmissible antigens be capable of inducing a similar autoimmune state in an appropriate, genetically susceptible experimental animal; 4) that the disease state induced in the animal mimic that of its human counterpart; 5) that the autoimmune pathology be transmissible to a genetically identical experimental animal by means of serum or lymphocytes; and 6) that the autoimmune pathology be preventable or suppressible by means of further inoculations of one of the transmissible antigens alone, or a monoclonal antibody or T-cell line. These criteria, I believe, will stimulate the elucidation of human autoimmune diseases and help to replace the clearly artificial animal models upon which research is currently done with ones that mimic conditions more likely to result in autoimmunity in human populations.

X. RETHINKING IMMUNOLOGICAL SELF

But what, to return to our central mystery, does all of this tell us about self, non-self, and autoimmunity? I find this question no easier to answer now than I did when I began this essay; but I find, too, that my thinking has changed in several ways. I began this essay assuming that there was a single phenomenon called autoimmunity. I am no longer sure that this is a valid conceptual framework. It might be more useful (only time will tell) to divide the phenomena of autoimmunity into three subcategories, corresponding to the observations of Ehrlich and Morgenroth, in one case; Metalnikoff, in the second; and Donath and Landsteiner, in the third [60]. As Ehrlich and Morgenroth found, true autoimmunization is clearly very rare and appears to be limited experimentally to "hidden" antigens such as nerve tissue or the lens of the eye, and then only in presence of repeated stimulation or an adjuvant. As Metalnikoff observed, production of autoantibodies is common but rarely if ever leads to active autoimmunity due to as yet unknown regulators of complement activation. There are major discoveries waiting to made here. And as Donath and Landsteiner found, isoimmunization is relatively common, but only leads (in my view) to active autoimmunity under the multiple antigenic, chemically complementary, conditions detailed in the dual antigen theory outlined above for EAE and AIDS dementias. I would therefore like to suggest that almost all cases of what we call autoimmunity are in fact a specific and unusual subset of what would better be termed cross-reactive isoimmunity leading to internal competition between the elements of the immune system itself, and that true autoimmunity, induced by autoimmunization, may be extremely rare even among those diseases currently encompassed by the term autoimmunity. In short, I am suggesting that the presence of autoantibodies and auto-reactive T-cells does not constitute sufficient evidence of autoimmunity, and may, under many circumstances, represent a perfectly healthy immune response.

The confrontation between philosophy, theory, and data outlined here requires other changes in how we view the concept of immunological self, particularly in light of the clearly important insights provided by prevention and suppression experiments of experimental autoimmunity. It now seems quite clear that the immunological history of an organism determines its susceptibility to autoimmunity. It follows that immunological self is determined not solely by genes, nor even by the expression of self-antigens (idiotypes and antiidiotypes, though these things clearly play a part), but by the organism's history of prior antigenic stimuli as well. This fact is clearly demonstrated by the prevention experiments outlined above. The only way in which self can be defined by the history of immunological responses, however, is to assume one of two things. First, it is possible that clones activated by an antigenic stimulus do not exist prior to that stimulus (that is to say, that the clones are actually generated by some as yet undiscovered form of template mechanism rather than by the current mechanism of pre-fabricated clones). In this case, the antigen would redefine immunological self by bringing into existence a clone that would be new both in its genetic and protein structure. The other possibility is that preexisting clones have no immunological status as self or non-self until they proliferate. In other words, by inducing cloning, the antigen increases the number of cells belonging to a particular lymphocyte clone to some level necessary to become recognized as self. This second possibility also implies that until this activation occurs, the immune system is not aware of its own existence. Both of these conclusions are odd, to say the least.

Another odd conclusion that I have arrived at is that autoimmunity results from the immune system attacking itself. Once again, this sort of autoimmunity seems to be possible only if the unexpanded clones of the immune system are not yet recognized as self; if the immune system does not, for some reason, monitor its own components as it does the rest of the body and "foreign" antigens; or if these clones do not, in fact, exist immunologically or perhaps even developmentally until the immune reaction itself begins. And so we encounter yet more paradoxes. If, in fact, the immune system normally defines self by deleting or inactivating all self-reactive clones and only self-reactive clones, then the immune system *must* be aware of its own existence and its own constitution as it is generated. Moreover, if many antigens are chemically complementary, or if the immune system must merely take into account the possibility of such complementary antigens existing, then by necessity, many clones would carry determinants on their cell surfaces that would be complementary to other clones. Theoretically, then, every clone becomes potentially self-reactive to some other clone through antiidiotypy.

Probability dictates that at some point these complementary clones would meet and presumably activate each other. It is difficult to understand why spontaneous autoimmunity would not be a common or even necessary occurrence under those circumstances.

How, then, can the current dual antigen theory of autoimmunity, requiring internecine warfare within the immune system itself, be compatible with Burnet's clonal selection theory or with Jerne's network theory of clonal selection control? I frankly find this question impossible to answer at present. The paradox I am left with is why the immune system allows antiidiotypy. Should not the immune system eliminate or inactive not only all self-reactive clones, but all clones bearing complementary idiotypes? Should it not, in other words, preclude at least one half of all the possible clonal types? But if it did so, would not such widespread inactivation paralyze the immune system? Clearly there are large and important questions concerning immunological self-recognition that remain unanswered, and the paradoxes raised by autoimmunity will be central to their solution.

At present, the only way out of the dilemmas posed above seems to be a resort to some sort of template theory of clonal production, such that the idiotype of any particular antibody is not actually generated and inserted into the proper genetic sequence until an antigen is present. As outrageous as this suggestion may sound, it is important to remember that all current theories of clonal production are no more than that: theories. The clones certainly exist, but no one has yet isolated the chemical machinery that supposedly mixes and matches the genes to create the hypervariable regions that characterize clonal diversity. Indeed, the discovery of such cellular machinery would certainly overthrow much of what we now consider to be dogma concerning genes, since it would permit lymphocytes the unique role of actually scrambling and even creating new genetic sequences. That is certainly an idea with many implications for the concept of self! Personally, I do not think that it is any more revolutionary to suggest that cellular machinery may exist in lymphocytes that actually tailors the idiotype to the antigen, and then inserts the tailored gene sequence into the proper site in the immunoglobin gene sequence. [61] Certainly, such a mechanism makes the whole immune system much simpler and many of the present puzzles disappear. For example, rather than having a preexisting immunological sense of self that determines all future antigenic reactions, the functional capacity of the immune system (and thus both the set of excluded idiotypes and the immunological sense of self) would develop as a consequence of the history of the interactions between the organism and its environment. Adding an historical component to the concept of immunological self would also provide a time-keeping mechanism (determined by which antigen is encountered first) so that the different responses to the same antigen in inducing, preventing, and suppressing autoimmunity that were so puzzling when described above, become natural consequences of continuous immunological development. Most importantly, the immune system would no longer need to waste huge amounts of energy producing and monitoring huge numbers of clones that are never activated.

The phenomena of autoimmunity, in short, provide a fascinating case study in the ways in which concepts of self evolve both in response to internal observations and theorizing, and in response to concepts of individuality that are being developed concomitantly in other areas of science. This history suggests that developments in autoimmunity (or in any area of biology) will necessarily be most fruitful in the future when the particular subject is considered with relation to the widest possible view of concomitant developments in science. The best we can say at present is, to quote Irun Cohen, "self and world are no longer absolutely distinct" [62]. The relationship between the immune system and the body, and the body and its environment are, at present, no clearer than the relationship between brain and mind. Immunological self is somehow determined not only by the genes we inherit but the antigens we encounter and when we encounter them, and is not until this relativistic systems approach replaces the current trend toward genetic reductionism that the unsolved problems will become accessible.

I believe that there is another lesson to be learned from studies of autoimmunity, as well, both with regard to immunology in particular and to concepts of self and nonself in general. In the past, whenever a problem as profound as autoimmunity has stymied the scientific community for as long a period as a century, it has invariably turned out in retrospect that the anomalies involved have provided the impetus for the next revolution in that field. Thus, despite a fairly recent assertion on the part of Neils Jerne that we are "awaiting the end of immunology" [63]—by which he meant that nothing but details remain to be filled in—I believe just the opposite. In every historical instance in which an eminent scientist has made such a rash statement about the "closing" of his field, the revolutionary investigators who overthrew the old dogma had already begun their work. I believe that such a revolution is underway now, and that the solution to the problem of autoimmunity viewed within the broadest context of the problems of self and nonself will provide the lever that allows the old dogma to be shouldered aside.

Science progresses by recognizing outstanding anomalies, defining new paradoxes, and then redefining its questions to make these anomalies addressable [64]. Science never progresses by basking in its all too limited successes. And so science itself is like the immune system: it is defined both by what it knows (self) and by what it does not (nonself); and the self part of science will be redefined only by a dialectical interaction with the nonself. Philosophy, theory, and experiment will all have their roles in this redefining process.

Michigan State University

NOTES

1. Doyle, A.C. *The memoirs of Sherlock Holmes*. New York: Pebbles Press, c. 1975.
2. Silverstein, A.M. *A History of Immunology*. San Diego: Academic Press, 1989; Cruse, J.M., Whitcome, D., and Lewis, R.E. Autoimmunity — historical perspective. Concepts Immunopathol. 1: 32–71, 1985; Eyquem, A. L'Immuno-pathologie de l'auto-immunisation. Archs. Inst. Pasteur, Tunis. 58: 281–295, 1981.
3. Ehrlich, P. and Morgenroth, J. Zur Theorie der Lysinwirkung. Berl. Klin. Wochenschr. 36: 6–21, 1899 Üeber Haemolysine (5 part paper). Berl. Klin. Wochenschr. 36: 481–493, 1899; 37: 453–466; 681–696, 1900; 38: 251–268, 569–601, 1901. Reprinted and translated in Ehrlich, P. F *The Collected Papers of Paul Ehrlich*, Vol. 2. New York: Pergamon, 1957.
4. Cruse, J.M., Lewis, R.E., Jr. Contemporary concepts of autoimmunity. Concepts Immunopathol. 1: 1–31, 1985. Beutner, E.H., Chorzelski, T.P., Binder, W.L. Nature of autoimmunity: pathologic versus physiologic responses and a unifying concept. In Beutner, E.H., Chorzelski, T.P., and Bean, R. *Immunopathology of the Skin* 2nd ed. New York, Wiley, 1979, pp. 147–180.
5. Ehrlich, P. and Morgenroth, J. Berl. Klin. Wochenschr. 28: 251–268, 1901; *Collected Papers*, p. 253.
6. Silverstein, 1989 (note 2) pp. 162–163, 190–213; Goltz, D. Das Donath-Landsteiner Haemolysin. Die Entstehung eines Mythos in der Medizin des 20. Jahrhunderts. Clio. Med. 16: 193–211, 1982.
7. Metalnikoff, S. Etudes sur la spermatoxine. Ann. Inst. Pasteur, Paris. 14: 577–603, 1900; Ehrlich and Morgenroth, 1901, 251 and *Collected Papers*, p. 253n.
8. Donath, J. and Landsteiner, K. Üeber paroxysmale Haemaglobinuric. Muench. med. Wschr. 51: 1590–1593, 1904; Silverstein, op. cit. 1989, 190–213; Goltz, op. cit. 1982.
9. Silverstein, op. cit. 1989 (note 2) 160–189; Cruse, Whitcome, Lewis, op. cit. 1985 (note 2).
10. Glynn, L.E., Holborow, E.J. *Autoimmunity and Disease* Oxford: Blackwell, 1965; Mackay, I.R., Burnet, F.M. *Autoimmune Diseases* Springfield, IL: CC Thomas, 1963; Cohen, I.R. The self, the world, and autoimmunity. Sci. Am. 52–61, April 1988; Sinha, A.A., Lopez, M.T., McDivitt, H.O. Autoimmune diseases: the failure of self tolerance. Science 248: 1380 – 1388, 1990.
11. Silverstein, op. cit. 1989 (note 2) 172–175.
12. Weigle, W.O. Analysis of autoimmunity through experimental models of thryoiditis and allergic encephalomyelitis. Adv. Immunol. 30: 159 – 273, 1980; Waksman, B.H. Experimental allergic encephalomyelitis and the "auto-allergic" diseases. Intern. Arch. Allergy Suppl. vol. 14, 1959.

13. Rivers, T.M., Sprunt, D.H., Berry, G.P. Observations on attempts to produce acute disseminated encephalomyelitis in monkeys. J. Exp. Med. 58: 39–57, 1933; Rivers, T.M., Schwentker, F.F. Encephalomyelitis accompanied by myelin destruction experimentally produced in monkeys. J. Exp. Med. 61, 689–705, 1935; Waksman, B.H., Adams, R.D. A histologic study of the early lesion in experimental allergic encephalomyelitis in the guinea pig and rabbit. Amer. J. Path. 41: 135–147, 1962.
14. Paterson, P.Y. Transfer of allergic encephlomyelitis in rats by means of lymph node cells. J. Exp. Med. 111: 119–136, 1960; Orgad, S., Cohen, I. Autoimmune encephalomyelitis: activation of thymus lymphocytes against syngeneic brain antigens *in vitro*. Science, 183: 1083–1085, 1974; Shiraki, H., Ohtani, S. Clinical and pathological features of rabies post-vaccinal encephalomyelitis in man. Relationship to multiple sclerosis and to experimental allergic encephalomyelitis in animals. In *Allergic Encephalomyelitis*, Kies, M.W. and Alvord, E.C., Jr., Eds. Springfield, Ill, Charles C. Thomas, 1959, p. 58; Waksman, B. H. Immunological aspects of multiple sclerosis (Forum) Res. Immunol. 140: 173 passim, 1990.
15. Freund, J., McDermott, K. Sensitization to horse serum by means of adjuvants. Proc. Soc. Exp. Biol. Med. 49: 548–553, 1942; Kabat, E.A., Wolf, A., Bezer, A.E. The rapid production of acute disseminated encephalomyelitis in rhesus monkeys by injection of heterologous and homologous brain tissue with adjuvants. J. Exp. Med. 85: 11–25, 1947; Voisin, G.A., Delaunay, A., Barber, M. Sur des lésions testiculaires provoquacutées chez le cobaye par iso- et autosensibilization. Ann. Inst. Pasteur 81: 48–72, 1951. Freund, J., Lipton, M.M., Thompson, G.E. Aspermatogenesis in the guinea pig induced by testicular tissue and adjuvants. J Exp. Med. 97: 711–726, 1953.
16. Freund, J. Mode of action of immunologic adjuvants. *Adv. Tuberc. Res. 7: 130 –148, 1956*; Finger, H. Possible mechanism of adjuvant action. *In: Pathogenesis and Etiology of Demyelinating Diseases*. Burdzy, K. and Callos, P., Eds. Basel: Karger, 1969, pp. 293–309; Paterson, P.Y. Autoimmune neurological disease: experimental animal systems and implications for multiple sclerosis. In: *Autoimmunity: Genetic, Immunologic, Virologic, and Clinical Aspects*. Talal, N., Ed. New York: Academic Press, 1977, pp. 644–692.
17. Gautier-Smith, P. Neurological complications of glandular fever (infectious mononucleosis). *Brain* 88, 323–334, 1965; Vortel, V., Plachy, V. Glial-nodule encephalitis associated with generalized cytomegalic inclusion body disease. Am. J. Clin. Pathol. 49: 319–324, 1968. Hanshaw, J. Cytomegalovirus and cerebral dysfunction.*Hosp. Practice* 5, 111–120, 1970; Heyworth, B., Ironside, A.G., Hewer, R.L., Leonard, J.C. *et al.* Polyneuritis associated with cytomegalovirus infections. Q. J. Medicine 40, 435–442, 1971; Talal, N. *Autoimmunity: Genetic, Immunologic, Virologic, and Clinical Aspects*, 1977; Lampert, P.W. Autimmune and virus-induced demyelinating diseases. Am. J. Pathol. 91: 176–197, 1978; Clyde, W.A. Neurological syndromes and mycoplasmal infections. *Arch. Neurol.* 37, 65–66, 1980; Dowling, P., Menonna, J., Cook, S. Cytomegalovirus complement fixation antibody in Guillain-Barré syndrome. Neurology (Minn.) 27, 1153–1156, 1977.
18. Sprent, J.F.A. Parasitism, immunity and evolution. In: *The Evolution of Living Organisms.* Leeper, G.W., Ed. Parkville, Victoria: Melbourne University Press, 1962, pp. 149–165.
19. Zabriskie, J.B. Mimetic relationships between group A streptococci and mammalian tissues. Adv. Immunol. 7: 147–188, 1967; Thornes, C.J., Morris, J.A. Common epitopes between mycobacterial and certain host tissue antigens. Clin. exp. Immunol. 61: 323–328, 1985; Heijne, G., Uhlén, M. Homology to region X from staphylococcal protein A is not unique to cell surface proteins. J. Theor. Biol. 127: 373–376, 1987; Anderson, D.C., van Schooten,

W.C.A., Barry, M.E., Janson, A.A.M., Buchanan, T.M., de Vries, R.R.P. A *Mycobacterium leprae*-specific human T-cell epitope cross-reactive with an HLA-DR2 peptide. Science. 242: 259-261, 1988; Barthomomaeus, W.N., O'Donoghue, H., Foti, D., Lawson, M., Shellam, G.R. Multiple autoantibodies following cytomegalovirus infection: virus distribution and specificity of autoantibodies. Immunology 64: 397–405, 1988; Fujinami, R.S., Oldstone, M.B.A. Amino acid homology between the encephalitogenic site of myelin basic protein and virus: mechanism for autoimmunity. Science 230: 1043–1045, 1985; Jahnke, U., Fischer, E.H., Alvord, E.C. Sequence homology between certain viral proteins and proteins related to encephalomyelitis and neuritis. Science 229: 282–284, 1985; Fujinami, R.S., Nelson, J.A., Walker, L., Oldstone, M.B.A. Sequence homology and immunologic cross-reactivity of human cytomegalovirus with HLA-DR beta chain: a means for graft rejection and immunosuppression. J. Virology 62: 100–105, 1988; Srinivasappa, J., Saegusa, J., Prabhakar, B.S., Gentry, M.K. *et al.* Molecular mimickry: frequency of reactivity of monoclonal antiviral antibodies with normal tissues. J. Virology 57: 397–401, 1986; Capron, A., Dessaint, J.P., Capron, M., Ouma, J.H., Butterworth, A.E. Immunity to schistosome: progress toward vaccine. Science 238: 1065–1071, 1987.

20. Krueger, J.M., Pappenheimer, J.R., Karnovsky, M.L. Sleep-promoting effects of muramyl peptides. Proc. Natl. Acad. Sci. USA 79: 6102–6106, 1982; Masek, K., Kadlecova, O., Raskova, H. Brain amines in fever and sleep cycle changes caused by streptococcal mucopeptide. Neuropharmacology 12: 1039–1047, 1973; Masek, K., Kadkecova, O., Petrovicky, P. The effect of some bacterial products on temperature and sleep in rats. Z. Immun-Forsch. 149: 273–282, 1975; Root-Bernstein, R.S., Westall, F.C. Sleep factors: do muramyl peptides activate serotonin binding sites? Lancet i: 653, 1983; Root-Bernstein, R.S., Westall, F.C. Serotonin binding sites. I. Structures of sites on myelin basic protein, LHRH, MSH, and ACTH. Brain Res. Bull. 12: 425–436, 1984; Karnovsky, M.L. Muramyl peptides in mammalian tissues and their effects at the cellular level. Fed. Proc. 45: 2557–2560, 1986; Silverman, D.H.S., Imam, K., Karnovsky, M.L. Muramyl peptide/serotonin receptors in brain-derived preparations. Peptide Res. 2: 338–344, 1989; Kaydalov, A.A., Utkin, Y.N., Androva, T.M., Tsetlin, V.I., Ivanov, V.T. Muramyl peptides bind specifically to rat brain membranes. FEBS Lett. 248: 78–82, 1989.
21. See note 19, especially: Jahnke, Fischer and Alvord, 1985; Fujinami and Oldstone, 1985.
22. Litvak, A.M., Sands, I.J., Gribel, H. Encephalitis complicating measles. Am. J. Dis. Child. 65: 265–273, 1943; Fenichel, G.M. Neurological complications of immunization. Ann. Neurol. 12: 119–128, 1982; Wilson, G.S. *The Hazards of Immunization*. London: The Athlone Press, 1967.
23. Jaworski, M. A., Severini, A., Mansour, G., Hennig, K. *et al.* Inherited diseases in North American Mennonites: Focus on Old Colony (Chortitza) Mennonites. *Am. J. Med. Genetics* 32: 158–168, 1989; Cruse and Lewis, 1985 (note 4); Todd, J.A., Acha-Orbea, H., Bell, J.I. *et al.* A molecular basis for MHC class II-associated autoimmunity. Science 240: 1003–1009, 1990.
24. Navia, B.A., Jordan, B.D., Price, R.W. The AIDS dementia complex: I. Clinical features. Ann. Neurology 19: 517–524, 1986; Navia, B.A., Cho, E.-S., Petito, C.K., Price, R.W. The AIDS dementia complex: II. Neuropathology. Ann. Neurol. 19: 525–535, 1986; Nunn, P.P., McAdam, K.P.W.J. Mycobacterial infections and AIDS. Brit. Med. Bull. 44: 801–813, 1988; Petito, C.K., Navia, B.A., Cho, E.-S., Jordan, B.D., George, D.C., Price, R.W. Vacuolar myelopathy pathologically resembling subacute combined degeneration in patients

with the acquired immunodeficiency syndrome. N. Engl. J. Med. 312: 874–879, 1985; Ryder, J.W., Croen, K., Kleinschmidt-DeMasters, B.K., Ostrove, J.M., Straus, S.E., Cohn, D.L. Progressive encephalitis three months after the resolution of cutaneous zoster in a patient with AIDS. Ann. Neurol. 19: 182–188, 1986; Shearer, G.M. Immune suppression and recognition of class II (Ia) antigens: a possible factor in the etiology of acquired immune deficiency syndrome. In *AIDS: The Epidemic of Kaposi's Sarcoma and Opportunistic Infections*, Friedman-Kien, A.E. and Laubenstein, L.J., Eds. New York: Masson, 1984, pp. 169–172; Welch, K., Findkbeiner, W., Alpers, C.E. *et al.* Autopsy findings in acquired immune deficiency syndrome. JAMA 252: 1152–1159, 1984; Reichert, C.M., O'Leary, T.J., Levens, D.L. *et al.* Autopsy pathology in the acquired immune deficiency syndrome. Am. J. Pathol. 112: 357–382, 1983; Neidt, G.W., Schinella, R.A. Acquired immunodeficiency syndrome. Clinicopathologic study of 56 autopsies. Arch. Pathol. Lab. Med. 109: 727–734, 1985.

25. Reviewed in Bergstrand, H. Immunochemistry of encephalitogenic protein. In: *Immunochemistry of Proteins* Atassi MZ, Ed. Vol. 1. New York: Academic Press, 1977, pp. 315–370; Weigle, 1980 (note 12); Westall, F.C., Root-Bernstein, R.S. An explanation of prevention and suppression of experimental allergic encephalomyelitis. Mol. Immunol. 20: 169–177, 1983; Davison, A.N., Cuzner, M.L., Eds. *The Suppression of Experimental Allergic Encephalomyelitis and Multiple Sclerosis*. London: Academic Press, 1980; Ben-Nun, A., Wekerle, H., Cohen, I.R. The rapid isolation of clonable antigen-specific T lymphocytes lines capable of mediating autoimmune encephalitis. Eur. J. Immunol. 11: 195–199, 1981; Mitsuzawa, E., Yasuda, T., Tamura, N., Ohtani, S. Pretreatment with antigen or adjuvant, and the preventability of allergic encephalomyelitis induced by basic protein with synthetic muramyl dipeptide in comparison with Freund's complete adjuvant. Jpn. J. Exp. Med. 54: 97–107, 1984.
26. Silverstein, op. cit. 1989 (note 2) 214ff.
27. See note 13.
28. Field, E.J., Caspary, E.A. Lymphocyte sensitization: an *in vitro* test for cancer? Lancet ii: 1337–1371, 1970; Coates, A., McKay, I.R., Crawford, M. Immune protection against experimental autoimmune encephalomyelitis: optimal conditions and analysis of mechanism. I. Cell. Immunol. 12: 370–381, 1974; Waksman, F.J., Fritz, R.B., Hinrichs, D.J. The presence of antigen-reactive cells during the induction, recovery, and resistance phases of EAE. Cell. Immunol. 48: 34–42, 1980.
29. Bergstrand, H., Kallen, B. On the statistical evaluation of the macrophage migration inhibition assay. Scand. J. Immunol. 2: 173–187, 1973; Freund, op. cit. 1956 (note 16); Mitchison, N.A. Induction of immunological paralysis in two zones of dosage. Proc. Roy. Soc. Lond. 161: 275–292, 1964/65.
30. Bergstrand and Kallen, op. cit. 1973 (note 29); Ben-Nun *et al.*, op. cit. 1981 (note 25); Spitler, L.E., von Muller, C.M., Fudenberg, H.H., Eylar, E.H. Experimental allergic encephalitis: dissociation of cellular immunity to brain protein and disease production. J. Exp. Med. 136: 156–174, 1972; Lisak, R.R., Zweiman, B. Immune responses to myelin basic protein in mycobacterial-induced suppression of experimental allergic encephalomyelitis. Cell. Immunol. 14: 242–254, 1974; Bergstrand, H. Immunochemistry of encephalitogenic protein. In: *Immunochemistry of Proteins* Atassi, M.Z., Ed. New York: Academic Press, 1977, pp. 315–370; Lyman, W.D., Kadish, A.S., Raine, C.S. Experimental allergic encephalomyelitis in the guinea pig: variation in peripheral blood lymphocyte

responsiveness to myelin basic protein during disease development. Cell. Immunol. 63: 409–416, 1981. Waksman, F.J., Fritz, R.B., Hinrichs, D.J.The presence of antigen-reactive cells during the induction, recovery, and resistance phases of EAE. Cell. Immunol. 48: 34–42, 1980. The only cross-reactivity between myelin basic protein and mycobacteria reported —a weak skin response in guinea pigs—was by Vandenbark, A.A., Burger, D.R., Vetto, R.M. Cell-mediated immunity in experimental allergic encephalomyelitis: cross-reactivity between myelin basic protein and mycobacteria antigens. Proc. Soc. Exp. Biol. Med. 148: 1233–1236, 1975. However, the same authors also reported that lymphocyte cultures form mycobacteria-inoculated animals did not respond at all to myelin basic protein.

31. Ellouz, F., Adam, A., Ciorvaru, R., Lederer, E. Minimal structural requirements for adjuvant activity of bacterial peptidoglycan derivatives. Biochem. Biophys. Res. Commun. 59: 1317–1325, 1974; Nagai, Y., Akiyama, K., Suzuki, K. *et al.* Structural specificity of synthetic peptide adjuvant for induction of experimental allergic encephalomyelitis. Cell. Immunol. 35: 158–167, 1978; Westall, F.C., Robinson, A.B., Caccam, J., Jackson, J.M., Eylar, E.H. Essential chemical requirements for induction of allergic encephalomyelitis. Nature 229: 22–24, 1971; Carelli, C., Audibert, F., Gaillard, J., Chedid, L. Immunological castration of male mice by a totally synthetic vaccine administered in saline. Proc. Natl. Acad. Sci. USA 79: 5392–5395, 1982. Testicular fractions are also effective: Glynn, L.E., Holborow, E.J. *Autoimmunity and Disease*. Oxford: Blackwell, 1965, pp. 306–319.

32. Root-Bernstein, R.S. Structure of a serotonin and LSD binding site on myelin basic protein. J. Theor. Biol. 100: 373–378, 1983; Root-Bernstein, R.S. "Molecular sandwiches" as a basis for structural and functional similarities of interferons, MSH, ACTH, LHRH, myelin basic protein, and albumins. FEBS Lett. 168: 208–212, 1984; Root-Bernstein and Westall, op. cit. 1984 (note 20).

33. Freund and Lipton, op. cit. 1948; Kabat *et al.*, op. cit. 1947 (note 15).

34. Shaw, C.M., Alvord, E.C., Fahlberg, W.J., Jr., Kies, M.W. Adjuvant-antigen relationships in the production of experimental allergic encephalomyelitis in guinea pigs. J. Exp. Med. 115: 169–179, 1962. Young, J.D., Tsuchiya, D., Geier, M. *et al.* The encephalitogenic dose-response in guinea pigs of the tryptophan region of myelin basic protein. Immunol. Commun. 3: 219–226, 1974; Nagai *et al.*, op. cit. 1978 (note 31).

35. Salk, J., Romine, J.S., Westall, F.C., Wiederholt, W.C. Myelin basic protein studies in experimental allergic encephalomyelitis and multiple sclerosis: a summary with theoretical considerations of multiple sclerosis etiology. In: *Experimental Allergic Encephalomyelitis and Multiple Sclerosis*. Davison, A.N. and Cuzner, M.L., Eds. London: Academic Press, 1980, pp. 141–156; Field, E. J. Immunological treatment for multiple sclerosis. Lancet i: 1272, 1989.

36. Carelli *et al*, 1982 (note 31); Dyck, R.J., Thomas, P.K., Lambert, E.H., Bunge, R., eds. *Peripheral Neuropathy* Philadelphia: WB Saunders, 1984; Waxman, B.H., Raymond, D.A., Mansmann, H.C. Experimental study of diphtheric polyneuritis in the rabbit and guinea pig. J. Exp. Med. 105: 591–614, 1957; Weigle, 1980 (note 12); Brostoff, S., Burne, H.P., Lampert, P., Eylar, E.H. Nature, New Biology 235: 210–214, 1972; Wacker, W.B., Donoso, L.A., Kalsow, C.M., Yankeelov, J.A. Jr., Organisciak, D.T. J. Immunol. 119: 1949, 1977; Kozak, Y., Faure, J.P., Ardy, H., Ursi, M., Thillaye, B. Ann. Immunol. (Institute Pasteur) 129C: 73–88, 1978.

37. Root-Bernstein, R.S. Amino acid pairing. J. Theor. Biol. 94: 885–895, 1982; Root-Bernstein, 1984 (note 32); Root-Bernstein and Westall, op. cit. 1984 (note 20).

Root-Bernstein, R.S. *Discovering*. Cambridge, MA.: Harvard University Press, 1989, pp. 195–196, 269–272, 309.

38. Westall and Root-Bernstein, op. cit. 1983 (note 25); Root-Bernstein, R.S., Westall, F.C. Complementarity between antigen and adjuvant in the induction of autoimmune diseases — a dual antigen hypothesis. J. Infer. Deduct. Biol. 1.5: 1–25, 1986.
39. Ibid.; Westall, F.C., Root-Bernstein, R.S. Cause and prevention of postinfectious and postvaccinal neuropathies in light of a new theory of autoimmunity. Lancet ii: 251–252, 1986; Root-Bernstein, R.S. Multiple antigen mediated autoimmunity (MAMA) in AIDS: a possible model for postinfectious autoimmunity. Res. Immunol. (Institute Pasteur), 141, 815–838, 1990.
40. Westall and Root-Bernstein, op. cit. 1983 (note 25).
41. Ibid.
42. Sinha, Lopez, and McDivitt, op. cit. 1990 (note 10).
43. Root-Bernstein, R.S., Yurochko, F., Westall, F.C. Clinical suppression of experimental allergic encephalomyelitis by muramyl dipeptide "adjuvant." Brain Res. Bull. 17: 473–476, 1986; Killen, J., Root-Bernstein, R.S., Lallouette, P., unpublished data.
44. See figure 1 of Suckling, A. J., Kirby, J. A., Rumsby, M.G. Characterization by acid alpha-naphthyl acetate esterase staining of the spinal cord cellular infiltrate in the acute and relapse phases of chronic relapsing experimental allergic encephalomyelitis. *Prog. Brain Res.* 59: 317–322, 1984; Hickey, W.F., Kimura, H. Perivascular microglial cells of the CNS are bone marrow-derived and present antigen in vivo. Science 239: 290–292, 1988.
45. Westall, F.C., Root-Bernstein, R.S. Serotonin-binding peptides bind muramyl dipeptide. Abst. 170 *Soc. Neurosci. Abst.* 11: 1985; Root-Bernstein, R.S., Westall, F.C. Serotonin binding sites. II. Muramyl dipeptide binds to serotonin binding sites on myelin basic protein, LHRH, and MSH-ACTH 4–10. Brain Res. Bull. 25: 827–841, 1990.
46. This issue of the cause of kuru is a sensitive one. Carlton Gadjusek received a Nobel Prize for the discovery of the disease, which is usually attributed to a "slow virus" (see Gadjusek, D.C., Gibbs, C.J., Jr., Alpers, M. Experimental transmission of a kuru-like syndrome to chimpanzees. Nature, 209: 794–796, 1966.) Despite claims by Stanley Prusiner to have discovered this "slow virus" (see Prusiner, S.B. Prions and neurodegenerative diseases. N. Engl. J. Med. 317: 1571–1581, 1987), his putative causative agent, "prions," are highly controversial because they are supposedly composed only of protein (without DNA or RNA) and are also found genetically encoded in normal brain tissue. These findings suggest that these "slow viruses" or "prions" are not infectious, and, in fact, all experimental protocols require direct inoculation of brain material into the recipient animal. Thus kuru may actually be an autoimmune reaction similar to EAE. Certainly the fact that Gajdusek's group succeeded in transmitting a Gerstmann-Straussler-like syndrome (another disease putative caused by a prion) to chimpanzees (again by inoculation) argues against there being an infectious origin for the disease, since this syndrome is known to be inherited in a manner consistent with a Mendelian dominant gene. (See Goodfield, J. *Quest for the Killers.* Boston: Birkhauser, 1985, ch. 1, especially p. 43). Notably, retrovirologist Peter Duésberg has recently called into question the existence of the entire class of so-called "slow virus" agents since none has ever been isolated (see Deusberg P. Retroviruses as carcinogens and pathogens: expectations and reality. Cancer Res. 47: 1199–1220, 1987).
47. Pollock, T.M., Morris, J.A. Seven year survey of disorders attributed to vaccination in northwest Thames region. Lancet i: 753–757, 1983; Eylar, E.H. Antibody to chicken P2 protein in swine flu-associated Guillain-Barré syndrome. Fed. Proc. 43: 1991, 1983;

Gendelman, H.E., Wolinsky, J.S., Johnson, R.T., Pressman, N.J. *et al.* Measles encephalomyelitis: lack of evidence of viral invasion of the central nervous system and quantitative study of the nature of demyelinization. Ann. Neurol. 15: 353–360, 1984.

48. Root-Bernstein, R.S. How scientists really think. Persp. Biol. Med. 32: 471–488, 1989; Root-Bernstein *Discovering* 1989 (note 37).
49. Root-Bernstein, op. cit. 1990 (note 39).
50. Welch *et al.*, op. cit. 1984; Reichert *et al.*, op. cit. 1983; Niedt and Schinella, op. cit. 1985 (note 24).
51. Hurtenbach, U., Shearer, G.M. Germ cell-induced immune suppression in mice. Effect of inoculation of syngeneic spermatozoa on cell-mediated immune responses. J. Exp. Med. 155: 1719–1729, 1982; Mathur, S., Goust, J.-M., Williamson, H.O. *et al.* Cross-reactivity of sperm and T-lymphocyte antigens. Am. J. Reprod. Immunol. 1: 113–118, 1981; Richards, J.M., Bedford, J.M., Witkin, S.S. Immune response to allogenic insemination via the rectum in the rabbit. Fed. Proc. 42: 1334, 1983; Witkin, S.S., Sonnabend, J. Immune responses to spermatozoa in homosexual men. Fertil. Steril. 39: 337–342, 1983; Mavligit, G.M., Talpaz, M., Hsia, F.T. *et al.* Chronic immune stimulation by sperm alloantigens. Support for the hypothesis that spermatazoa induce immune dysregulation in homosexual males. JAMA 251(2): 237–241, 1984; Mathur, S., Goust, J.-M., Williamson, H.O. *et al.* Cross-reactivity of sperm and T-lymphocyte antigens. Am. J. Reprod. Immunol. 1: 113–118, 1981; Marcus, Z.H., Lunenfeld, B., Weissenberg, R., Lewin, L.M. Immunosuppressant material in human seminal fluid. Inhibition of blast transformation and of NK activity by seminal fluid patients of a male infertility clinic. Gynecol. Obstet. Invest. 23: 54–59, 1987; Williamson, J.D. Semen polyamines in AIDS pathogenesis. Nature 310: 103, 1984.
52. Lewis, J.H. The antigenic relationship of the alcohol-soluble fractions of brain and testicle. J. Immunol. 26: 473–478, 1934.
53. Root-Bernstein, op. cit. 1990 (note 39).
54. Cohen, op. cit. 1988 (note 10); personal observations of the author.
55. Fowler, W.M. *Hematology* 2nd ed., New York: Paul B. Hóeber, 1949, pp. 230–232.
56. Weisman, H.F., Bartow, T., Leppo, M. *et al.* Soluble human complement receptor type 1: in vivo inhibitor of complement suppressing post-ischemic myocardial inflammation and necrosis. Science 249: 146–151, 1990.
57. Silverstein, op. cit. 1989, pp. 167–169 (note 2); Fleck, L. *Genesis and development of a scientific fact.* Bradley, F. and Trenn, T.J., Transl. Chicago: University of Chicago Press, 1979.
58. Witebsky, E., Rose, N.R., Terplan, K., Paine, J.R., Egan, R.W. Chronic thyroiditis and autoimmunization. JAMA 164: 1439–1447, 1957.
59. Cruse, Whitcomb and Lewis, op. cit. 1985, pp. 34–35 (note 2).
60. Cruse and Lewis, op. cit. 1985; Beutner, Chorzelski and Binder, op. cit. 1979 (note 4).
61. Root-Bernstein, R.S. Protein replication by amino acid pairing. J. Theor. Biol. 100, 99–106, 1983; Mekler, L.B. Antibody formation. Nature 215, 481–483, 1967; Cook, N.D. The case for reverse translation. J. Theor. Biol. 64, 113–135, 1977.
62. Cohen, op. cit. 1988 (note 10).
63. As quoted by Burnet, F.M. *Changing Patterns: An Atypical Autobiography* Melbourne: Heinemann, 1968, pp. 11–12.
64. Root-Bernstein, op. cit. 1990, pp. 44–47 (note 46); Kuhn, T.S. *The Structure of Scientific Revolutions.* Chicago: University of Chicago Press, 1959.

EDITOR'S COMMENTS TO FOSTER AND SARKAR

The papers by Foster and Sarkar are essentially a dialogue and to complement each other, comprehensively review the current argument of whether the environment induces genomic responses in bacteria. The basic issue dates at least to the early 19th century. Although Lamarck is often credited with having been the first major proponent of the idea of organic evolution, he is probably most often associated with the infamous notion of acquired heredity (broadly accepted at the time) presented in *Zoological Philosophy* (1809). The first part of his case was correct, when he observed that "the environment affects the shape and organization of animals," not directly, but by the sequence of "altered need" leading to "changed functions." Clearly his notions of how this evolution might be affected, and inherited, pre-dated Darwinian selection and our modern Synthetic Theory, where we understand the environment serving only to select given genetic fitnesses from an ever changing genomic library. However as a result of recent studies, evidence for a Neo-Lamarckian mechanism for evolutionary change is now being considered; Foster and Sarkar summarize and critically examine this recent research to present the case of an adaptive genome in a balanced perspective.

In regards to our theme concerning the origins of the Self, it would appear appropriate to reconsider the integrity of the genome as the inviolate parameter of Self-definition. We are not considering the case of random somatic mutation, but the issue of a *responsive* gene. The paradigm of information proceeding from the gene to the protein and thereby determining the organism's structural definition may in fact miss a truly dialectical bidirectionality of informaton transfer. Mechanisms of the genome shuffling its components, as in the case of McClintock's early studies with maize, and more recently in the assembly of immunoglobulin, is a phenomenon qualitatively different than a directed mutant response to the environment of a starved bacterium. Even those who suggest that such mutagenesis exists, usually postulate an underlying random process by which variance originates. (It is important to recognize that this discussion more radically extends the notion of the responsive immune genome, where a genetic library is shuffled to present an accommodating antibody for antigen binding.) If the biologic Self includes the criteria of purposeful reaction, then perhaps this most basic response may well extend to the very foundation of the organism's identity, that is its genetic constitution. Or, as Joshua Lederberg has said in a different context, "Mutate or die". Thus the Neo-Lamarckian studies discussed by Foster and Sarkar are central to our concern.

PATRICIA L. FOSTER

DIRECTED MUTATION IN *ESCHERICHIA COLI*: THEORY AND MECHANISMS

INTRODUCTION

For a haploid unicellular organism, such as *Escherichia coli*, that reproduces asexually by binary fission, the concept of "self", or more appropriately, "individual", may be indistinguishable from the concept of organism. Controversy has arisen in the past about whether and how such creatures maintain themselves as a species since every new mutant that appears could, theoretically, give rise to a clone of unique descendants. Indeed, based largely upon protein electrophoresis patterns it has been estimated that the world-wide population of *E. coli* consists of only 10^2–10^3 such clones [1]. This result implies that *E. coli* in the wild rarely engage in chromosomal recombination, and that the genetic exchanges that take place via extrachromosomal elements leave chromosomal genes largely unchanged [1, 2]. (However, see Reference 3 for an alternative interpretation).

In a recent paper, Rayssiguier *et al.* [4] demonstrated that methyl-directed mismatch repair is a major barrier to recombination between *E. coli* and *Salmonella typhimurium*, which are about 20% divergent in their DNA sequences. Methyl-directed mismatch repair is a multienzyme system that recognizes and initiates the repair of error-containing DNA (see below). By binding to the many mismatches in heteroduplex DNA, the mismatch repair system also aborts recombination. It is estimated that this system could prevent recombination between DNA differing in sequence by only 2% (M. Radman, personal communication); thus it may be an effective mechanism for genetic isolation and subsequent speciation [4]. Given an average spontaneous mutation rate of 3×10^{-3} mutations per round of DNA replication per genome (J.W. Drake, personal communication), of which about 30% occur at silent sites [2], it would appear that even in the absence of selection an *E. coli* cell could give rise to a new species in about 10^8 generations. In natural populations with an estimated $10^2 - 10^3$ generations per year [2], 10^8 generations could, for a very lucky bacterium, correspond to only about 10^5 years. Of course, most *E. coli* face nearly instantaneous extinction of their clones.

It is generally accepted that most spontaneous mutations in bacteria arise as errors made during DNA replication. Normal error frequencies for replica-

Alfred I. Tauber (ed.), Organism and the Origins of Self, 213–234.

tion in *E. coli* are 10^{-9} to 10^{-10} per base [5]. This overall accuracy results from four processes. First, DNA polymerase preferentially inserts the base that is complementary to the template base, achieving an error rate of 10^{-4} – 10^{-5} [6]. Second, if the polymerase inserts an incorrect base, the mispaired terminus is immediately degraded by an exonuclease that is associated with the polymerase (the proofreader). By forcing repolymerization, proofreading increases the accuracy of replication 10–200 fold, depending on the specific mispair [7]. Third, several separate enzyme systems prevent or correct mispairs, resulting in a further 10–1000 fold reduction in error rate [8, 9].

The most powerful of the post-replication error-correction pathways is methyl-directed mismatch repair, mentioned above. Because the details of this system will be important to a subsequent section, it is worth describing here (see Reference 10 for a comprehensive review). In *E. coli*, adenines in the sequence GATC are methylated by DNA adenine methylase, the product of the *dam* gene. Methylation by the *dam* methylase lags behind replication about 10% of a generation [11], resulting in a segment of newly replicated DNA that is methylated only on the template strand. The mismatch repair system recognizes mispaired bases, cuts the unmethylated strand at a GATC site, and initiates the degradation and resynthesis of the unmethylated strand with the methylated strand serving as template. Thus, corrections are made in favor of the preexisting strand, resulting in net error reduction. However, in the absence of methylation (e.g., in a *dam*$^{-}$ mutant), either strand can be cut and degraded, and the error-containing strand is as likely to serve as template as the correct strand. The result is a net increase in mutation rate since approximately 50% of the errors made during replication will be "fixed" into the genome.

Mutations that disrupt any of the four processes discussed above elevate spontaneous mutation rates 10^2 – 10^5 fold [8]. That the fidelity of replication has evolved to be an optimum, not a maximum, was suggested by the discovery of mutations in bacteriophage T4 that lower the spontaneous mutation rate by increasing the accuracy of DNA replication [12]. Subsequently, Cox and Gibson [13] have shown that mutator strains of *E. coli* outcompete wildtype strains in chemostat experiments, demonstrating that high mutation rates can be advantageous in some situations. A theoretical model by Leigh [14] predicts selection for high mutation rates in fluctuating environments. However, it is curious that, in contrast to the case of T4, only one antimutator affecting spontaneous mutation rates has been characterized in *E. coli*. The antimutator phenotype is due to a mutation in *purB*, which encodes adenylosuccinate lyase, an enzyme in the adenosine biosynthetic pathway [15]. The mechanism by which this defect inhibits spontaneous mutation is still unknown.

In addition to simple replication errors, which mainly result in base substitution and frameshift mutations, major genomic rearrangements also occur spontaneously or are mediated by mobile genetic elements. The evolutionary role of mobile genetic elements, such as insertion sequences (which contain recognition sequences and a transposase gene) and transposons (which also carry other genes, such as those encoding antibiotic resistances), in promoting such genomic changes may be significant. Indeed, Campbell [16] has argued that certain classes of rearrangements can only be mediated by transposable elements. In chemostat experiments, transposon-containing cells can outcompete non-transposon-containing cells due to the mutator effect of the transposon [17]. It is interesting that the methylation state of GATC sequences, which directs the mismatch repair system, also regulates the movement of some transposons; genetic rearrangements associated with these elements are increased 10–200-fold in *dam* cells (reviewed in Reference 18). Thus, the *dam* methylase, by design or accident, has a large impact on at least two sources of genomic variability. A cell that wished to regulate genetic variation might do well to regulate the activity of its *dam* methylase, although there does not appear to be any evidence that such a mechanism exists.

In the neo-Darwinian view, random mutations in the DNA due to the mechanisms outlined above are the source of variation upon which natural selection acts. The evidence upon which this view rests was obtained in the 1940's and 50's by Luria and Delbrück, the Lederbergs, Newcombe, and others (see the chapter of this volume by Sarkar). These scientists established beyond a doubt that, in *E. coli*, mutations arise before the population is subjected to selection. Hence, mutations must occur randomly in the sense that they can arise without regard to their utility. However, it the ensuing 40 years it was largely forgotten that these experiments did not eliminate the alternative hypothesis — mutations might also arise adaptively, i.e., in response to the selective agent, if the selective agent were not lethal to non-mutants. Indeed, Delbrück made this point in 1946 [19], but it was not until recently that the issue has been reopened [20, 21].

Recognizing that the early studies were not definitive, and realizing that the results of certain studies of mutation in *E. coli* could not easily be explained by the random mutation hypothesis [22] (also see below), Cairns *et al.* [21] undertook a reexamination of the issue. They found that under certain non-lethal conditions, the only mutations that accumulate in a population of bacteria under selective pressure are those that are advantageous [21]. The phenomenon has been named "directed mutation" (since the selective conditions appear to direct which mutations appear) in contrast to "random mutation" (in which the mutations arise without regard to utility). Not adverse to controversy, Cairns *et al.* [21] pointed out that these results could mean that cells have a mechanism for "the inheritance of acquired characteristics".

The controversy that indeed ensued has pitted neo-Darwinism against neo-Lamarckism, as if *E. coli* 's only *raison d'etre* was to validate or invalidate current evolutionary theory. But *E. coli* is an organism that must survive and propagate itself, and to do so it must interact with its environment to its own advantage. As pointed out by Cairns [22], there is no intrinsic reason why the simple expedient of testing a phenotypic change for utility before it becomes permanently fixed in the genome could not have evolved, at least in this organism. This assertion does not imply that a similar mechanism has or could evolve in organisms in which phenotype and genotype are not so closely linked. The intent of this chapter is to refocus the controversy about directed mutation on the biology of *E. coli*. It is only from study of the organism, not from theoretical considerations, that an understanding of the phenomenon will come. I shall first try to provide a conceptual precedent for directed mutation by describing how *E. coli* can and does control its mutation rate. I shall then discuss the evidence for directed mutation, some recent findings, and my current thoughts on the underlying mechanisms.

THE SOS RESPONSE OF *ESCHERICHIA COLI*

DNA damage by ubiquitous environmental agents, such as UV light and oxygen, is a common hazard to bacteria. When suffering substantial DNA damage, *E. coli* responds by increasing the expression of approximately 20 genes scattered throughout the genome (see Reference 23 for a comprehensive review). Known as the "SOS response", the induction of these genes is under the coordinate control of the *recA* and *lexA* genes. RecA protein normally mediates homologous recombination, but when complexed with single-stranded DNA and ATP, it is activated, probably via a conformational change, to a state known as RecA*. RecA* facilitates the autoproteolytic cleavage of several other proteins, among which is LexA, the common repressor of the SOS genes. The identity of the activating signal *in vivo* has not been definitively established, but it is likely to be the single-stranded DNA that occurs when replication forks are stalled at DNA lesions. These regions of single-stranded DNA would activate RecA to perform its proteolytic feat, resulting in the inactivation of LexA and the rapid induction of the SOS proteins. LexA is also the repressor of itself and of RecA. This refinement allows the response to rapidly amplify when DNA damage is massive and to immediately shut down after damage is repaired, but to be insensitive to small levels of DNA damage or to single-stranded DNA produced during normal cell processes.

While the genetic regulation of the SOS response is well understood, the biochemical activities of many of the SOS proteins are obscure. As might be expected, several of the SOS genes code for DNA repair enzymes. Thus *E. coli* mutants that are globally defective for the SOS response are exquisitely sensitive to DNA damage. Less expectedly, such mutants are also immutable by a wide variety of DNA damaging agents. The conversion of many (but not all) types of DNA lesions into mutations requires the active participation of at least three SOS proteins: RecA, UmuD, and UmuC. Just as RecA* turns on the SOS response by facilitating the cleavage of LexA, it also turns on the mutagenic processing of DNA lesions by facilitating the cleavage of UmuD [24-26], generating its active form, UmuD* [25]. RecA also plays a further role in mutagenesis, possibly as part of a RecA*UmuD*UmuC complex [27]. Genetic evidence suggests that all three proteins are involved in modifying the DNA replication machinery so that it can replicate past DNA lesions [25, 28, 29]. A popular current model is that RecA inactivates the proofreading function of DNA polymerase, allowing the insertion of a base opposite a DNA lesion [30-34], and that UmuD*C stabilizes the polymerase so it can continue replication from this aberrant terminus [35, 36]. However, this model does not explain all the phenomenology associated with SOS mutagenesis.

The mutations produced by SOS mutagenic processing occur almost exclusively at the sites of DNA lesions [37–39]. Either the modified replication machinery is able to replicate undamaged DNA accurately, or the error-correction enzymes can eliminate the occasional mistakes that the modified polymerase makes at normal DNA bases. Even the modest increase in spontaneous mutations that results from genetic induction of the SOS response in the absence of DNA damage can be explained by mutagenic processing of cryptic DNA lesions [40]. In addition, the spectrum of mutations that is generated by SOS processing is determined by the nature of the DNA lesion and the DNA sequence in which the lesion occurs (reviewed in Reference 39). Thus, SOS induction does not result in a general mutagenic state.

Inactivation of UmuDC strongly inhibits mutagenesis but results in only a modest increase in sensitivity to killing by DNA damage [23]. This result implies that mutations are not an inevitable consequence of DNA damage. Indeed, many species of bacteria are poorly mutable by UV, apparently because they lack fully functional equivalents of UmuDC [41, 42]. It appears that the degree to which bacteria respond to DNA damage by converting DNA lesions into mutations is an evolutionary choice. Some species may have evolved to wait for accurate repair pathways to correct DNA damage before replication, whereas others may have evolved to rapidly proceed with DNA replication and cell division despite the mutational load generated by replicating a damaged template. As discussed above, the ability to generate genetic variation in the face of environmental change can result in increased fitness. Thus, the SOS system may have evolved, in part, for the purpose of producing such variation. UmuDC-like activities are found on a variety of naturally occurring plasmids [43–45], suggesting that the selective advantage of these proteins may fluctuate over time. Thus, some bacterial species may be non-mutators, some occasional mutators, and some, like *E. coli*, "professional mutators" [46].

In effect, SOS processing represents a method by which a population of bacteria can generate genetic variation in response to adverse conditions. SOS mutagenesis differs from directed mutation in that, as far as we know, it does not discriminate among genes. In this sense it is a blind response to the environment. Nonetheless, SOS mutagenesis could result in an acceleration of evolution within a population subjected to a hostile environment. Such an acceleration would be even more pronounced if cells had a way to preferentially produce mutations appropriate to the environmental conditions. The next section presents the evidence that *E. coli*, at least, has such a mechanism. If the hypothesis of directed mutation is true, no matter by what molecular mechanism it is achieved, evolutionary theory will have to be adjusted to allow for the preferential production of variants with increased fitness.

THE EVIDENCE FOR DIRECTED MUTATION

Cairns *et al.* [21] observed that when a certain strain of *E. coli* that cannot metabolize lactose (Lac$^-$) due to a mutation in the *lacZ* gene (which codes for β-galactosidase, the enzyme that cleaves lactose) was plated out on medium in which the only source of carbon was lactose, lactose metabolizing (Lac$^+$) revertants appeared and continued to appear for up to a week (at which time the plate contained the maximum number of Lac$^+$ colonies supportable). These revertants did not accumulate in the absence of lactose, nor did cells with mutations in another, unrelated gene, accumulate in the presence of lactose.

The continued appearance of revertants was itself *prima facie* evidence that the mutations leading to the Lac^+ phenotype were occurring on the selective plates, but Cairns *et al.* [21] also demonstrated that the distribution of mutations from parallel cultures showed a significant departure from the Luria-Delbrück distribution, the hallmark of pre-selection mutation (see the chapter of this volume by Sarkar). As Cairns *et al.* [21] pointed out, post-selection mutation could not have been demonstrated by the classical studies of spontaneous mutation since the selections employed were for antibiotic or viral resistances, which immediately kill non-mutant cells.

A second example of directed mutation discussed by Cairns *et al.* [21] and Hall [47] is activation of cryptic genes. Natural isolates of *E. coli* are normally unable to metabolize certain uncommon sugars, for example the β-glucosides salicin, arbutin, and cellobiose. With prolonged incubation in the presence of such substrates, however, mutants arise that can utilize one or more of these sugars. Hall *et al.* [48] has proposed that the expression of such operons is disadvantageous when the substrate is not present. Thus, in a slowly fluctuating environment, the operons would be maintained in the population in three alternative states: active, cryptic (silent but activatible by mutation), and irreversibly inactivated (due to the passive accumulation of additional mutations in the silent genes). The particular state of an operon in a population would depend on the time since the last period of selection for utilization of the substrate.

Hall [47, 49] has investigated two examples of cryptic genes—*ebgA*, which encodes a β-galactosidase, and *bglF*, which encodes the salicin transport protein—in which activation required two genetic changes. In both cases, mutations were required in the structural gene (to produce activity) and in a regulatory gene (to allow expression of the structural gene). Yet with prolonged incubation in the presence of substrate, the required double mutants arose spontaneously at frequencies far in excess of what would be expected from the rate of each mutation alone. For example, the mutations that activate *ebg* occur during normal growth at rates of 10^{-9} (*ebgA*) and 10^{-8} (*ebgR*) per generation; thus the double mutant should appear at a rate of 10^{-17} per generation. Yet, the double mutant appeared within aging colonies at a frequency of about 10^{-9} per cell. Even allowing for cryptic growth within a colony, not enough generations could take place to account for the frequency at which the double mutant appeared [49]. Hall [47] further showed that the event which restored *bglF* to functionality, the excision of an insertion sequence, occurred in 1–10% of the cells after 8–12 days on salicin-containing plates, whereas the rate of excision during normal growth was $< 2 \times 10^{-12}$ per generation. Reconstruction experiments showed that in the absence of the second mutation (in *bglR*), the excision mutants had no selective advantage on the salicin-containing plates. Furthermore, the excision did not accumulate if the cells were maintained in the absence of salicin, and cells with mutations in a second, unrelated gene did not accumulate in the presence of salicin.

A third example of directed mutation given by Cairns *et al.* [21] also involved movement of a transposable element. Shapiro [20] studied a strain of *E. coli* in which a portion of the bacteriophage Mu, which is a transposon, was inserted between the regulatory region of the arabinose operon and the *lacZ* gene. This strain, designated Lac(Ara)$^-$, is Lac$^-$ because of transcriptional and translational stops in the inserted DNA, and Ara$^-$ because *araB*, required to utilize arabinose, has been interrupted. Certain deletions of the Mu DNA fuse *lacZ* in frame with *araB* so that the resultant mutant can utilize lactose if arabinose is present as an inducer [designated Lac(Ara)$^+$]. These rearrangements are under control of the phage transposase, since the presence of a second, non-defective Mu prophage in the cell inhibits the appearance of Lac(Ara)$^+$ cells [20].

Shapiro [20] found that after growth of the Lac(Ara)$^-$ strain under non-selective conditions, the frequency of Lac(Ara)$^+$ cells among the population was $<10^{-10}$. However, Lac(Ara)$^+$ colonies began to appear several days after the Lac(Ara)$^-$ cells were plated on medium containing arabinose and lactose. The rate of appearance of Lac(Ara)$^+$ colonies went up sharply with time until a maximum of about 100 colonies a day, after which the rate declined. Using this strain, Cairns *et al.* [21] further demonstrated that even when the cells were maintained in stationary phase, Lac(Ara)$^+$ cells did not accumulate unless both arabinose and lactose were present.

MODELS OF DIRECTED MUTATION

From the examples given above, Cairns *et al.* [21] drew the conclusion that "populations of bacteria, in stationary phase, have some way of producing (or selectively retaining) only the most appropriate mutations". It is worthwhile dividing this statement into two parts: (1) bacteria, in stationary phase, produce mutations; and, (2) only the most appropriate mutations are produced, or, selectively retained.

The first half of the conclusion is equivalent to stating that mutations can occur in the absence of normal DNA replication (i.e., replication that leads to cell division). This conclusion was also reached by Ryan *et al.* [50], who, in a number of experiments, showed that mutations to histidine prototrophy occurred in stationary phase cells that were neither dividing nor engaging in cell turnover to any appreciable extent. Furthermore, Arber *et al.* [51], Shapiro [20], and Hall [52] have suggested that cells in stationary phase enter into a program that increases their genetic variability. That such a program might exist is not unreasonable since it is known that about 30 proteins are induced upon starvation, some of which are required for survival [53]. Cells undergoing starvation are not quiescent, but are engaged in a process of slow degradation of some cellular components and synthesis of others [54]. In addition, as discussed above, *E. coli* are able to increase genetic variability when the need arises. Thus, this part of the argument is reasonable and precedented. However, the question remains how such variability might be generated in stationary phase when normal DNA replication is not occurring.

In the subsequent discussion, it is necessary to distinguish between the generation of variation and the production of mutations. A mutation is a heritable change in the DNA sequence and a mutant is a cell that has such a change. The phenotype of a cell is what is measurable or observable, i.e., growth rate, amino acid requirements, etc., whereas the genotype is the particular set of DNA sequences which the cell contains. The acquisition of a mutation does not necessarily confer a new phenotype, but it is usually assumed that the phenotype of a cell is an expression of its genotype. However, this is not a requirement, as a phenotype can arise from some other source of variation, although such a phenotype may not be stable. To express this distinction, Cairns *et al.* [21] used the term variant to mean any change that gives rise to a new phenotype even if that change is not in the DNA. In Table I are listed some of the sources of variation in *E. coli's* informational molecules, and estimates of the frequencies at which such variants can occur.

Cairns *et al.* [21] suggested that inaccurate transcription could be the source of variation in stationary phase; a cell under stress might produce variant messenger RNAs which could then be reverse transcribed into DNA and recombined or replicated into the genome. The original radicalness of this idea has been somewhat ameliorated by the subsequent discovery that at least some strains of *E. coli* do have reverse transcriptase [55, 56]. However, Cairns *et al.* [21] also noted that the production of variant regions of DNA, for example, by gene duplication, could also achieve the same result. Stahl [57] proposed that a non-dividing cell may replicate small regions of its DNA during stationary phase, but may do so inaccurately because its methyl-

TABLE I
Sources of variation

Level	Source	Frequency
DNA	Replication errors	10^{-10} to 10^{-9} per base (a)
	Endogenous and induced DNA lesions	10^{-7} depurinations per purine per hour (b)
	Movement of mobile genetic elements	10^{-7} per cell (c)
	Genomic rearrangements	10^{-4} to 10^{-3} duplications per cell (d)
RNA	Transcriptional errors	10^{-4} per base (e)
Protein	Translational errors	10^{-4} per codon (f)

[a] Drake, J.W. 1969. Comparative rates of spontaneous mutation. *Nature (London)* 221:1132.
[b] Lindahl, T., and Nyberg, B. 1972. Rate of depurination of native deoxyribonucleic acid. *Biochem*. 11:3610-3618.
[c] Schaaper, R.M., Danforth, B.M., and Glickman, B.W. 1986. Mechanisms of spontaneous mutagenesis: An anlysis of the spectrum of spontaneous mutation in the *Excherichia coli lacI* gene. *J. Molec. Biol.* 189:272–284.
[d] Anderson, R.P., and Roth, J.R. 1981. Spontaneous tandem genetic duplication in *Salmonella typhimurium* arise by unequal recombination between rRNA (rnn) cistrons. *Proc. Natl. Acad. Sci. USA* 78:3113–3117.
[e] Rosenberger, R.F., and Foskett, G. 1981. An estimate of the frequency of in vivo transcriptional errors at a nonsense codon in *Escherichia coli*. *Molec. Gen. Genet.* 183:561–563.
[f] Parker, J. 1989. Errors and alternatives in reading the universal genetic code. *Microbiol. Rev.* 53:272–298.

directed mismatch repair system is inefficient under starvation conditions. Similarly, Ryan *et al.* [50] had suggested that occasional repair synthesis during stationary-phase might give rise to the mutations that he observed. Davis [58] hypothesized that transcribed regions of DNA, which are at least partially single stranded, are inherently vulnerable to attack by normal cellular and environmental agents, resulting in an accumulation of DNA lesions. [This idea was also proposed by Fitch [59] who specifically predicted the occurrence of directed mutation.] Davis [58] further speculated that changes in DNA topology upon starvation could trigger the movement of transposable elements.

The second half of the conclusion drawn by Cairns *et al.* [21] has, to a much greater degree, provoked the wrath of neo-Darwinists as it can be regarded as a mechanism for the inheritance of acquired characteristics. One model that Cairns *et al.* [21] proposed can be called "specific". After producing variant RNA transcripts, as described above, the cell could have a mechanism by which each transcript is identified with the particular protein it encodes. If a protein were successful, the cell would reverse transcribe only the RNA transcript that coded for the successful protein and immortalize only that mutation in its genome. This model is as close to Lamarckian as possible without violating any principles of molecular biology. However, a number of other models postulated by Cairns *et al.* [21] and others have few Lamarckian overtones (see Table II). Starting again with RNA variants, Cairns *et al.* [21] proposed a "non-specific" model by which, upon receiving an energy boost from a successful protein, all RNA transcripts extant at the time would be reverse transcribed so that, with some probability, the successful mutation would be incorporated into the genome. If DNA is the starting point, this model is conceptually the same as Stahl's [57]—DNA mutated on the transcribed strand would produce variant proteins, one of which may be successful, allowing the cell to replicate its DNA and fix the mutation into the genome before it is corrected by mismatch repair or by recombination.

TABLE II
Models of directed mutation

Informational flow from the protein product to the gene.
Cairns *et al.* (a): reverse transcription of specific transcripts.
The production of random but transient variation that is immortalized only upon success.
Stahl (b): persistence of errors in the DNA due to slow repair.
Cairns *et al.* (a): production of variant transcripts or DNA copies.
Hall (c): a subpopulation of cells enters a hypermutable state.
Biased mutation of the relevant gene.
Davis (d): transcriptional bias.

[a] Cairns, J., Overbaugh, J., and Miller, S. 1988. The origin of mutants. *Nature (London)* 335:142–145.
[b] Stahl, F.W. 1988. A unicorn in the garden. *Nature (London)* 335:112–113.
[c] Hall, B.G. 1990. Spontaneous point mutations that occur more often when they are advantageous than when they are neutral. *Genetics* 126: 5–16.
[d] Davis, B.D. 1989. Transcriptional bias: A non-Lamarckian mechanism for substrate-induced mutations. *Proc. Natl. Acad. Sci. USA* 86:5005–5009.

The essence of the above models is that the source of variation is temporary and correctable. For example, in the absence of success, the variant RNA transcripts would be eventually degraded by normal cell processes, or the variant DNA regions would be corrected in favor of the preexisting methylated strand by mismatch repair. A mutation would become permanent only upon success, and mutations leading to non-useful phenotypes would not accumulate. Thus, the bulk of the cells' genomic DNA would be protected while only "a subset of their informational macromolecules [would be subjected] to the forces of natural selection" [21]. The Davis model [58] achieves the same selectivity—genes whose transcription is being induced by the selective agent would be preferentially mutated—but, since the model incorporates no mechanism to reverse the damage, at the cost of potential irreversible loss of some genetic information. An even more costly alternative, explicitly proposed by Hall [52], is that in a population experiencing starvation, a small proportion of the cells enter an hypermutable state in which (genomic) DNA is damaged and/or misrepaired. This state can only be exited if a cell is successful in producing a useful mutation; if unsuccessful, the cell dies. Although the consequences of failure are more extreme for the individual, this model achieves the same end for the population, i.e., the fixation of useful mutations while the bulk of the genomic DNA is unchanged. What is of further interest about these models is that each can account, with some degree of success, for the phenomenon of directed mutation. Yet none, from the most radical to the most prosaic, requires anything more than an extension of currently understood principles of molecular and cellular processes.

A reductionist can consider three, and only three, levels at which variation might take place—DNA, RNA, and protein. Each of these are subject to random errors in their synthesis, assaults from environmental and cellular agents, and specific modifications (methylations, phosporylations, etc.) to suit the needs of the current moment. Since the phenotypes of mutants that occur by directed mutation are clearly heritable, the variations that gave rise to the phenotypes must proceed from, or ultimately result in, changes in the DNA. The "central dogma" of molecular biology holds that information flows from nucleic acid to protein [60], and the discovery of reverse transcriptase provided the mechanism for a RNA to DNA link [21, 61]. Indeed, Reanney [62] has suggested that, because of the high error rate of reverse transcriptase, genetic diversity could be generated by periodically passing genetic information through a RNA intermediate. It is only the possibility of information transfer from protein to RNA (or DNA) that has no conceptual framework in currently molecular biology. Not even Cairns *et al.* [21] proposed that changes arising de *novo* in a protein might end up encoded in the DNA (a truly Lamarckian concept). Even the dichotomy between "random" and "directed" mutation is false as every mechanism proposed is based on an initial random event, thus at some level preserving the neo-Darwinian canon [22].

I have presented these models at length to illustrate that the theory of directed mutation certainly does not threaten the current paradigm, and can probably be incorporated happily into neo-Darwinism (see the chapter in this volume by Sarkar). However, as discussed in the next section, subsequent research has cast doubt on several of these models.

ALTERNATIVE MODELS AND ADDITIONAL EVIDENCE

In Table III are summarized several of the alternative models that have been proposed to explain directed mutation via mundane phenomena. Much of the criticism directed Cairns *et al.* [21] focused on the evidence that mutants are actually arising on the selective plate in the absence of cell division. The question of cryptic growth on the plate was dismissed by Cairns *et al.* [21] because the population of bacteria that was accumulating Lac^+ mutants was not also accumulating an unselected mutation, which would be expected if the cells were generating random mutations by normal DNA replication. [Note, however, that Danchin [46] suggested that different mutations may not respond in the same way to starvation conditions]. Subsequently, using rever-

TABLE III
Trivial explanations of directed mutation

The non-mutant population of cells can grow under the selective conditions.
During prior growth under non-selective conditions, certain classes of mutants do poorly (a-c).
Certain classes of mutants arise before selection but grow slowly after selection.
Full expression of the mutant phenotype requires multiple mutations (c).

[a] Charlesworth, D., Charlesworth, B., and Bull, J.J. 1988. Scientific correspondence. *Nature (London)* 336:525–525.
[b] Tessman, I. 1988. Scientific correspondence. *Nature (London)* 336:527–527.
[c] Lenski, R.E., Slatkin, M., and Ayala, F.J. 1989. Mutation and selection in bacterial populations: Alternatives to the hypothesis of directed mutation. *Proc. Natl. Acad. Sci. USA* 86:2775–2778.

sion of a frameshift mutation in *lacZ*, we [72] have shown that the number of viable cells actually measured on selective plates cannot account for the number of mutants that arise unless the mutation frequency per cell division increases by 2-3 orders of magnitude after plating. This result does not formally exclude the possibility that some cells are dying and being cannibalized by others. However, since it takes 3-5 cell deaths to allow one cell to double [63], cannibalism would lead to a decline of the population on the plate, whereas the rate of appearance of new mutants is almost constant for several days after plating [21, 72]. In addition, Hall [52] has used reconstruction experiments to show that the maximum number of cell generations possibly attributable to gross cell turnover fails by at least an order of magnitude to account for the number of revertants of a *trpB* mutation that appear after selection. Thus, either mutation rates per cell division are orders of magnitude higher when cells are under selection, or mutation rates under these conditions are independent of cell division.

Several respondents to Cairns *et al.* [21] pointed out that if mutants have a lower fitness than non-mutants before selection, the resultant distribution of mutants after selection will deviate from the Luria-Delbrück distribution towards the Poisson distribution [64, 65]. The most comprehensive of these responses was by Lenski *et al.* [66] who presented a mathematical model for the resulting distribution. The essence of the argument can be illustrated by its extreme—if every newly-born mutant is unable to multiply until the selective agent is present, the resulting distribution will be Poisson (see the chapter in this volume by Sarkar). However, this hypothesis only addressed the distribution of Lac^+ colonies seen immediately (48 hours), not the continued appearance of Lac^+ colonies for days or weeks after plating. The latter could be explained by the occurrence of a class of slow-growing mutants. However, neither Hall [52] nor we [72] have detected such a class. Lenski *et al.* [66] presented a two-step mutation model that could also explain the late appearance of Lac^+ cells. A mutant with an intermediate phenotype occurs before selection and grows slowly on the substrate, increasing its population size but not producing detectable colonies. Among this population, second mutations occur that yield a fast growing phenotype, and these mutants produce colonies after a delay that is determined by the growth rate of the first mutant and the mutation rate of the first to the second. The problem with this model is that the second mutation rate has to be unrealistically high (10^{-5} per cell per generation); with such a high mutation rate, it would be very difficult to isolate the intermediate mutant.

The Lac^- phenotype of the original strain used by Cairns *et al.* [21] was due to a nonsense mutation, which causes protein synthesis to be prematurely terminated. Nonsense mutations can be reverted in two different ways: the nonsense codon can be mutated to a sense codon that codes for an amino acid (a true reversion); and, a transfer RNA can be mutated to read the nonsense codon and insert an amino acid (a suppressor mutation). Both types of mutation occur by single base changes. Depending on the particular amino acid inserted, mutants of both classes may have various abilities to metabolize lactose, and thus various growth rates after selection. However, only suppressor mutants would be expected to be less fit during growth prior to selection since the mutant tRNA would cause some suppression of the normal translational termination of other proteins.

With this nonsense mutant, post-selection mutation was greatly enhanced if the strain also carried a deletion of a region of the chromosome (the *uvrB-bio* region) that is well removed from the *lac* region. The gene or genes missing in this strain influence mutation in complex ways, but when plated on minimal medium with lactose, the deletion mutant continues to produce Lac^+ revertants at a greater rate and for a longer time than the wild-type strain (Reference 21, Fig. 2). By any hypothesis to explain directed mutation by the selective survival or growth of one class of mutant, the deletion mutation would be predicted to bias the frequency of one or more of the events described above. However, we (P.L. Foster and J. Cairns, unpublished results) have found that over several days after plating the *uvrB-bio* deletion strain on lactose medium, both true and suppressor mutants continue to appear, and in about the same proportion as in the wild type strain. Thus, both classes of mutation are subject to directed mutation and both are enhanced by the *uvrB-bio* deletion. While not definitive, these results place severe constraints on the alternative models.

The post-selection appearance of second-site mutations (i.e., in a gene that does not code for the enzymatic function being selected for) has implications for other, more interesting models of directed mutation as well. As pointed out by Davis [58], such a result is difficult to explain by the transcriptional bias hypothesis since the tRNA genes would not be specifically induced by the selective conditions. This result also makes the specific reverse transcription model proposed by Cairns *et al.* [21] unlikely. With a suppressor mutation, the amino acid change that results in the functional protein is not encoded in either the RNA or the DNA sequence specifying the protein.

Hall [52] has also obtained evidence against a specific mechanism for directed mutation. Among a population of Trp$^+$ revertants appearing after selection, he found a 1500-fold higher than expected number of second, unselected mutations. This is the predicted result of all the non-specific models—whether DNA or RNA is the source of variation, upon success all variants extant in the cell would have some probability of being fixed in the genome.

Recently Mittler and Lenski [67] also have suggested that starvation increases genetic variability. Using the Lac(Ara)$^-$ fusion strain mentioned above, they found that when the cells were kept for many days under starvation conditions, Lac(Ara)$^+$ cells could accumulate in the absence of arabinose and lactose. This result is in direct conflict with that of Cairns *et al.* [21], who found no such fusions under comparable conditions unless arabinose and lactose were present. One possible explanation for the difference is that, in the Mittler and Lenski experiments, cells were under additional selection to delete the Mu DNA. Indeed, under starvation, survival of Lac(Ara)$^+$ cells was significantly better than survival of the Lac(Ara)$^-$ parent cells [67]. Why the cells were subjected to this additional selective pressure in Mittler and Lenski's hands but not in Cairns *et al.* 's is a bit mysterious. Nevertheless, Mittler and Lenski [67] have confirmed the general results of Arber *et al.* [51], Shapiro [20], Cairns *et al.* [21], and Hall [47], that the probability of movement of an insertion element increases sharply with time when the cells are deprived of an energy source. Further, the experiments reported to date suggest that genomic rearrangements mediated by insertion elements are several orders of magnitude more sensitive to starvation conditions than are simple point mutations.

Since the publication of the paper by Cairns *et al.* [21], directed mutation has been observed in *E. coli* using reversion of frameshift mutations in *lacZ* [68, 72], missense and frameshift mutations in the *trp* operon [52, 72], an X-ray-generated mutation in *metB* [69], and mutations affecting membrane proteins [70]. Yet not all mutations that could, in principle, occur after selection display directed mutation, at least in otherwise wild-type cells [21, 68, Foster and Cairns, unpublished results]. Furthermore, one or more genes in the *uvrB-bio* region of the chromosome specifically affects post-selection mutation, not mutations occurring during growth under non-selective conditions [21, Cairns, personal communication]. Indeed, the original antimutator isolated by Geiger and Speyer [15] may be of this class since it also affects only mutations that arise under non-lethal selection.

CONCLUSIONS

Table IV summarizes what is known to date about directed mutation. The most important conclusions are that the phenomenon appears to be general, i.e., not restricted to certain class of mutations or genes, and under genetic control. It is, nonetheless, dependent on some unknown property of the mutational target observed. These characteristics must be accounted for by any hypothesis that attempts to explain directed mutation with mundane phenomena or trivial artifacts. We can be fairly certain that the process is initiated by a random change in the genetic material, be it DNA or RNA. The most conservative model is that under stress an *E. coli* cell can, either by design or accident, transiently increase its genetic variability, thus enhancing its chances of surviving with a more or less intact genome. Additional explication of the molecular mechanism underlying directed mutation must await further experimentation.

The controversy surrounding directed mutation has illustrated that at least two basic postulates of evolutionary theory have to be reexamined. First, it cannot be assumed that the production of variation upon which natural selection acts is constant with time and place. Second, it also cannot be assumed

TABLE IV
Characteristics of directed mutation

Mutants with the appropriate phenotype arise continuously under non-lethal selection.
These mutants have a Poisson, not a Luria-Delbrück, distribution.
Under selection, only the mutants being selected for accumulate, and they do not accumulate in the absence of selection.
The rate of accumulation of new mutants can be linear with time, or, in some instances, can increase with time.
The mutations affected include base substitutions, frameshifts, and movement of insertion sequences.
The genes affected include those that code for enzymes involved in carbohydrate utilization and amino acid synthesis, tRNA suppressors, and membrane structural proteins.
The phenomenon is dependent on some property of the mutational target.
The phenomenon appears to be under genetic control.

that the phenotype that natural selection acts upon is necessarily stable and not transient. However, it remains to be determined if any organisms other than bacteria display directed mutation. To date, the closest analogy reported in an eucaryotic organism is in a plant—certain environmental conditions can induce phenotypic changes in a flax plant that are passed on to the next generation [71]. This response appears to be due to an inducible genetic variability affecting only a subset of the genome; mutated somatic cells with an advantageous phenotype then develop into a sport [71]. This example suggests that the ability to respond to environmental changes by producing genetic variants may be restricted to organisms that lack a clear separation between somatic and germ cell lines. Even so, as Cairns [22] has pointed out, at least one class of non-heritable somatic mutation, that giving rise to a cancerous cell, is of vital importance to the individual, if not the species.

ACKNOWLEDGEMENTS

Since the publication of Cairns *et al.*, [21] I have had numerous stimulating and sometimes impassioned discussions with both scientists and nonscientists about the results and the implications of the phenomenon of directed mutation. I would particularly like to acknowledge Eric Eisenstadt, John Drake, Maurice Fox, Edward Cox, Franklin Stahl, C. Mark Smith, Jeffrey Miller, and Sahotra Sarkar. And, of course, I thank John Cairns for being the ever provocative and inspiring person that he is. This work was supported by National Science Foundation grant DMB-8905004.

Boston University School of Public Health

REFERENCES

1. Selander, R.K., Caugant, D.A., and Whittam, T.S. 1987. Genetic structure and variation in natural populations of *Escherichia coli. In Escherichia coli and Salmonella typhimurium, Cellular and Molecular Biology*. J.L. Ingraham, K.B. Low, B. Magasanik, M. Schaechter, and H.E. Umbarger, editors. American Society for Microbiology, Washington, DC. 1625–1648.
2. Ochman, H., and Wilson, A.C. 1987. Evolutionary history of enteric bacteria. In *Escherichia coli and Salmonella typhimurium, Cellular and Molecular Biology*. J.L. Ingraham, K.B. Low, B. Magasanik, M. Schaechter, and H.E. Umbarger, editors. American Society for Microbiology, Washington, DC. 1649–1654.
3. Milkman, R., and Stoltzfus, A. 1988. Molecular evolution of the *Escherichia coli* chromosome. II. Clonal segments. *Genetics* 120: 359–366.

4. Rayssiguier, C., Thaler, D.S., and Radman, M. 1989. The barrier to recombination between *Escherichia coli* and *Salmonella typhimurium* is disrupted in mismatch-repair mutants. *Nature (London)* 342: 396–400.
5. Drake, J.W. 1969. Comparative rates of spontaneous mutation. *Nature (London)* 221: 1132.
6. Topal, M.D., and Fresco, J.R. 1976. Complementary base pairing and the origin of substitution mutations. *Nature (London)* 263: 285–289.
7. Fersht, A.R., Knill-Jones, J.W., and Tsui, W-C. 1982. Kinetic basis of spontaneous mutation: Misinsertion frequencies, proofreading specificities and cost of proofreading by DNA polymerases of *Escherichia coli*. *J. Molec. Biol.* 156: 37–51.
8. Cox, E.C. 1976. Bacterial mutator genes and the control of spontaneous mutation. *Ann. Rev. Genet.* 10: 135–156.
9. Choy, H.E., and Fowler, R.G. 1985. The specificity of base pair substitutions induced by the *mutL* and *mutS* mutators in *E. coli*. *Mutat. Res.* 142: 93–97.
10. Radman, M., and Wagner, R.E., Jr. 1986. Mismatch repair in *Escherichia coli*. *Ann. Rev. Genet.* 20: 523–538.
11. Ogden, G.B., Pratt, M.J., and Schaechter, M. 1988. The replicative origin of the *E. coli* chromosome binds to cell membranes only when hemimethylated. *Cell* 54: 127–135.
12. Drake, J.W., and Allen, E.F. 1968. Antimutagenic DNA polymerases of bacteriophage T4. *Cold Spring Harbor Symp. Quant. Biol.* 33: 339–344.
13. Cox, E.C., and Gibson, T.C. 1974. Selection for high mutation rates in chemostats. *Genetics* 77: 169–184.
14. Leigh, E.G. 1970. Natural selection and mutability. *Am. Natural.* 104: 301–305.
15. Geiger, J.R., and Speyer, J.F. 1977. A conditional antimutator in *E. coli*. *Molec. Gen. Genet.* 153: 87–97.
16. Campbell, A.N. 1983. Transposons and their evolutionary significance. In *Evolution of Genes and Proteins*. M. Nei, and R.K. Koehn, editors. Sinauer Associates, Sunderland, MA, 259–279.
17. Chao, L., Vargas, C., Spear, B.B., and Cox, E.C. 1983. Transposable elements as mutator genes in evolution. *Nature (London)* 303: 633–635.
18. Messer, W., and Noyer-Weidner, M. 1988. Timing and targeting: The biological functions of *dam* methylation in *E. coli*. *Cell* 54: 735–737.
19. Delbrück, M. 1946. Heredity and variations in microorganisms, *Cold Spring Harbor Symp. Quant. Biol.* 11: 154.
20. Shapiro, J.A. 1984. Observations on the formation of clones containing *araB-lacZ* cistron fusions. *Molec. Gen. Genet.* 194: 79–90.
21. Cairns, J., Overbaugh, J., and Miller, S. 1988. The origin of mutants. *Nature (London)* 335: 142–145.
22. Cairns, J. 1988. Scientific correspondence. *Nature (London)* 336: 527–528.
23. Walker, G.C. 1984. Mutagenesis and inducible responses to deoxyribonucleic acid damage in *Escherichia coli*. *Microbiol. Rev.* 48: 60–93.
24. Burckhardt, S.E., Woodgate, R., Scheuermann, R.H., and Echols, H. 1988. UmuD mutagenesis protein of *Escherichia coli*: Overproduction, purification, and cleavage by RecA. *Proc. Natl. Acad. Sci. USA* 85: 1811–1815.
25. Nohmi, T., Battista, J.R., Dodson, L.A., and Walker, G.C. 1988. RecA-mediated cleavage activates UmuD for mutagenesis: Mechanistic relationship between transcriptional derepression and posttranslational activation. *Proc. Natl. Acad. Sci. USA* 85: 1816–1820.
26. Shinagawa, H., Iwasaki, H., Kato, T., and Nakata, A. 1988. RecA protein-dependent

cleavage of UmuD protein and SOS mutagenesis. *Proc. Natl. Acad. Sci. USA* 85: 1806–1810.

27. Freitag, N., and McEntee, K. 1989. "Activated"-RecA protein affinity chromatography of LexA repressor and other SOS-regulated proteins. *Proc. Natl. Acad. Sci. USA* 86: 8363–8367.
28. Bridges, B.A. 1988. Mutagenic DNA repair in *Escherichia coli* XVI. Mutagenesis by ultraviolet light plus delayed photoreversal in *recA* strains. *Mutat. Res.* 198: 343–350.
29. Dutreix, M.P., Moreau, P.L., Bailone, A., Galibert, F., Battista, J.R., Walker, G.C., and Devoret, R. 1989. New *recA* mutations that dissociate the various RecA protein activities in *Escherichia coli* provide evidence for an additional role for RecA protein in UV-mutagenesis. *J. Bacteriol.* 171: 2415–2423.
30. Villani, G., Bouteux, S., and Radman, M. 1975. Mechanisms of ultraviolet induced mutagenesis: Extent and fidelity of *in vitro* DNA synthesis on irradiated templates. *Proc. Natl. Acad. Sci. USA* 78: 3037–3041.
31. Fersht, A.R., and Knill-Jones, J.W. 1983. Contribution of 3' → 5' exonuclease activity of DNA polymerase III holoenzyme from *Escherichia coli* to specificity. *J. Molec. Biol.* 165: 669–682.
32. Lu, C., Scheuermann, R.H., and Echols, H. 1986. Capacity of RecA protein to bind preferentially to UV lesions and inhibit the editing subunit (epsilon) of DNA polymerase III: A possible mechanism for SOS-induced targeted mutagenesis. *Proc. Natl. Acad. Sci. USA* 83: 619–623.
33. Foster, P.L., and Sullivan, A.D. 1988. Interactions between epsilon, the proofreading subunit of DNA polymerase III, and proteins involved in the SOS response of *Escherichia coli*. *Molec. Gen. Genet.* 214: 467–473.
34. Foster, P.L., Sullivan, A.D., and Franklin, S.B. 1989. Presence of the *dnaQ-rnh* divergent transcriptional unit on a multicopy plasmid inhibits induced mutagenesis in *Escherichia coli*. *J. Bacteriol.* 171: 3144–3151.
35. Bridges, B.A., and Woodgate, R. 1985. Mutagenic repair in *Escherichia coli* : Products of the *recA* gene and of the *umuD* and *umuC* genes act at different steps in UV-induced mutagenesis. *Proc. Natl. Acad. Sci. USA* 82: 4193–4197.
36. Battista, J.F., Nohmi, T., Donnelly, C.E., and Walker, G.C. 1989. Role of UmuD and UmuC in UV and chemical mutagenesis. In *Mechanism and Consequences of DNA Damage Processing*. E. Friedberg, and P. Hanawalt, editors. Liss, New York. 455 – 459.
37. Foster, P.L., Eisenstadt, E., and Cairns, J. 1982. Random components in mutagenesis. *Nature (London)* 299: 365–367.
38. Miller, J.H. 1982. Carcinogens induce targeted mutations. *Cell* 31: 5–7.
39. Eisenstadt, E. 1988. SOS mutagenesis in *Escherichia coli* occurs primarily, perhaps exclusively, at sites of DNA damage. In *DNA Replication and Mutagenesis*. R.E. Moses, and W.C. Summers, editors. American Society for Microbiology, Washington, DC. 397–402.
40. Miller, J.H., and Low, K.B. 1984. Specificity of mutagenesis resulting from the induction of the SOS system in the absence of mutagenic treatment. *Cell* 37: 675–682.
41. Ghosh, S.K., Panda, D.K., and Das, J. 1989. Lack of *umuDC* gene functions in *Vibrio cholerae* cells. *Mutat. Res.* 210: 149–156.
42. Sedgwick, S.G., and Goodwin, P.A. 1985. Differences in mutagenic and recombinational DNA repair in enterobacteria. *Proc. Natl. Acad. Sci. USA* 82: 4172–4176.
43. Molina, A.N., Baburdrl, M., Tamaro, M., Venturini, S., and Monti-Bragadin, C. 1979.

Enterobacteriacea plasmids enhancing chemical mutagenesis and their distribution among incompatability groups. *FEMS Lett.* 5: 33–37.
44. Upton, C., and Pinney, R.J. 1980. Expression of eight unrelated Muc^+ plasmids in eleven DNA repair deficient *Escherichia coli* strains. *Mutat. Res.* 112: 261–273.
45. Strike, P., and Lodwick, D. 1987. Plasmid genes affecting DNA repair and mutation. *J. Cell Sci.* 6 (Suppl.): 303–321.
46. Danchin, A. 1988. Scientific correspondence. *Nature (London)* 336: 527–527.
47. Hall, B.G. 1988. Adaptive evolution that requires multiple spontaneous mutations. I. Mutations involving an insertion sequence. *Genetics* 120: 887–897.
48. Hall, B.G., Yokoyama, S., and Calhoun, D.H. 1983. Role of cryptic genes in microbial evolution. *Molec. Biolog. Evolution* 1: 109–124.
49. Hall, B.G. 1982. Evolution on a petri dish. In *Evolutionary Biology. Vol. 15*. M.K. Hecht, B. Wallace, and G.T. Prance, editors. Plenum Press, New York. 85–150.
50. Ryan, F.J., Okada, T., and Nagata, T. 1963. Spontaneous mutation in spheroplasts of *Escherichia coli. J. gen. Microbiol.* 30: 193–199.
51. Arber, V., Iida, S., Jutte, H., Caspers, P., Meyer, J., and Hanni, C. 1978. Rearrangements of genetic material in *Eschericia coli* as observed on the bacteriophage P1 plasmid. *Cold Spring Harbor Symp. Quant. Biol.* 43: 1197–1208.
52. Hall, B.G. 1990. Spontaneous point mutations that occur more often when they are advantageous than when they are neutral. *Genetics* 126: 5–16.
53. Groat, R.G., Schultz, J.E., Zychlinsky, E., Bockman, A., and Matin, A. 1986. Starvation proteins in *Escherichia coli*: Kinetics of synthesis and role in starvation survival. *J. Bacteriol.* 168: 486–493.
54. Koch, A.L. 1971. The adaptive responses of *Escherichia coli* to a feast and famine existence. *Adv. Microb. Physiol.* 6: 147–217.
55. Lim, D., and Maas, W.K. 1989. Reverse transcriptase-dependent synthesis of a covalently linked, branced DNA-RNA compound in *E. coli* B. *Cell* 56: 891–904.
56. Lampson, B.C., Sun, J., Hsu, M-Y., Vallejo-Ramirez, J., Inouye, S., and Inouye, M. 1989. Reverse transcriptase in a clinical strain of *Escherichia coli* : Production of branched RNA-linked msDNA. *Science* 243: 1033–1038.
57. Stahl, F.W. 1988. A unicorn in the garden. *Nature (London)* 335: 112–113.
58. Davis, B.D. 1989. Transcriptional bias: A non-Lamarckian mechanism for substrate-induced mutations. *Proc. Natl. Acad. Sci. USA* 86: 5005–5009.
59. Fitch, W.M. 1982. The challenges to Darwinism since the last centennial and the impact of molecular studies. *Evolution* 36(6): 1133–1143,
60. Crick, F.H.C. 1957. On protein synthesis. *Symp. Soc. Exp. Biol.* 12: 138 –163.
61. Temin, H.M. 1971. The protovirus hypothesis: Speculations on the significance of RNA-directed DNA synthesis for normal development and carcinogenesis. *J. Natl. Cancer Inst.* 46: iii.
62. Reanney, D. 1984. Genetic noise in evolution?. *Nature (London)* 307: 318–319.
63. Postgate, J. 1967. Viability measurements and the survival of microbes under minimum stress, *Adv. Microb. Physiol.* 1: 1–23.
64. Charlesworth, D., Charlesworth, B., and Bull, J.J. 1988. Scientific correspondence. *Nature (London)* 336: 525–525.
65. Tessman, I. 1988. Scientific correspondence. *Nature (London)* 336: 527–527.
66. Lenski, R.E., Slatkin, M., and Ayala, F.J. 1989. Mutation and selection in bacterial popula-

tions: Alternatives to the hypothesis of directed mutation. *Proc. Natl. Acad. Sci. USA* 86: 2775–2778.

67. Lenski, R.E., and Mittler, J.E. 1990. New data on excisions of Mu from *E. coli* MCS2 cast doubt on directed mutation hypothesis. *Nature (London)* 344: 173–175.
68. Levin, B.R., Gordon, D.M., and Stewart, F.M. 1990. Is natural selction the composer as well as the editor of genetic variation? (in press).
69. Hall, B.G. 1989. Selection, adaptation, and bacterial operons. *Genome* 31: 265–271.
70. Benson, S.A. 1988. Scientific correspondence. *Nature (London)* 336: 21–22.
71. Cullis, C.A. 1987. The generation of somatic and heritable variation in reponse to stress. *Am. Natural.* 130: s62–s73.
72. Cairns, J., and Foster, P.L. 1991. Adaptive reversion of a frameshift mutation in *Escherichia coli. Genetics* (in press).

SAHOTRA SARKAR

LAMARCK *CONTRE* DARWIN, REDUCTION *VERSUS* STATISTICS: CONCEPTUAL ISSUES IN THE CONTROVERSY OVER DIRECTED MUTAGENESIS IN BACTERIA [1]

I. INTRODUCTION

The purpose of this paper is to examine some conceptual aspects of the controversy over the possibility of directed mutagenesis in bacteria that has erupted since the publication of some provocative results by Cairns, Overbaugh and Miller [2]. The paper by Foster in this volume takes up more "empirical" issues, surveys the experimental literature, and offers occasionally different interpretations of the results. The conceptual issues that are important here occur at least at two levels, the first of which is, in a sense, *metaphysical* and the second, *epistemological*. *First*, the possibility of directed mutagenesis challenges the core of the current orthodox framework of evolutionary theory. Thus the sense in which mutations can indeed be "directed" is of considerable foundational importance to evolutionary theory. To the extent that such foundational issues are "metaphysical," in the sense that they concern the most general and universal underlying features of the world explored by science, these conceptual issues properly belong to metaphysics. *Second*, much of the evidence on which the current controversy thrives is statistical evidence about the number of mutant bacteria. Experimental methods which rely on such evidence are non-reductive in the sense that they attempt to understand what occurs at a lower level—for instance, that within a bacterial cell—by making observations at a higher level—in the example, that of the cell. The interesting question is whether such non-reductive strategies can resolve the controversy being considered here. This issue is epistemological because it concerns the powers and limitations of different scientific methods. It is particularly interesting in the present context because the role played by reductionism in molecular biology remains a matter of considerable controversy [3].

Section II of this paper attempts to explicate the nature of "neo-Darwinism" and "neo-Lamarckism" and to put them in their proper conceptual contexts. In particular it attempts to distinguish various senses of "neo-Lamarckism" which will ultimately be relevant to the characterization of various potential positions that might be adopted in this controversy.

Alfred I. Tauber (ed.), Organism and the Origins of Self, 235–271.

Though it briefly touches on the long and variegated history of the persistent disputes between neo-Darwinians and neo-Lamarckians, it is by no means even a representative summary of this history, let alone a complete one. Section III gives a short account of the early history of the actual problem of concern here, namely, the nature of the origin of bacterial mutants. The account is intended as a critical one in the sense that both the experiments performed and their interpretations are carefully appraised. Section IV is a very short account of the new experimental work on the topic which forms part of the new controversy over such mutants. A more complete review will be found in the contribution by Foster in this volume.

Section V takes up, in detail, the question of the interpretation of all the experimental results, that is, those described or mentioned in the preceding two sections. Given the various interpretive possibilities encountered, it argues that mere statistical data is unlikely to be able to resolve the controversy: the discovery of the appropriate mechanisms of mutagenesis will be necessary for that purpose. It then analyzes various suggested, but as yet unestablished, mechanisms that might account for the empirical data. The conceptual framework developed in Section II is utilized to clarify the various issues that arise. It is also suggested that there is a sense in which neo-Darwinism and neo-Lamarckism might peacefully co-exist. Section VI turns to the issue of reduction. It uses a model of reduction, based on explaining wholes in terms of their parts by invoking only physically characterized interactions, to argue that the type of research that was found to be most likely to resolve this controversy, that is, the search for mechanisms, is fundamentally reductionist [4]. In Section VII the various conclusions drawn during the course of this paper are reiterated. A final important, but often ignored, moral is drawn: certain metaphysical and epistemological questions, such as those mentioned above, can be answered, and perhaps only be answered, experimentally.

II. NEO-DARWINISM AND NEO-LAMARCKISM

The dominant intellectual current within contemporary evolutionary biology is neo-Darwinism, an interpretation of evolution in which the emphasis is on natural selection acting on random variation. What constitutes evolution by natural selection is easy enough to explicate. As formulated by Richard Lewontin in 1970, a population of individuals evolves by natural selection if three criteria are satisfied [5]. *First* there must be phenotypic variation: "Different individuals in a population have different morphologies, physiologies, and behaviors." [6] Were all individuals identical with respect to all properties that interact with their environment, there would either be complete stasis or change due to some process rather than natural selection. *Second*, these different phenotypes, or sets of properties of the individual that interact with the environment, must have, associated with them, different fitnesses, that is, "different rates of survival and reproduction." [7] Here these rates and, consequently, fitness itself is to be construed probabilistically, that is, the fitness of an individual is to be construed as the probability of it leaving offsprings in the next generation. *Third*, fitnesses must be heritable. The fitnesses of offsprings of an individual must depend on the fitness of the individual itself in that environment [8].

The notion of randomness that is appropriate for evolutionary biology, however, is quite murky [9]. The considerations adduced here will use as weak a notion of randomness as possible. It will only be assumed that the type of variation that arises through a mutation of a gene is random if and only if the probability of its occurrence in an environment has no correlation with the fitness of the phenotype induced by it in that environment. Thus, though certain chemical mutagens or types of radiation are known to induce entire classes of mutations with a fairly predictable frequency, these mutations would still be regarded as random because the sort of correlation just mentioned does not exist. Once the randomness of a mutation is construed in this fashion, all mutations that are *undirected* by the environment are obviously random. Such a weak notion of randomness is being used here for two reasons. *First*, in any analysis, it is generally better to assume as little as possible so as not to make the success of the analysis depend at all on dispensable features. *Second*, no stronger notion of randomness in evolutionary biology seems to be empirically warranted. Certainly the possibility of enhancing mutation rates in general through the action of radiation and chemical mutagens shows that not all mutation events are equiprobable at the physical level.

Though random variation and natural selection are being emphasized here as the dominant aspects of the conceptual framework of neo-Darwinism there are also other features of neo-Darwinism that are often regarded as being quite important. In fact, from some points of view, it is these other features that are central to the neo-Darwinian orthodoxy. Some see neo-Darwinism as requiring that *all* evolutionary change be caused by natural selection [10]. For such points of view, the existence of other mechanisms of evolutionary change such as allometry, or differential rates of growth of different parts of an organism, or pleiotropy or multiple, partly non-selective, effects of genes, or even neutral mutations at the molecular level, constitute a challenge to neo-Darwinism. Others view neo-Darwinism as requiring that evolution proceeds by continuous slight variations. Indeed, in the debate between the biometricians (who claimed to be following Darwin strictly) and the Mendelians at the turn of this century, it was this feature of neo-Darwinism that was considered central [11]. These other features are not emphasized here because they are routinely incorporated into the neo-Darwinian framework without any perceptible rupture of its conceptual framework. It is largely uncontroversial today that mechanisms other than natural selection are operative in evolutionary change though opinion is divided on how important these other features may be [12]. Further, the reduction of biometrical insights to Mendelian genetics by Haldane, Fisher and Wright in the 1920's, which largely created the current version of neo-Darwinian orthodoxy, has already shown one way in which debates over the size of variations might be resolved [13]. Yet, the neo-Darwinian framework has largely survived.

The two claims that are being construed here to be central to neo-Darwinism, that is, that natural selection is operative, and that mutations are random, are independent. Each could be violated without the other. Mutations might occur at random but all mutants might have identical fitnesses. Natural selection would thus not be operative. Similarly, natural selection could be operative in a population but the mutations that occur in it might not be random. This is the position that corresponds to that of neo-Lamarckism but, in order to specify neo-Lamarckism exactly enough to evaluate its claims as an alternative to neo-Darwinism, three other distinctions are necessary.

First, neo-Darwinism is being construed here to require that *all* mutations are random. This usage is standard. Neo-Lamarckism could potentially be analogously construed as requiring that *all* mutations are not random (or even directed which, as will be discussed below, is a stronger claim). Such a construal of neo-Lamarckism, which might be called "strong neo-Lamarckism," would be uninteresting since it is quite trivially false: that at least *some* mutations are random is uncontroversial. Neo-Lamarckism will, therefore, be construed here as only requiring that some mutations are not random. If a distinction between the two ways of construing "neo-Lamarckism" is needed, this weaker notion might be called "weak neo-Lamarckism." Weak neo-Lamarckism then just amounts to a denial of the universalization over *all* mutations that is part of the standard construal of neo-Darwinism. Indeed, historically, neo-Lamarckian positions have usually been construed in this fashion [14]. Only during the height of the justly infamous Lysenko affair did most of Lysenko's defenders suggest that all mutations had to be specifically induced by the environment and would not, therefore, be random [15]. In the discussion below, unless otherwise indicated, "neo-Lamarckism" will be taken to be "weak neo-Lamarckism."

Second, neo-Lamarckism, as it is usually construed, and as it will be used here, actually requires something stronger than the assertion that mutations are not all random. Non-random mutations could either enhance or decrease the fitness of the associated phenotypes in that environment. Neo-Lamarckism requires the former. It is precisely in this sense that neo-Lamarckism requires that some mutations are *directed*. The sense of direction that is involved here is that mutations occur, or at least occur more frequently, in those environments in which the associated phenotype has an enhanced fitness. Thus there is, basically, a positive correlation between the mutation rate and the fitness of the associated phenotype.

Third, it is necessary to distinguish carefully between two ways in which a directed mutation might be defined. A strong definition would be one that requires that a mutation be considered "directed" *if and only if* it occurs (or occurs more frequently) in an environment where its associated phenotype has an enhanced fitness [16]. A weaker definition would be one that dropped the "only if" clause: a mutation is then considered "directed" if it occurs (or occurs more frequently) in the fitness-enhancing or "selective" environment as characterized above [17]. In the context of particular mutations, the former corresponds to strong neo-Lamarckism and the latter to weak neo-Lamarckism, since the former excludes any random (or selectively neutral) occurrence of these mutations. Only the latter will be used here because of the considerations given during the discussion of the first point above and because it suffices to capture the disagreements that characterize the controversy over directed mutagenesis in bacteria.

"Neo-Lamarckism," when construed in this fashion, makes no reference to what has historically been a fairly standard conception of the position during debates between neo-Darwinians and neo-Lamarckians, namely, that of the "inheritance of acquired characteristics." There are two reasons for this. *First*, in the current controversy regarding the possibility of directed mutagenesis in bacteria, which is the immediate concern of this paper, there is usually no reference to the inheritance of acquired characteristics [18]. Construing the issues involved in terms of "directed mutations" thus makes it easy to remain faithful to the debates. *Second*, the "inheritance of acquired characteristics" is a concept too vague to serve any useful purpose here [19]. Interpreted broadly enough, the "inheritance of acquired characteristics" can include so much that it becomes trivially true. In some plants, such as flax, environmentally induced changes in the genome are routinely inherited [20]. These clearly should be considered as the inheritance of acquired characteristics. In some sense, even radiation induced damage to chromosomes is an "acquired" characteristic that is obviously inherited. What is missing from such examples, however, is that they leave open the question of the correlation of these variations with the fitnesses of the associated phenotypes. It is exactly the last issue that is of interest in the controversy being examined here. Thus to construe it in terms of the inheritance of acquired characteristics is unilluminating whereas the characterizations of neo-Lamarckism given above capture the spirit of the disputes [21].

The terms "neo-Darwinism" and "neo-Lamarckism" obviously refer to Darwin and Lamarck and are intended to capture some part of their views. However, the terms as explicated above (and those explications are consistent with how the terms are usually used) have very significant differences with the views that Darwin and Lamarck actually held. Lamarck suggested in 1809 that the "a more frequent and continuous use of any organ gradually strengthens, develops and enlarges that organ"; disuse has the opposite effect; and that such changes are inherited [22]. An animal thus has the property of being able to induce variations that, in some sense, enhance its ability to survive in a particular environment. There is thus some resonance between Lamarck's views and the explication of "neo-Lamarckism" given above but it cannot be suggested that this explication constitutes a reconstruction of Lamarck's actual views.

Exactly fifty years after Lamarck, Darwin published the *Origin*, in which he argued that variations usually arose at random [23]. However, Darwin did not exclude the possibility of some variations arising through the "use and disuse" mechanism of Lamarck or through direct environmental induction [24]. Moreover, under the pressure of various objections to his theory on the grounds that there could not have been enough time for evolution to have taken place purely through natural selection acting upon random variations, he began to emphasize such mechanisms over random variation in later editions of the *Origin* [25]. The conception of neo-Darwinism that excludes all possibilities of variation except random mutations is largely due to Weismann [26]. The term "neo-Darwinism" seems to have been coined to describe Weismann's views by one of his principle opponents, Romanes, who tried to maintain Darwin's more "pluralist" approach to the sources of variation [27]. Thus, once again, while there is some consonance between "neo-Darwinism," as used above, and Darwin's own views, the explication is not to be taken as one of Darwin's actual views. In fact, while these historical associations are interesting, it should be emphasized that what truly matters in the adoption of the given analyses of "neo-Darwinism" and "neo-Lamarckism" is their utility for conceptual clarification and not whether they accurately reflect historical developments [28].

III. THE 'CLASSIC' EXPERIMENTS

During the late 1920's and 1930's, considerable biological attention shifted to micro-organisms such as viruses, thanks to the research programs initiated by Delbrück and others [29]. Out of these developments came such crucial events as the formation of the Phage Group and Lederberg's discovery of bacterial recombination which would both play a critical role in the emergence of molecular biology. At this point it was already known that certain viruses, called "bacteriophages" or "phages" preyed upon bacteria that were "sensitive" to them. It was also known that when such viruses and bacteria were grown together, occasionally, strains of bacteria that were "resistant" or immune to the viruses arose. Apparently, some bacteria had been somehow transformed from virus sensitivity to virus resistance and this transformation was heritable. The cause of this transformation, however, remained unknown and subject to considerable controversy [30]. D'Herelle, one of the discoverers of phage, was among those who argued that the presence of the virus induced this transformation while Gratia and Burnet, among others, argued that the transformation in the bacteria arose before exposure to the virus [31].

The former position conforms to neo-Lamarckism, the latter to neo-Darwinism as these terms have been explicated in the last section. If it is necessary that the virus be present in order to induce the observed transformation in a bacterium, then there is a correlation between the occurrence of this transformation and the environment of the bacterium. Further, this transformation, by permitting the bacteria which would otherwise fall prey to the virus to survive and reproduce, obviously enhanced the fitness of the bacteria. Thus the transformation is clearly neo-Lamarckian. If, however, the transformation occurred spontaneously in the sense that the presence of the virus was irrelevant to it, then the occurrence of such transformations would have no correlation with the presence of virus. Further, assuming that these transformations did not affect the fitness of the bacteria in some unknown way other than permitting virus resistance, there would also be no correlation with the fitness of any associated phenotype in the environment. Thus they would be random in the sense explicated in Section II. Therefore, this position is genuinely neo-Darwinian.

The canonical attempt to resolve this dispute was through the use of the so-called "fluctuation test" devised by S. Luria and M. Delbrück [32]. The basic idea was not new though Luria and Delbrück seem to have arrived at it independently. As early as 1934, Yang and Bruce White had observed that there were wide fluctuations in the occurrence of "rough" variants of *V.*

cholerae, that is, variants which formed morphologically rough colonies, and had concluded that the occurrence of a large number of variant colonies signified the spontaneous, that is, random occurrence of the variation and its subsequent clonal expansion during growth [33]. They did not develop the idea mathematically and their work seems to have been largely ignored [34]. Luria and Delbrück are responsible for developing the same insight and exploiting it to its full in order to establish the neo-Darwinian position noted above.

If the transformation of bacteria from virus sensitivity to virus resistance occurs as a result of variations induced by the presence of the virus, then there is a finite probability that such an event occurs in each bacterium after exposure to the virus. Subsequently, this variation is inherited. Luria and Delbrück called this possible outcome the hypothesis of "acquired hereditary immunity." [35] If the transformation of bacteria from virus sensitivity to virus resistance occurs as the result of a spontaneous variation, then variants arise at random during the growth of the bacteria. The number of variants arising from a particular initial event resulting in a variant grows exponentially with time because variants geometrically give rise to more variants in the next genera-

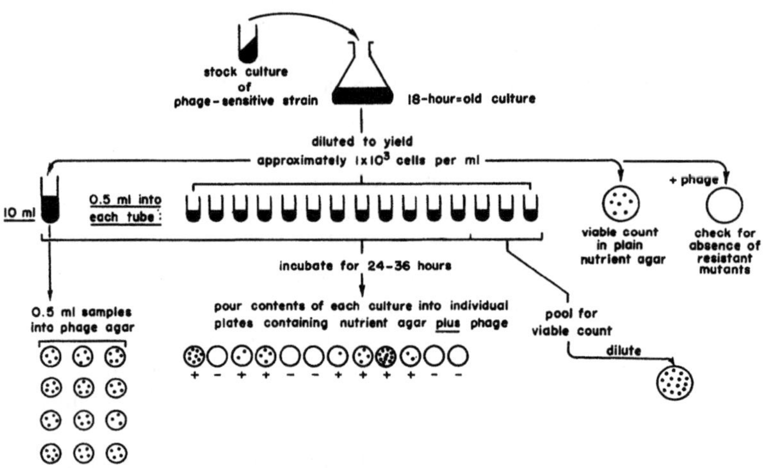

Fig. 1. The Fluctuation Test. A single bacterial cell from a stock culture is taken and incubated. Sister cultures are then grown in parallel, each from a single cell, in different test-tubes. These are then plated on the selective medium (phage-sensitive) agar. Mutants showing phage resistants are the only cells that can form colonies. Thus the number of colonies on a plate equals the number of resistant cells in the corresponding test-tube before plating. © W.B. Saunders. From Braun, W. *Bacterial Genetics*, W.B. Saunders and Company.

tion by normal cell division. Luria and Delbrück called this possibility the hypothesis of "mutation" [36]. They note with care, however, that they were not claiming that such mutations were the same, or even very strongly similar, to what were commonly called "mutations" in higher organisms [37].

They then devised the so-called "fluctuation test" to distinguish between these two hypotheses (see Fig. 1). A number of "sister" cultures of bacteria are grown to a specified density in liquid media in different test-tubes, ideally from a single ancestral cell that is known to be sensitive to the virus, though in practice from a few such cells that are assumed to be identical in their genotype. The number of resistant bacteria in each culture after exposure to the phage is then determined by mixing each culture (or a fixed fraction of it) with an excess of phage on a single test plate. Sensitive bacteria die while the resistant ones form colonies that are visually recognizable on the plate. The number of such colonies on a particular plate is, therefore, equal to the number of resistant bacteria in the corresponding culture (or the fraction of it that was plated).

Under the hypothesis of acquired hereditary immunity, the resistant mutant bacteria arose because of the finite probability that a bacterium would be

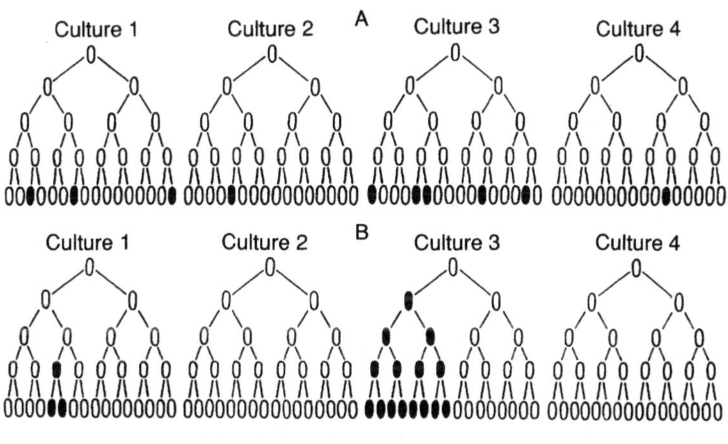

Fig. 2. The Poisson (A) and Luria-Delbrück (B) distributions. The ovals represent bacteria growing clonally. Empty circlesrepresent non-mutants; black circles represent mutants. Note how the Poisson distribution arises when mutations occur only after exposure to the selective environment. The variance is equal to the mean. The Luria-Delbrück distribution has one "jack-pot" culture in which a mutation occurred in the first generation and the number of mutants grew with the clone. © W.H. Freeman and Company. From Stent, G.S. 1971. *Molecular Genetics*, W.H. Freeman and Company.

induced to resistance by exposure to the phage. The distribution of the number of mutants would, therefore, follow the well-known Poisson distribution with a mean equal to the product of that probability and the number of bacteria in the culture (see Fig. 2). Since the distribution is Poisson, the mean would be equal to the variance. Under the hypothesis of mutation, however, the number of mutants in a culture would vary depending on how early in the history of that culture the mutation had occurred. The mutations that occurred early would be represented by an enormous number of mutants due to exponential growth. These are usually called "jackpot" cultures. Because of the presence of such "jackpot" cultures the "Luria-Delbrück" distribution has a much greater variance-to-mean ratio than the Poisson (for which this ratio is 1). Thus the fluctuations in the number of mutants would be much greater in this case—hence the name "fluctuation test."

Assuming that the mutant and non-mutant strains of bacteria grow at the same rate before exposure to the phage, which occurs if they had the same fitness, Luria and Delbrück managed to derive a formula for this variance which showed that it is much greater than 1. However, they could neither give an explicit expression for the distribution nor specify a procedure for calculating the probability that a given culture would have a certain number of mutants. Lea and Coulson first solved the latter problem by finding an iterative procedure for calculating these probabilities [38]. Recently, Stewart, Gordon and Levin have given a systematic treatment of this distribution under a wider set of assumptions than initially conceived by Luria and Delbrück, such as differential reproductive rates of mutant and non-mutant strains and inefficient plating techniques [39]. Their procedure permits the computation of the actual distribution expected in each of these cases. No explicit analytic formula for the distribution has yet been found.

Luria and Delbrück applied the fluctuation test to the action of the phage T1 on a strain of the bacterium, *E. coli*, that was normally sensitive to it. The distribution for the number of mutants obtained in "sister" cultures had a far higher variance to mean ratio than 1. They concluded, on the basis of this statistical test, in favor of the hypothesis of mutation. Between 1943 and 1949 the Luria-Delbrück experiment was repeated (in the course of other work) by Newcombe, and also by Demerec and Fano, for the same system [40]. In fact, the last two authors systematically applied fluctuation analysis to the case of resistance to phages T3 to T7 by *E. coli*. Witkin similarly studied the interaction of *E. coli* with radiation while Ryan studied its histidine independence [41]. Demerec applied fluctuation analysis to the resistance of *Staphylococcus aureus* to penicillin; Oakberg and Luria to its resistance to sulphathiazole; and

Demerec, later, to its resistance to streptomycin [42]. Ryan, Schneider and Ballentine applied the method to the study of uracil independence of *Clostridium septicum*; Alexander and Reidy to streptomycin resistance of *Hemophilus influenza*; and Curcho to the tryptophan independence of *Eberthella typhosa* [43].

In general, the evidence was taken to show that the observed distributions corresponded to the Luria-Delbrück distribution within the limits of experimental error. Sometimes, however, the deviations were obvious as in the case of Newcombe who then invoked phenotypic lag though only Ryan, in the cases of the interaction of *E. coli* with radiation and its mutations to histidine independence, argued that the deviations were significant [44]. However, in the former of Ryan's examples, Witkin and Newcombe were explicitly satisfied with the usual interpretation that the deviations were not significant [45]. In the case of mutations of *Clostridium* to uracil independence, Ryan, Schneider and Ballentine observed that the mutation occurred in spite of the presence of uracil in the environment, and thus was not directed, but that the data did not permit an assumption that the mutations were random in the sense that the distribution of mutants followed the Luria-Delbrück distribution [46].

In fact, it was Ryan who struck a further slightly discordant note in the growing consensus, among those using fluctuation analysis, in favor of the neo-Darwinian hypothesis that bacterial mutations occurred at random [47]. He studied mutations of two strains of bacteria from Lac^-, a phenotype which does not have the ability to digest lactose to Lac^+, a phenotype which does [48]. In both strains, there was some divergence from the expected Luria-Delbrück distribution but Ryan considered only one of these significant enough to report [49]. However, he did not interpret this divergence to suggest that any non-random process was occurring. He argued that the assumptions of Luria and Delbrück were too simplistic and seems to have been suggesting that factors like differential fitness of mutants and non-mutants in the original medium or phenotypic lag (the latter of which had already been shown by Armitage to lead to some divergence from the Luria-Delbrück distribution) might be operative [50].

However, not all were convinced by fluctuation analysis. Partly this must have been because of the abstract mathematical nature of the argument. The most strident critics, however, were those who, like Hinshelwood, held that there were no "genes" in bacteria analogous to genes in higher organisms even as late as 1957 [51]. According to Hinshelwood, bacterial "genetics" could be accounted for in terms of complex patterns of chemical reaction networks and the response of these to environmental stimuli was responsible for transformations like the one from virus sensitivity to virus resistance. The large variations observed in the fluctuation analysis experiments were to be explained on grounds of uncontrolled environmental differences between the various fluctuation test tubes, a claim initially made on experimental grounds by Eriksen, and subsequently repeated by Hinshelwood and his collaborators [52].

Cavalli and Mitchison attempted to provide an independent statistical argument for the spontaneous origin of the mutants by studying secondary characteristics of the mutants, that is characteristics other than the one directly selected for in the fluctuation tests [53]. Such characteristics would presumably differ in descendants of mutants from different mutations, that is in different test tubes of the fluctuation experiment if the hypothesis of mutation was correct [54]. Further all mutants within a test tube, having then arisen from the same mutation, would be very likely to have such characteristics to the same degree. Indeed, they found a strong statistical correlation for such characteristics for mutants within a single test tube and much greater variation between test tubes. This strongly suggested the clonal nature of the mutants as would be expected if they arose from single spontaneous mutations in each test tube. Cavalli also argued that the absence of a correlation between the number of mutants detected by the fluctuation test and the average resistance of the mutants (the secondary characteristic) further suggested that the environmental factors were irrelevant since it seemed likely that the same environmental factors would be relevant to both [55].

Meanwhile Newcombe carried out a different experiment on *E. coli* bacteria normally sensitive to the phage T1 [56]. Bacteria sensitive to phage were incubated for a few hours on several plates in order to allow them to grow. After this, on half the plates, the cells were redistributed, more or less uniformly, using a little salt solution. The other half of the plates were left undisturbed. Phage was then sprayed on to all the plates and the plates were incubated again. Under the hypothesis of acquired hereditary immunity, the respreading should have no effect on the appearance and number of resistant colonies on the plates since the mutants would only arise due to interaction with the phage after exposure. Under the hypothesis of (spontaneous) mutation, however, the respreading would make a difference. Mutants would have arose in clones which would have remained as part of the same colony before respreading. Respreading would spatially redistribute clones whose members would then form separate colonies. Thus the plates in which respreading occurred would be expected to have more colonies than the plates in which it did not occur. The experimental results provided striking confirmation of this expectation.

Newcombe recommended his method for the demonstration that mutations arose at random over the fluctuation test because the experimental method seemed to be more "direct." [57] Undoubtedly, this directness helped establish that result. However, the most decisive experiments were two equally direct experiments involving "indirect selection," where the mutants were obtained without any prior exposure to the selective media of any of their ancestors. These experiments, involving the use of "replica plating" and "sib-selection" in liquid media were performed by Lederberg and his collaborators [58]. This work involved some brilliant experimental designs and will be described in some detail, here, because of their role in establishing the orthodox neo-Darwinian position.

Replica plating permits the direct transfer of a pattern of bacterial growth on one plate to several others. A piece of velveteen is pressed against the plate containing the original bacteria. The pattern of growth on the plate is thus transferred on to the velveteen. The imprinted velveteen can then be used to inoculate, serially, several plates. Thus several replicas of the original pattern are obtained on these plates. It would be the case that if these new plates contained phage, bacterial colonies at corresponding sites would show phage resistance on them. Were resistance being induced by the phage, resistant colonies would be randomly distributed in each of these new plates. It turned out that resistance occurred at corresponding sites, which established that it had arisen in the colonies at these sites prior to exposure to the phage. This result already provided quite conclusive evidence that the mutation from phage sensitivity to phage resistance was occurring spontaneously. However, Lederberg and Lederberg went on to extend it even further to select mutants that had never been exposed to phage, as will shortly be described.

Lederberg and Lederberg carried out this test not only for phage resistance of bacteria but also for resistance to streptomycin. In the case of phage resistance, they also managed to use replica selection to isolate resistant variants without their, or their ancestors, ever having been exposed to phage. Replicas from the original plate were imprinted on one plate containing phage and another without it. A colony from the latter plate whose site corresponded to a resistant one in the former plate was selected. Cells from this colony were then allowed to grow and plated again. Replicas were then plated onto plates with and without phage. Iteration of this procedure permitted purification of the resistant line. Thus a phage-resistant strain of bacteria was "indirectly" selected without its ever having been exposed to phage.

Results obtained from replica plating do not permit quantitative analysis. However, Cavalli-Sforza and Lederberg initiated a scheme that permits quantitative investigation of the problem by devising a method of indirect selection, usually called "sib selection," using liquid media. If one starts out with a culture in a liquid medium where a resistant cell is so rare that only 1 in 10 test tubes would contain a resistant cell, this sample can be detected by exposure of a fraction of its contents to the selective medium. Discarding the other 9 test-tubes leads to a 10-fold enhancement of the concentration of resistants in the population. Dilution and repeated iteration of this procedure permits, in principle, the isolation of a pure line of resistants that had no contact with the selective media [59]. Not only that, the extent of the gradual enrichment of resistants could be precisely calculated from the hypothesis that all mutations were spontaneous. Cavalli-Sforza and Lederberg carried out this experiment for resistance to streptomycin and chloramphenicol. For streptomycin resistance, they found full agreement between theory and data (within the limits of the statistical power of their experiment). For chloramphenicol resistance, a large amount of statistical error did not permit detailed comparison of expected and observed enrichments. However, the data was consistent with all mutants arising spontaneously.

There are three possible criticisms of these experiments. All three are particularly relevant to the resurgence of the controversy over the possibility of directed mutagenesis in bacteria during the last three years and will, therefore, be discussed in some detail. The *first* criticism, which has been emphasized by Stahl, is that the selective media used were lethal to non-mutants [60]. Thus, these experiments do not truly explore the possibility that environmentally directed mutations might occur, especially if they occur at a slow enough rate. Exceptions to this general practice were the experiments of Ryan, but those experiments were precisely the ones that showed some significant deviation from the Luria-Delbrück distribution [61]. Thus the experimental data of the 1950's might not have been quite as conclusive as it then seemed [62].

Second, all these experiments seem to have been designed to test whether *all* mutations were random or whether all mutations were directed. The possibility that some mutations could be directed while others, perhaps most, were random seems not to have been taken seriously. In other words, the experiments do not seem to concern themselves with what was proposed to be (weak) neo-Lamarckism in Section II. Luria and Delbrück explicitly assumed that one of two extreme hypotheses hold: all mutations are spontaneous (random) or all are directed, as if each of these is the negation of the other [63]. The fluctuation test as devised, and used by them, only distinguished these two cases. Indeed, until the work of Lea and Coulson in 1949, there was no procedure to calculate the Luria-Delbrück distribution and any observation that the variance of an experimentally obtained distribution was significantly greater than its mean was taken to be indicative that the Luria-Delbrück distribution was obtained [64]. Further, the work of Armitage in 1952 made it even more difficult to be more precise about when the Luria-Delbrück distribution holds because deviations from it could already be attributed to factors other than the occurrence of directed mutations [65]. This point will be extremely important below. Even the quantitative sib-selection experiments of Cavalli-Sforza and Lederberg is flawed in the sense that it, too, would fail to distinguish between any except the extreme hypotheses of all spontaneous or all directed mutations [66]. Since the proportion of non-mutants decreases asymptotically in these experiments with every iteration of the dilution procedure, the statistical limits of the experiment would not permit the identification of additional mutants arising from the non-mutants after exposure to the selective medium, whether or not the selective medium was lethal to the non-mutants. Finally, the replica plating experiments and Newcombe's experiment are simply not quantitative in their nature and can be of no help in deciding this issue [67].

Third, and least important, there is a further difficulty with the design of these experiments, though it can be easily avoided with some secondary choices. *Directly*, all that they test is whether certain mutations occur *before* or *after* exposure to certain media. The choice whether the medium is selective is made independently of the experimental design. The experiments discussed chose media that are trivially selective. However, what is truly of significance in order to distinguish the neo-Lamarckian and neo-Darwinian hypotheses is the response of the bacteria precisely to selective media. In practice, this difficulty was avoided by ensuring that the medium used was trivially selective for the mutants. However, this does not permit consideration of subtle fitness differences that might be involved. This will also be of some importance below.

IV. RECENT EXPERIMENTS [68]

The first new experimental evidence that suggests the occurrence of directed mutations came from Shapiro [69]. Shapiro studied a strain of *E. coli* in which the regulatory part of the arabinose operon was separated from the terminal portion of the lac operon by a short segment of the transposable phage Mu which contained signals for the termination of transcription of DNA to RNA. When the intervening phage segment is deleted, cells of this strain can digest and grow on lactose provided that arabinose is also present to induce transcription. This phenotype is symbolized Lac (Ara)$^+$; the phenotype with an inability to do this is symbolized Lac (Ara)$^-$. Shapiro observed that when this strain of bacteria, still containing the intervening phage sequence between the parts of the ara and lac operons, was plated in media containing minimal arabinose and lactose, Lac (Ara)$^+$ mutants, which can grow and form colonies on this medium, occur only after a delay of about three days. However, if these plates are seeded with strains of pre-existing Lac (Ara)$^+$ cells, they grow and form colonies immediately. It appears from this experimental result that the formation of the Lac (Ara)$^+$ mutants, presumably by the deletion of the intermediate phage sequence, did not occur before the exposure of the original strain to arabinose and lactose, and occurred because of some interaction with the environment afterwards. Moreover, since the Lac (Ara)$^+$ mutants are selected for in this environment because of their ability to digest lactose, this is a case when the mutation is one that enhances fitness. This result is thus potentially a case of neo-Lamarckian mutagenesis.

Cairns, Overbaugh and Miller have presented two sets of experiments designed to show that some directed mutations occur in bacteria [70]. *First*, they investigated the problem of the origin of mutants in bacteria using the ability to digest lactose as the phenotypic marker somewhat along the lines initiated by Ryan [71]. They studied two strain of *E. coli*. The first had a mutation in the structural gene of the lac operon that prevented the production of the enzyme that digests lactose. The second contained an additional mutation, a deletion in the so-called "*uvrB-bi*" region of the bacterial chromosome. The distribution of mutants for these that "reverted" to a form that could digest lactose (the so-called "revertants") in a minimal lactose medium apparently turned out to have less variance than that was expected from the Luria-Delbrück distribution. Further, the deviation was much more significant for the *second* than the first. Cairns, Overbaugh and Miller interpreted this result to suggest that both random mutations before exposure to lactose and directed mutations after such exposure were occurring. This interpretation was based on the observation that the distribution obtained was consistent with a Poisson component "added on" to a Luria-Delbrück distribution as obtained from other strains. Further, the distribution of the mutants appearing after exposure to lactose, according to them, followed a Poisson distribution, though they do not actually perform any statistical analysis. Further, that the second strain had a much greater deviation indicated an influence of environmental factors.

In a *second* set of experiments, Cairns, Overbaugh and Miller replicated the observations of Shapiro just discussed and extended them further. First they showed that strains of *E. coli* that were Lac (Ara)$^-$, not because of the inserted phage sequence, but because of a point mutation in either the lac or the ara operon, showed no delay in producing the revertant Lac (Ara)$^+$ colonies. They interpret this to argue that there must be some special property of the intervening phage sequence (at that position) that permits its deletion (and thus reversion to Lac (Ara)$^+$) in the presence of lactose and arabinose. They also report a series of experiments in liquid cultures that showed the same effect.

Meanwhile Hall has experimented with a strain of *E. coli* that required two mutations in the β-glucosidase (*bgl*) operon in order to have the ability to digest salicin [72]. One of these (here referred to as the first mutation) is the excision of an intervening sequence of nucleotides from the structural gene that codes for the protein that actually participates directly in the digestion of salicin. The other (here referred to as the second mutation) is one in the regulatory gene of the operon. Note that neither mutation by itself confers the ability to digest salicin. Hall found that the first (excision) mutation rate in a strain with the second mutation was very low in the absence of salicin. Yet, Hall found that when this strain of bacteria with neither mutation is grown in agar rich in salicin, the first (excision) mutation rate is greatly enhanced. This observation is particularly odd because the first (excision) mutation, by itself, does not confer any ability to digest salicin [73]. One possibility is that the intermediate genotype, that is, the one with the first (excision) but not the second mutation, enhances fitness in some way other than the ability to digest salicin. Hall tested this possibility by growing mixtures of cells with or without the first (excision) mutation, but not the second mutation, in an environment containing salicin. No difference in fitness associated with the intermediate genotype was observed. Though Hall does not explicitly interpret these results as favoring directed mutation, he notes the "striking similarity" between his observations and those of Cairns, Overbaugh and Miller discussed above.

Levin, Gordon and Stewart have repeated and generalized the experiments of Cairns, Overbaugh and Miller, though for a different strain of *E. coli* and different culture media [74]. They report deviations from the Luria-Delbrück distribution similar to those observed by Cairns and his collaborators. However, they generally interpret their results to suggest that the mutations had already occurred in the fluctuation test tubes prior to plating and the colonies had appeared late after plating due to "physiological processes" associated with their interaction with high densities of a stationary parental population of non-mutants. Meanwhile, Mittler and Lenski have repeated the experiments of Shapiro and Cairns, Overbaugh and Miller on the Lac (Ara)$^+$

mutation on *E. coli* [75]. They verified the reported observations but ran a control that casts serious doubts on the interpretation favored by Cairns, Overbaugh and Miller. While the mutant colonies in question were present in media with lactose and arabinose, they found a comparable number of Lac $(Ara)^+$ colonies produced even in plates where both were absent. These colonies were detected by spraying plates with lactose and arabinose after different intervals of time during which these sugars had been absent. Similar results were obtained with liquid cultures. They interpret these findings to show that the appearance of Lac $(Ara)^+$ mutants does not require the presence of lactose and arabinose but seems do depend on the physiological state of the cells. However, it is not clear yet whether these results can be repeated [76].

V. INTERPRETATION

The results briefly touched upon in the last sections show that the experimental situation is still quite uncertain. However, this uncertainty is the result not so much of disagreements about the observations themselves as about their interpretation. These interpretive problems arise, to a very large extent, because of the three limitations, discussed at the end of Section III, in the design of the experiments that established the orthodox neo-Darwinian position regarding bacterial mutagenesis and which remain equally problematic for most of the experiments mentioned in the last section. The first of those limitations is not particularly important at this stage since the current experiments all study the response of non-mutant bacteria in media which are selective for the mutants but not lethal to the non-mutants. The third problem was that those experiments only directly test whether certain mutations occur before or after exposure to certain media and not whether there is any correlation between these mutations and the fitness of the resultant phenotype in that medium. The requirement that a medium be selective is not, in any direct sense, part of the experimental designs. Thus these experiments do not help in resolving questions of the sort raised by Mittler and Lenski [77]. In practice, as in the experiments described in Section III, the medium is chosen to be trivially selective. The second limitation is the most interesting and needs to be discussed in greater detail. In discussing it, attention will, however, be confined to fluctuation analysis since that is the only method that is relevant to the new experiments.

The source of the trouble is that virtually any factor operative in cultures that influences the rate of colony formation shifts a Luria-Delbrück distribution to one with less variance which can be interpreted as being a shift in the direction of the Poisson. This makes it extremely difficult, if not impossible, to use the fluctuation test even indirectly to argue that some directed mutations are occurring by showing such deviations from the Luria-Delbrück distribution as Cairns, Overbaugh and Miller have attempted [78]. In other words, this unfortunate property of the Luria-Delbrück distribution is precisely what limits the potential use of the fluctuation test to decide whether *some* directed mutations might be occurring. The effect of various factors on the Luria-Delbrück distribution is best seen by exploiting the general procedure for computing these distributions given by Stewart, Gordon and Levin [79]. They have computed the distributions under a variety of circumstances. These include differential fitness of mutant and non-mutants, a mutation rate dependent on the nutritional state of the cell, and the possibility that plating efficiency is less than 1. In all cases, there is a shift from the Luria-Delbrück distribution to one with less variance, which can be interpreted as a shift in the direction of the Poisson and, therefore, as a mixture of the two distributions. Thus, the type of observation reported by Cairns, Overbaugh and Miller can be accounted for without recourse to the hypothesis of directed mutation. Some simple qualitative arguments demonstrate these effects. These will now be given for: (i) differential fitness of mutants and non-mutants; (ii) phenotypic lag; (iii) plating efficiency less than 1; and also (iv) for multiple mutations and selection [80].

The effect of differential fitness of mutants and non-mutants [81]: If the fitness of the mutant is lower than that of the non-mutant in the original medium, that is, not the medium in which the mutant is being selected for, then the distribution of mutants as observed during a fluctuation analysis experiment will be shifted from a Luria-Delbrück one in the direction of a Poisson distribution. This is easy enough to motivate in an extreme case. Consider the situation when the fitness of the mutant is so low that it is not even viable for one generation. Then the distribution obtained, even if mutants only arose at random, would be a Poisson one, because the only mutants that would be observed through colony formation after plating on the selective medium would be the ones that arose during the last generation. In less extreme cases, this tendency towards the Poisson is just less extreme [82]. This can be shown using the iterative procedure developed by Stewart, Gordon and Levin as has been mentioned above [83]. The same effect is also seen in a simulation reported by Lenski, Slatkin and Ayala where they showed that the distribution obtained from a model with spontaneous and some directed mutation could be mimicked by one with differential fitnesses [84]. However, the differential fitness model needs to assume a much higher spontaneous mutation rate than the directed mutation model. This, too, can be seen intuitively. Since the mutants have a lower fitness, they grow more slowly than the non-mutants in the original medium. Thus to achieve the same concentrations as in the directed mutation model (where mutants and non-mutants have the same fitness), when mutants have lower fitness, they have to arise more frequently. Further, other mutants must also arise to compensate for those apparently arising after exposure to the selective environment in the directed mutation model.

The effect of phenotypic lag: Phenotypic lag is observed when a mutant phenotype is not expressed for some generations after a mutation occurs. This is best illustrated by the following example: suppose that a mutation makes some functionally necessary protein abnormal; suppose also that the functionally necessary protein is present in quantities much greater than strictly necessary in the cytoplasm when the mutation occurs. Then, for some generations of cell division, the resultant cells would continue to have functional proteins in them in sufficient quantities to preclude loss of function because of the division of the cytoplasm of the original cell. The phenotypic change—loss of the function associated with the protein—would only be observed late. Phenotypic lag obviously explains the late appearance of mutants. The reason why it seems to shift the Luria-Delbrück distribution in the direction of the Poisson, however, is a little complicated. Such a shift occurs only when the selective medium favors the mutants and the mutants are relatively rare, as is true in the cases being investigated experimentally. Mutants grow faster than non-mutants, and this process disproportionately affects the relative concentration of mutants to non-mutants in those cultures where mutants are rare and has very little effect on jackpot cultures. Thus, the variance of the distribution decreases.

The effect of plating efficiency being less than 1: It is easy to see why this would also decrease the variance of a distribution. Assume that the loss due to non-ideal plating efficiency is the same for all cultures. Assume also that this loss is proportional to the total number of cells, mutant or non-mutant. Then the proportional loss of mutants in a jackpot culture is the same as in a culture with few or no mutants. However, the number of mutant cells lost in a jackpot culture is much greater than in cultures with few or no mutants. This decreases the variance of the distribution, thus shifting a Luria-Delbrück distribution in the direction of the Poisson.

The effect of multiple mutations and selection: Lenski, Slatkin and Ayala have also proposed, and simulated, a different model which gives rise to a similar distribution of eventual mutants as the one obtained from the directed mutation model [85]. In this model, g_0, the initial non-mutant which cannot grow on the selective media first undergoes a mutation to g_1, which can only grow slowly on that media, and then to g_2 which grows rapidly on that media and whose distribution is the one that is measured. Colonies of g_2 appear late because of the slow growth of the intermediate genotype, g_1. Thus, though the late appearance of the colonies would have been inconsistent with the measured growth rate of g_2, the discrepancy can be explained by invoking the slow rate of growth of g_1. The source of this delay is, of course, the slow growth of the second genotype. The result of such a process would be that the initial results from the fluctuation test, which uses colony counts, would show less variance than is actually warranted. However, Levin, Gordon and Stewart have failed to detect such multiple mutation events (though they did not systematically search for them) [86].

The variant interpretive possibilities of the same fluctuation analysis data that has just been summarized highlight the point that fluctuation analysis alone cannot resolve the dispute over whether some directed mutations occur in bacteria, that is, whether there is some neo-Lamarckian mutagenesis. Any observation of a deviation from the Luria-Delbrück distribution can be explained by invoking subsidiary interactions such as those of differential fitnesses, phenotypic lag or inefficient plating while remaining strictly within the neo-Darwinian framework. Of course, hypothesizing such subsidiary interactions consists of making additional empirical claims and these, in turn, require experimental support. However, the kind of experiment these require must, in most cases, go beyond mere statistical analysis of the number of mutants and non-mutants [87]. Those favoring directed mutations, meanwhile, can experimentally rule out the existence of the subsidiary interactions posited by the neo-Darwinians. If literally all such explanations are ruled out, and it is shown that a certain mutation occurs (or occurs more frequently) if the original strain is exposed to a media in which the phenotype associated with the mutation has enhanced fitness, then this evidence can be taken to suggest the plausibility of the neo-Lamarckian position.

However, given the variety of neo-Darwinian alternatives that have already been posited, such a strategy seems unlikely to generate easy acceptance of directed mutations. What the neo-Lamarckian needs to do, in order to convince the skeptic, is to find *mechanisms* for directed mutations. This will involve both theoretical speculation—imagining physically plausible mechanisms for directed mutagenesis—and experimental corroboration. Cairns, Overbaugh and Miller have already begun the former process and have suggested three possible mechanisms [88]. Stahl and Davis have suggested others [89]. Some of these mechanisms have already been experimentally ruled out or, at least, been made quite implausible [90]. They are still discussed here in order to show how the possibility of directed mutagenesis can quite easily and plausibly be modelled at the molecular level [91]. All of these merit some further discussion because they illustrate that there might be nothing genuinely mysterious, or even fundamentally new, about some neo-Lamarckian processes [92]. In the discussion of these mechanisms the distinction of "variant" strands of DNA or RNA and "mutations" is maintained scrupulously since it will be important later. "Variant" strands of DNA are simply alternative strands (from that which is initially part of the genome) that arise through a variety of mechanisms such as failure of the repair of errors or the action of the environment when a single-stranded segment of DNA is exposed to it. Similarly, variant strands of RNA can also arise by a variety of mechanisms including transcription errors. Such variants might or might not get expressed. Only when such a variant gets incorporated into the genome when the cell divides has a "mutation" taken place.

Cairns, Overbaugh and Miller discuss three scenarios [93]. The first of these will be called a "specific reverse transcription" mechanism. When somewhat generalized from the specific description of this mechanism given by them, it operates along the following lines: the cell produces a highly variable set of messenger RNA molecules (possibly through processes like inaccurate transcription). Of these, one codes for the functionally best protein. This messenger RNA molecule would then be reversely transcribed into DNA. Thus a variant DNA molecule would arise by an obviously directed mutation; as the cell then divided, this variant DNA molecule would give rise to a mutation in the sense that it would become part of the genome. This mechanism requires three things: (i) a special organelle for reverse transcription such as the gag-pol complex of retroviruses which contains reverse transcriptase; (ii) a method by which messenger RNA molecules are monitored for performance of proteins coded for, that is, some kind of a feedback loop back from protein to nucleic acid; and (iii) some interaction from this feedback process to the organelle in which reverse transcription would take place [94].

The second scenario presented by them will, in analogy to the first, be called a "non-specific reverse transcription" mechanism: in this mechanism, a highly variable set of messenger RNA molecules, after being produced, would be routinely reverse transcribed into DNA. However, reverse transcription would only be initiated when a cell began growing. This would amount to a positive feedback loop that would ensure that the variant sequence that permitted the cell to grow out of a situation of stress would be the one that was reverse transcribed. This model also requires two things: (i) the existence of some reverse transcriptase in the cell; and (ii) a regulatory mechanism that initiates reverse transcription when cell growth begins. Though this mechanism would be less efficient than the last one, it is a little more plausible for two reasons: (i) it assumes less specificity before reversely transcription begins; and (ii) the regulatory mechanism that correlates reverse transcription and cell growth does not seem too improbable. The third scenario will be called a "variant DNA copies" mechanism: here the DNA itself is assumed to give rise to mutant copies and positive feedback is assumed from the best mutant. No other detail was specified.

Stahl presents another possible scenario which might be best referred to as the "correction enzyme cannibalization" mechanism [95]. In normal cells, routine errors during replication are normally corrected by mis-match repair enzymes. In starving cells these enzymes might well be cannibalized for energy. In that case variant strands of DNA can easily occur during the repair synthesis of DNA in a non-dividing cell. Should one fortuitously code for a protein that digests some substrate, the cell replicates the DNA as, perhaps, it divides. Thus the variant strand in question is stabilized as a mutation. The positive feedback from protein to DNA is thus mediated by cell division. Davis has presented yet another scenario which will be called the "transcriptional bias" mechanism [96]. According to this mechanism, single stranded DNA, exposed during transcription routinely gives rise to variant strands. Should one of the variants be such that an inducer, when some little energy is available, can initiate transcription, it does so. Thus lactose would initiate transcription when a DNA strand corresponding to Lac^+ arises. This direct induction of transcription would constitute the positive feedback from the environment to the DNA. Now, as the cell with this variant has the ability to digest lactose, energy becomes available cell growth takes place and the mutation is stabilized. Thus indirect feedback from the protein to the DNA occurs.

There are three important points to note about these mechanisms. *First*, all the mechanisms easily satisfy the requirements of (weak) neo-Lamarckism [97]. In each of these cases the mutation rate for a certain mutation is enhanced if the environment, that is, the medium, is selective, that is, it is one in which the associated phenotype has a higher fitness. The associated phenotype is the presence of a protein capable of digesting a particular substance and thereby providing the energy necessary for growth which obviously increases the fitness of the cell. It might, however, be argued that, for example, in the "correction enzyme cannibalization" and the "transcriptional bias" mechanisms, that the underlying variations are occurring spontaneously (randomly) and are only being selected for by the environment. However, even here, selection occurs within a generation, and is a selection of *variants*, not *mutations*, unlike the usual neo-Darwinian processes.

However, because of this underlying selection of variants, these mechanisms illustrate very well the *second* important point that needs to be noticed, namely, that there might not be any serious conflict between neo-Darwinism and neo-Lamarckism. In this sense, at least, there might be *peaceful coexistence* between neo-Darwinism and neo-Lamarckism in such circumstances. The mutations are arising in a directed fashion. However, in the "correction enzyme cannibalization" and the "transcriptional bias" mechanisms, the underlying genesis of variants is, *formally*, neo-Darwinian in the sense that there is random variation and natural selection, although that selection is extremely rapid. However, it is not completely clear whether the same can be said about the mechanisms hypothesized by Cairns, Overbaugh and Miller [98]. The "specific" and "non-specific reverse transcription' mechanisms do postulate the existence of new organelles and processes. How they fit with neo-Darwinism can only be fruitfully investigated if such entities are discovered and the mechanisms specified. The conclusion arrived here might seem surprising given that, in Section II, neo-Lamarckism was explicated as the denial of neo-Darwinism. However, there is no confusion here because the formally neo-Darwinian process is occurring at the level of the genesis of variants whereas the neo-Lamarckian effect is only seen at the level of the occurrence of mutagenesis. Thus, all that is being suggested here is that, in some circumstances, directed mutagenesis can take place because basically random changes that can be incorporated in the genome occur and, they are subsequently so incorporated because the environment is fiercely selective for them in the sense that only one variant is viable [99]. *Third*, all these mechanisms rely on random processes to some extent. However, this still does not preclude the occurrence directed mutagenesis simply because the randomness is at the level of the genesis of variants. However, it is conceivable that there could be models of directed mutagenesis that do not rely at all on random genesis of variants. Should such models be formulated, and then subsequently found to capture what occurs in nature, they would present greater difficulties for the potentially peaceful coexistence of neo-Darwinism and neo-Lamarckism that is being suggested here.

VI. REDUCTION

It was argued in the last section that it is unlikely that fluctuation analysis alone would be able to resolve the controversy about the possibility of some directed mutagenesis in bacteria. Further, it was argued that the resolution of that controversy would have to await the discovery and specification of the mechanisms that account for the experimental data. It will now be argued that the search for such mechanisms constitute *reductionist* research. This point is important for two reasons. *First* the role of reduction in molecular biology has been extremely controversial [100]. In applying physical (including chemical) techniques to biology, both at the theoretical and the experimental levels, molecular biologists are obviously contributing to a physical understanding of biological processes. This makes it quite plausible to suggest that, in some sense, molecular biologists are reducing biology to physics and chemistry. However, such a suggestion, however plausible, is problematic. There are two main reasons for this: (i) some of the techniques used such as statistical analyses, for example, fluctuation analysis or sib selection, as will be argued below, are definitely non-reductionist; and (ii) there is remarkable disagreement among philosophers as to what constitutes reduction and it is certain that at least some of the models of reduction that have been proposed by philosophers fail to capture the nature of research or explanation that is involved in molecular biology [101]. *Second*, Cairns, Overbaugh and Miller have explicitly connected the possibility of the occurrence of directed mutations in bacteria with a purported failure of a reductionist account of contemporary research in molecular biology. "The early triumphs of molecular biology strongly supported the reductionists," they argue. However, "when we come to consider what mechanism might be the basis for the forms of mutation described in this paper we find that molecular biology has, in the interim, deserted the reductionist. Now, almost anything seems possible. In certain systems, information freely flows back from RNA into DNA; genomic instability can be switched on under conditions of stress, and switched off when stress is over; and instances exist where cells are able to generate extreme variability in localized regions of the genome. The only major category of informational transfer that has not been described is between proteins and the messenger RNA (mRNA) molecules that made them. If a cell discovered how to make that connection, it might be able to exercise some choice over which mutations to accept and which to reject." [102] They go on to describe the scenarios described in the last section.

In discussing questions related to that of reductionism, two issues need to be precisely formulated. *First*, since there are many questions in whose answers reductionsm is at stake, the actual question that is being considered needs to be precisely specified. For example the question whether the research strategy of molecular biology is uniformly reductionist is distinct from the question whether the explanations afforded by it are reductionist. The two might receive different answers as will be seen below. The discussion here will focus on the questions whether fluctuation analysis (or other statistical analyses at the same "level," that is, of individual bacteria), and the search for mechanisms, are reductionist as experimental methods in the sense whether they seek reductionist explanation. Only a few remarks will be made about the question whether the structure of explanations in molecular biology is reductionist and the two questions will be systematically kept distinct. These questions are each *epistemological*. *Ontological* questions regarding whether some entity is "formed" of, or "consists entirely" of other entities or types of entities will be completely ignored because they do not seem to be particularly interesting in this context.

Second, since there are a large number of models of reduction, it needs to be stated clearly which model is being used [103]. The model of reduction adopted here is due to Sarkar, based to a large extent on the earlier work of Kauffman and partly on Wimsatt [104]. Reductionist explanation is construed here as (i) an explanation based on the constituent parts of the entity under investigation; and (ii) an explanation that only posits those interactions between these parts that have a "physical warrant" in the sense of being derivable from physical (or chemical) theory or being corroborated by experiments involving only inanimate matter. It is assumed in this model that the explanation afforded satisfies any criteria required for its adequacy as a scientific explanation. What is of interest, here, is whether it satisfies the two additional requirements noted above to be reductionist explanations. The basic intuition, which is incorporated in the first of these requirements, is that the explanation is of a whole in terms of its parts. This intuition is not the same as that of an "upper-level" theory or phenomenon being explained in terms of more general "lower-level" theories which lay behind many of the classic accounts of reduction [105]. The intuition incorporated in the second requirement is that the interactions between the parts invoked in the explanation are physical in the sense that they are known on the basis of physical theory or experimentation.

Explanation of observed distributions of mutant cells by invoking mechanisms of mutagenesis *within* a cell obviously satisfies the first of these two criteria since the explanation of the behavior of a cell is being given at the level of the nucleic acids and the proteins inside it. However, to the extent that the explanation invokes the notion of "information," in the sense that it continuously refers to "codes," "translation," and so forth, it initially appears to violate the second criterion because the properties of biological information are obviously neither derivable from physical theory nor learnt from the type of experiment that is envisioned in the statement of that criterion. However, "information," as used here is not a technical concept and, indeed, has never received sufficient explication in biology to become such a concept: it is, in fact, quite dispensable in the considerations that are at stake here. Once the mechanism for a directed (or, for that matter, a random) mutation is specified, as in the case of the examples discussed in the last section, the concept of information is redundant and the explanation can be given in terms of molecular interactions and reaction rates. It is customary to adhere to the usual terminology only for the sake of clarity. Thus, the explanation of the occurrence of mutations, directed or random, by the invocation of mechanisms of mutagenesis satisfies both requirements of reductionist explanation. Thus, the type of research that is involved in searching for these mechanisms is also fundamentally reductionist. Assuming that such explanations in terms of mechanisms are eventually obtained, both the epistemological questions mentioned above receive the same answer in favor of reductionism.

In contrast, fluctuation analysis or other such statistical analyses are not reductionist in their approach. By fluctuation analysis, conclusions about sub-cellular processes are drawn from statistical examination at the cellular level which, in the case of bacteria, is the organismic level. In effect, Luria and Delbrück were doing for bacteria what Mendel did for peas [106]. Properties of the whole were neither being explained nor being investigated by examining properties of parts. In fact, fluctuation analysis turns the investigation the other way around. Thus the first criterion in the model of reduction specified above is violated. Such research strategies are, therefore, clearly non-reductionist. Moreover, it is not even clear that the results obtained from fluctuation analysis (or other such statistical analyses) would constitute explanation, even if these results were unequivocal with respect to their interpretation. In a sense, all that would have been obtained is a description of what must be happening at the sub-cellular level (which is, of course, extremely important) but not an explanation. Consequently, the question whether the results of such analyses permit reductionist explanation cannot be fruitfully asked.

The failure of fluctuation analysis and the necessity for the search for mechanisms to resolve the controversy of directed mutagenesis in bacteria underscores the importance of reductionist research in molecular biology. In fact, contrary to the claims of Cairns, Overbaugh and Miller, this controversy further illustrates the increasing success of reductionist research in this field [107]. This need not have been so: many of the founders of molecular biology were not reductionists. Delbück and Stent, motivated by the views of Bohr, hoped that the laws of physics might fail at a certain level that would be discovered through biological research [108]. "Complementary" biologica laws would be discovered. The research strategy of Delbrück and his followers reflected these non-reductionist hopes, at least to some extent, as the design of the fluctuation test makes apparent [109]. The explanations they hoped to find would require laws beyond the domain of physics and chemistry, that is, the laws governing the behavior of inanimate matter. Such explanations, too would not have been reductionist.

It is precisely this hope that has not been realized. Explanations offered by molecular biology in the 1950's and 1960's were routinely reductionist [110]. Thus, the issue of reductionism in the questions regarding research strategy and the structure of explanations was resolved differently in this case. This is precisely why the questions needed to be kept separate in the first place. Since the 1960's, moreover, molecular biology has progressively become more reductionist in its research strategies. It has generally been successful at explanation and these explanations have continued to be routinely reductionist. Few would doubt that, in the process, a deeper understanding of some fundamental biological phenomena has been obtained.

VII. CONCLUSION

This paper has attempted to establish two sets of claims. *First,* it has attempted to clarify the differences between neo-Darwinism and neo-Lamarckism by proposing a certain (weak) explication of "neo-Lamarckism" that would permit serious consideration of neo-Lamarckian hypotheses in the sense that such hypotheses would not be trivially false or uninteresting. It has then used this notion of neo-Lamarckism to give criteria which bacterial mutagenesis would have to satisfy in order to be neo-Lamarckian. It has also examined possible mechanisms of mutagenesis to see whether they satisfy these criteria. It has also suggested that, in some cases, neo-Darwinist and neo-Lamarckist hypotheses do not stand in serious conflict. Nothing substantive depends on whether the term "neo-Lamarckist" is used to describe the relevant positions. Those who object to the use of the term for ideological reasons, or for the connotations of the Lysenko affair that it still often carries, might well prefer to used a myriad of terms like "directed mutations." What is important is that the criteria for distinguishing neo-Darwinist and conflicting positions are clearly explicated: that is the substantive issue.

Second, it has been argued that the type of research that is likely to resolve the controversy under discussion is reductionist in the precise sense of reduction discussed in Section V. It has also been argued that the type of explanation that would be obtained, if the research strategy suggested succeeds, would also be reductionist. These arguments should not, however, be construed to suggest either that reductionist research strategies are useful, or even viable, in all contexts or that reductionist explanations can always be obtained. What research strategy is appropriate in which context is something that can only be decided by a detailed consideration of many factors, including the nature of the questions asked, the complexities of the entities being investigated, and the theoretical and experimental resources available. Both types of strategy have had long and illustrious careers in biology: in molecular biology, however, the reductionist currently has the upper hand and it is probably reasonable to suggest that this situation will continue for the foreseeable future. Whether all phenomena can eventually be afforded reductionist explanation, whether in biology or elsewhere, is also an open question. Evolutionary explanations routinely violate reductionist criteria by appealing to composite systems that affect fitness [111]. It would be rash to say that all such explanations will eventually receive reductionist support. It would be equally rash to assume they never will. Only further work can decide this question.

The issues in the first set of claims are metaphysical since they involve the most fundamental aspects of the framework of evolutionary theory, as was pointed out in Section II. To the extent that evolutionary theory, as it is formulated at present, is correct, these aspects of evolutionary theory reflect fundamental features of the natural world. Put in another way, the fundamental question simply is whether there is, even to a very limited extent, any direction in evolution. Perhaps it is precisely because the question can be formulated in this way, loaded with ideological and even theological connotations, that the behavior of a strain of bacteria, often under stress, has generated so much controversy. However, at this level, the question is of scant scientific interest. This is exactly why such broad formulations of the question have here been avoided. The issues in the second set of claims are epistemological in a straightforward sense: they are issues concerning the nature of scientific explanations and strategies of research and, therefore, strategies for obtaining further knowledge.

Both sets of issues will be resolved only by experience. It is obvious that whether any neo-Lamarckian processes exist is an empirical question to be settled by further experimentation. However, whether reductionist research is necessary or whether statistical analyses such as fluctuation analysis can resolve the controversy being discussed here is also an empirical question. The future history of this problem will determine the answer. Perhaps a truly important moral, and one which is not sufficiently recognized, that can be drawn from all these considerations is that metaphysical and epistemological questions of this sort can be, and perhaps only be, answered from experience. Perhaps this is truly the sense in which the interaction between philosophy and the world of experience, that is, the world described by science is dialectical: each progresses in response to the development of the other.

Boston University

NOTES

1. Thanks are due to John Cairns, Robert S. Cohen, James F. Crow, Patricia Foster, Stephen Jay Gould, David Hull, Joshua Lederberg, Richard Lenski, Heidi Lindquist, Bruce R. Levin, Richard Lewontin, Abner Shimony, John Maynard Smith, Alfred I. Tauber and William Wimsatt for helpful discussions and comments on earlier versions of parts of this analysis. Comments by members of the audience at a colloquium at Boston University, Spring 1990, where this analysis was presented, were also very helpful.
2. Cairns, J., J. Overbaugh, and S. Miller. The origin of mutants. *Nature* 335: 142–145, 1988.
3. Cairns, J. *et al.*, *op. cit.* in Note 2; Sarkar, S. Reductionism and molecular biology: a reappraisal. Ph. D. Dissertation. Department of Philosophy, University of Chicago, 1989.
4. Sarkar, S., *op. cit.* in Note 3; Kauffman, S.A. Articulation of parts explanation in biology and the rational search for them. *Boston Studies in the Philosophy of Science* 8: 257–272, 1972; Sarkar, S. Models of reduction and categories of reductionism. Forthcoming in *Synthese*.
5. Lewontin, R.C. The units of selection. *Annual Review of Ecology and Systematics* 1: 1–18, 1970.
6. Lewontin, R.C., *op. cit.* in Note 5, p. 1.
7. Lewontin, R.C., *op. cit.* in Note 5, p. 1.
8. Note that this explication of evolution by natural selection makes no mention of what the individuals of the population are. They could be molecules, organelles within a cell, cells, individuals (which in multi-cellular organisms would not be identical with cells), kins, groups, populations or species. What the appropriate units of selection are is a matter of considerable biological and philosophical controversy. In the case of bacteria, which is all that is directly of concern in this article, the levels of cells and individuals are identical thus partly avoiding this controversy. For details of this controversy, see Brandon, R. N. and R. N. Burian (eds.). *Genes, Organisms, Populations: Controversies Over the Units of Selection.* Cambridge, MA: MIT Press, 1984.
9. See Wimsatt, W.C. Randomness and perceived randomness in evolutionary biology. *Synthese* 43: 287–329, 1980, for an attempt to analyze the difficulties with the notion of randomness appropriate for evolutionary biology.
10. Kimura (Kimura, M. *The Neutral Theory of Molecular Evolution.* Cambridge: Cambridge University Press, 1983), in his more polemical moments seems to suggest such a viewpoint.
11. See Provine, W. *The Origins of Theoretical Population Genetics.* Chicago: University of Chicago Press, 1971, for details of this fascinating history.
12. See Maynard Smith, *J. Evolutionary Genetics.* Oxford: Oxford University Press, 1989, for an emphasis on natural selection and Lewontin's review (Lewontin, R.C.A natural selection. *Nature* 339: 107, 1989) for a critique.

13. Haldane, J.B.S. *The Causes of Evolution*. London: Harper & Brothers, 1932; Huxley, J. 1942. *Evolution: The Modern Synthesis*. London: George Allen & Unwin.
14. Not one of the figures discussed by Hull (Hull, D. Lamarck among the Anglos. In J.B. Lamarck, *Zoological Philosophy*. Chicago: University of Chicago Press, 1984, pp. xl – lxvi) in his colorful, though short, account of the history of neo-Lamarkism seems to have suggested that all mutations were directed.
15. For details of Lysenkoist claims, see Hudson, P.S., and R.H. Richens. *The New Genetics in the Soviet Union*. Cambridge: Imperial Bureau of Plant Breeding and Genetics, 1946. Among those at least partly sympathetic to Lysenko's claims, only Haldane (Haldane, J.B.S. Lysenko and genetics. *Science and Society* 4: 433–437, 1940) seems to have been fully clear that all that was needed to explain the kind of effects Lysenko claimed to be observing was that some mutations could be induced in such a fashion.
16. This strong notion of directedness has been championed by Lenski, R.E. Are some mutations directed? *Trends in Ecology and Evolution* 4: 148–150, 1989.
17. This weaker definition has been used by Sarkar, S. On the possibility of directed mutations in bacteria: statistical analyses and reductionist strategies. In *PSA 1990*. A. Fine, editor, 1990. Philosophy of Science Association, East Lansing. In press.
18. Many of the participants in the controversy would also reject the use of "neo-Lamarckism" on various grounds including vagueness. However, vagueness, at least, has been removed if the distinctions and definitions elaborated in the text are clear enough. For responses to some other objections to the use of "neo-Lamarckism," see Sarkar, S., *op. cit.* in Note 17.
19. However, for an important treatment of the possible inheritance of acquired characteristics where the concept is not used vaguely, see Jablonka, E., and M.J. Lamb. The inheritance of acquired epigenetic variations. *Journal of Theoretical Biology* 139: 69–83, 1989.
20. Cullis, C.A. The generation of somatic and heritable variation in response to stress. *American Naturalist* 130: S62–S73, 1987.
21. The reasons given for not using the notion of the inheritance of acquired characteristics are also those for not using another related distinction due to Mayr, namely, that between "hard" and "soft" inheritance (Mayr, E. *The Growth of Biological Thought*. Cambridge, MA: Harvard University Press, 1982), pp. 687–689). These choices, however, are made for the purpose of conceptual clarification. No suggestion is being made here that these other construals of the differences between neo-Darwinism and neo-Lamarckism can be avoided when the history of these disputes, especially during the first half of this century, is considered.
22. Lamarck, J.B. *Zoological Philosophy*. Chicago: University of Chicago Press, 1984, p. 113.
23. Darwin, C. *On the Origin of Species*. London: John Murray, 1859.
24. See, for example, Darwin, C., *op. cit.* in Note 23, pp. 134–139. Mayr lists nine other examples (Mayr, E. Introduction. In C. Darwin, *On the Origin of Species*. Cambridge, MA: Harvard University Press, 1964, p.xxvi).
25. See Eiseley, L. *Darwin's Century*. New York: Doubleday Anchor, 1961, and Ruse, M., *The Darwinian Revolution: Science Red in Tooth and Claw*. Chicago: University of Chicago Press, 1979, for details of these developments.
26. See, for example, Weismann, A. *Das Keimplasma: Eine Theorie der Vererbung*. Jena: Gustav Fischer, 1892.
27. Romanes, G.J. *Life and Letters*. London: Longmans, Green, 1896.
28. In particular the unfortunate historical association between neo-Lamarckism and Lysenkoist positions during the Lysenko affair needs to be carefully avoided.

29. For details of this history, see Sarkar, S, *op. cit.* in Note 3, and Fischer, E.P., and C. Lipson. *Thinking About Science: Max Delbrück and the Origins of Molecular Biology.* New York: Knopf, 1988, and references therein.
30. Though the terms, "mutant" and "mutation" were routinely used for this transformation, it remained an open question whether these "mutations" were mutations of genes. In 1943, for example, Luria and Delbück carefully observe: "Naming such hereditary changes 'mutations' of course does not imply a detailed similarity with any of the classes of mutations that have been analyzed in terms of genes for higher organisms. The similarity may be merely a formal one (Luria, S.E., and M. Delbrück. Mutations of bacteria from virus sensitivity to virus resistance. *Genetics* 28: 491–511, 1943, p. 492)." Even in the late 1950's a few skeptics such as Hinshelwood (for example, in Dean, A.C.R., and C.N. Hinshelwood. Aspects of the problem of drug resistance in bacteria. In *Drug Resistance in Micro-Organisms.* G.E.W. Wolstenholme, and C.M. O'Connor, editors. London: J. & A. Churchill, 1957, pp. 4–24) would maintain that any reference to genes in bacteria would actually be a reference to complex chemical reaction networks.
31. d'Herelle, F. *The Bacteriophage and Its Behavior.* Baltimore: Williams and Wilkins, 1926; Gratia, A. Studies on the d'Herelle phenomenon. *Journal of Experimental Medicine* 34: 115–131, 1921; Burnet, F. M. "Smooth-rough" variation in bacteria in its relation to bacteriophage. *Journal of Pathology and Bacteriology* 32: 15–42, 1929.
32. Luria, S.E. and M. Delbrück, *op. cit.* in Note 30. According to Luria, he was led to the idea of the fluctuation test while watching the operation of slot machines during a faculty dance at the Bloomington Country Club in Indiana. If these machines were programmed to return money at random, the returns would form a Poisson distribution clustered around a mean and there would be virtually no jackpots. If, however, the machines were programmed so that they return occasional jackpots with many very tiny returns, the returns would fluctuate much more widely. The average return would be the same in both cases. The way in which the machines were programmed could only be determined by actually observing the actual distribution of the returns and not from the mean alone. Luria immediately applied this insight to the question of the distribution of mutants in bacteria. For further details see Luria, *S.E. A Slot Machine, a Broken Test Tube: An Autobiography.* New York: Harper and Row, 1984, pp. 74–79.
33. Yang, Y.N., and P. Bruce Wright. Rough variation in *V. cholerae* and its relation to resistance to cholera-phage (Type A). *Journal of Pathology and Bacteriology* 38: 187–200, 1934.
34. Luria, S. E. and M. Delbruck (*op. cit.* in Note 30) do not cite them. Neither do Newcombe, H. B. Origin of bacterial variants. *Nature* 164: 150–151, 1949, Lederberg, J., and E. Lederberg. Replica plating and indirect selection of bacterial mutants. *Journal of Bacteriology* 63: 399–406, 1952, or Cavalli-Sforza, L.L., and J. Lederberg. Isolation of preadaptive mutants by sib selection. *Genetics* 41: 367–381, 1956, which were critical papers in the establishment of the neo-Darwinian view. Cavalli-Sforza and Lederberg do list them in their bibliography in an earlier paper (Cavalli-Sforza, L.L., and J. Lederberg. Genetics of resistance to bacterial inhibitors. In *Symposium: Growth Inhibition and Chemotherapy.* Rome: Istituto Superiore di Sanita, 1953, pp. 108–142) but do not discuss them in the text. A recent article by Lederberg is responsible for drawing attention to the contribution of Yang and Bruce White (Lederberg, J. Replica plating and indirect selection of bacterial mutants: isolation of preadaptive mutants in bacteria by sib selection. *Genetics* 121: 395–399, 1989).
35. Luria, S.E. and M. Delbrück, *op. cit.* in Note 30, p. 493.

36. Luria, S.E. and M. Delbrück, *op. cit.* in Note 30, p. 493.
37. From this point, however, the distinction between "variation" and "mutation" and related terms will be ignored for the sake of convenience of expression.
38. Lea, D.E., and C.A. Coulson. The distribution of the number of mutants in bacterial populations. *Journal of Genetics* 49: 264–285, 1949. As a historical curiosity, it is worth noting that it is possible that Fisher might have solved the Luria-Delbrück distribution first. Crow (Crow, J.F., R.A. Fisher, a centenniel view. *Genetics* 124: 207–211, 1990) recalls that while he found the Luria-Delbrück argument convincing, he thought that the mathematical treatment "shoddy and confusing (Crow, J.F., *op. cit.*, p. 210)." Consequently, in 1946, he approached Fisher with the problem. Fisher "leaned back in his chair, thought for perhaps a minute, took a scrap of paper, and wrote a generating function (Crow, J.F., *op. cit.*, p. 210)." Crow, not understanding the formula yet, put that scrap of paper aside, intending to work on it later and then lost it!
39. Stewart, F., D. Gordon, and B. Levin. Fluctuation analysis: the probability distribution of the number of mutants under different conditions. *Genetics* 124: 175–185, 1990. There were many previous attempts to generalize the Luria-Delbrück distribution but none to this extent. For example, phenotypic lag is considered in Armitage, P. The statistical theory of bacterial populations subject to mutation. *Journal of the Royal Statistical Society B* 14: 1–40, 1952, and Koch, A.L. Mutation and growth rates from Luria-Delbrück fluctuation tests. *Mutation Research* 95: 129–143, 1982. The latter and Mandelbrot, B. A population birth-and-mutation process I: explicit distributions for the number of mutants in an old culture of bacteria. *Journal of Applied Probability* 11: 437–444, 1974 also consider differential fitnesses (growth rates), of the original and mutant strains. The effect of these factors on the distributions will be considered in Section 5.
40. Newcombe, H.B. Delayed phenotypic expression of spontaneous mutations in Escherichia coli. *Genetics* 33: 447–476, 1948; Demerec, M., and U. Fano. Bacteriophage-resistant mutants in Escherichia coli. *Genetics* 30: 119–136, 1945.
41. Witkin, E.M. Genetics of resistance to radiation in *Escherichia coli*. *Genetics* 32: 221–248, 1947; Ryan, F.J. On the stability of nutritional mutants of bacteria. *Proceedings of the National Academy of Sciences (USA)* 34: 425–435, 1948.
42. Demerec, M. Production of staphylococcus strains resistant to various concentrations of penicillin. *Proceedings of the National Academy of Sciences (USA)* 31: 16–24, 1945; Oakberg, E.F., and S.E. Luria. Mutations to sulfonamide resistance in *Staphylococcus aureus*. *Genetics* 32: 249–261, 1947; Demerec, M. Origin of bacterial resistance to antibiotics.*Journal of Bacteriology* 56: 63–74, 1948.
43. Ryan, F.J., L.K. Schneider, and R. Ballentine. Mutations involving the requirement of uracil in Clostridium. *Proceedings of the National Academy of Sciences (USA)* 32: 261–271, 1946; Alexander, H.E., and J. Leidy. Mode of action of streptomycin on type b Hemophilus influenzae. *Journal of Experimental Medicine* 85: 607–621, 1947; Curcho, M. de la G. Mutation to tryptophan independence in *Erbethella typhosa*. *Journal of Bacteriology* 56: 374–375, 1948.
44. Newcombe, H.B., *op. cit.* in Note 40; Ryan, F.J. Distribution of numbers of mutant bacteria in replicate cultures. *Nature* 169: 882–883, 1952.
45. Newcombe, H.B., *op. cit.* in Note 34; Witkin, E.M., *op. cit.* in Note 41.
46. Ryan, F.J. *et al.*, *op. cit.* in Note 43.
47. Ryan, F.J., *op. cit.* in Note 44.

48. Ryan, F.J., *op. cit.* in Note 44; Ryan, F.J. Adaptation to use lactose in *Escherichia coli. Journal of General Microbiology* 7: 69–88, 1952.This mutation is particularly relevant here because it is one of those studied by Cairns, J. *et al., op. cit.* in Note 2. In fact, that work used Ryan's studies as one of its starting points.
49. Ryan, F.J., *op. cit.* in Note 44.
50. Armitage, P., *op. cit.* in Note 39. Note that Ryan does not explicitly invoke phenotypic lag or any other specific mechanism.
51. Dean, A.C.R. and C.N. Hinshelwood, *op. cit.* in Note 30.
52. Eriksen, K.R. Studies on the mode of origin of penicillin resistant staphylococci. *Acta Pathologica et Microbiologica Scandinavica* 26: 269–279, 1949. However, Eriksen's experimental results were not particularly convincing because small samples were taken from the cultures which made the statistical tests inefficient as Cavalli-Sforza and Lederberg (1953 *op. cit.* in Note 34) point out. For Hinshelwood's position, see, for example, Hinshelwood, C.N. Chemistry and bacteria. *Nature* 166: 1089–1092, 1950; and Dean, A.C.R., and C.N. Hinshelwood. 1952. The resistance of *Bact. Lactis aerogenes* to proflavine (2: 8-diaminoacridine). I: the applicability of the statistical fluctuation test. *Proceedings of the Royal Society B* 139: 236–250, 1952.
53. Cavalli, L.L. Genetic analysis of drug-resistance. *Bulletin of the World Health Organization* 6: 185–206, 1952; Michison, D.A. The occurrence of independent mutations to different types of streptomycin resistance in *Bacterium coli. Journal of General Microbiology* 8: 168–185, 1953.
54. For example, Mitchison, D.A., *op. cit.* in Note 53, studied three types of strains of *Bacterium coli* which had differences in growth rates while being selected for streptomycin resistance by fluctuation analysis.
55. Cavalli, L.L., *op. cit.* in Note 53.
56. Newcombe, H.B., *op. cit.* in Note 34.
57. Newcombe, H.B., *op. cit.* in Note 34, p. 150.
58. For replica plating, see Lederberg, J. and E.M. Lederberg, *op. cit.* in Note 34; for sib-selection, see Cavalli-Sforza, L.L. and J.L. Lederberg, 1956 *op. cit.* in Note 34.
59. In practice Cavalli-Sforza and Lederberg (1956 *op. cit.* in Note 34) had to increase cell concentration in each cycle because of slow selection.
60. Stahl, F. Bacterial genetics: a unicorn in the garden. *Nature* 335: 112–113, 1988. Lederberg was certainly aware of this but seems not to have regarded it as particularly relevant (see Cavalli-Sforza, L.L. and J. Lederberg, 1953 *op. cit.* in Note 34, p. 121).
61. Ryan, F.J., *op. cit.* in Notes 41, 44 and 48.
62. This does not, of course, endorse in any way criticisms of the sort made by Hinshelwood, C.N., *op. cit.* in Notes 30 and 52.
63. Luria, S.E. and M. Delbrück, *op. cit.* in Note 30. Note that they do not explicitly state that one alternative is the negation of the other; they simply ignore intermediate possibilities.
64. Lea, D.E. and C.A. Coulson, *op. cit.* in Note 38.
65. Armitage, P., *op. cit.* in Note 39. The objection just noted in the text is independent of whether the last one, that of the selective media being lethal for non-mutants, is valid. Indeed, phenotypic lag had already been invoked by Newcombe, H.B., *op. cit.* in Note 34, even before Lea and Coulson had calculated the Luria-Delbrück distribution.
66. Cavalli-Sforza, L.L. and J. Lederberg, 1956 *op. cit.* in Note 34.
67. Lederberg, J. and E.M. Ledergerg, *op. cit.* in Note 34; Newcombe, H.B., *op. cit.* in Note 34.

Both experiments rely on visual recognition and make no quantitative arguments whatsoever. What they do show, however, is that *some* mutations were spontaneous (random).

68. The discussion in this section is not intended to be complete. For details, see the paper by Foster in this volume and Sarkar, S., *op. cit.* in Note 17.
69. Shapiro, J. Observations on the formation of clones containing *araB-lacZ* cistron fusions. *Molecular and General Genetics* 194: 79–90, 1984.
70. Cairns, J. *et al., op. cit.* in Note 2.
71. See Note 48.
72. Hall, B. Adaptive evolution that requires multiple spontaneous mutations I. *Genetics* 120: 887–897, 1988.
73. Lenski, R.E., *op. cit.* in Note 16, emphasizes the importance of this point.
74. Levin, B., D, Gordon, and F. Stewart. Is natural selection the composer as well as the editor of genetic variation? Forthcoming, 1990; Cairns, J. *et al., op. cit.* in Note 2.
75. Mittler, J.E. and R.E. Lenski. New data on excisions of Mu from E. coli MCS2 cast doubt on directed mutation hypothesis. *Nature* 334: 173–175, 1990; Shapiro, J., *op. cit.* in Note 69; Cairns, J. *et al., op. cit.* in Note 2.
76. See the contribution by Foster (this volume, pp. 213–234) for further detail on this point and, also, discussion of other experimental results of Hall.
77. Mittler, J.E. and Lenski, R.E., *op. cit.* in Note 75.
78. Cairns, J. *et al., op. cit.* in Note 2. Note that these authors interpret the shift as a Poisson component *added* on to a Luria-Delbrück distribution, but this is an *interpretation* and, therfore, subject to legitimate questioning.
79. Stewart, F. *et al., op. cit.* in Note 39.
80. The second and the last of these cannot be shown using the analysis of Stewart, F. *et al., op. cit.* in Note 39.
81. The importance of this case has been particularly emphasized by Charlesworth, D., B. Charlesworth, and J.J. Bull. Origin of mutants disputed. *Nature* 336: 525, 1988, and Tessman, I. Origin of mutants disputed. *Nature* 336: 527, 1988.
82. Cairns has replied to this objection by observing that there did not appear to be any difference in fitness between mutant and non-mutant types in his experiments (Cairns, J. Origin of mutants disputed. *Nature* 336: 527–528, 1988.)
83. Stewart, F. *et al., op. cit.* in Note 39.
84. Lenski, R.E., M. Slatkin, and F.J. Ayala. Mutation and selection in bacterial populations: alternatives to the hypothesis of directed mutation. *Proceedings of the National Academy of Sciences (USA)* 86: 2775–2778.
85. Lenski, R.E., M. Slatkin, and F.J. Ayala, *op. cit.* in Note 84; Lenski, R.E., M. Slatkin, and F.J. Ayala. Another alternative to directed mutation. *Nature* 337: 123–124, 1989.
86. Levin, B., D, Gordon, and F. Stewart, *op. cit.* in Note 74.
87. An obvious exception is the hypothesis of differential fitness.
88. Cairns, J. *et al., op. cit.* in Note 2. These mechanisms will be discussed in detail in the text.
89. Stahl, F., *op. cit.* in Note 60; Davis, B.D. 1989. Transcriptional bias: a non-Lamarckian mechanism for substrate-induced mutations. *Proceedings of the National Academy of Sciences (USA)* 86: 5005–5009, 1989.
90. For example, the mechanism suggested by Davis, B. D., *op. cit.* in Note 89, can be ruled out (in part due to the experimental reports of Davis himself).
91. See the contribution by Foster (this volume, pp. 213–234) for further discussion of these

mechanisms and the current state of the experimental research.

92. The description of the mechanisms given here are simplified, and sometimes generalized, versions of those proposed by the original authors. Though they are intended to remain faithful to the spirit of those versions, they might disagree in detail.
93. Cairns, J. *et al., op. cit.* in Note 2.
94. Cairns, J. *et al., op. cit.* in Note 2, require the same organelle to contain the reverse transcriptase and monitor the performance of the RNA molecules. Thus the interaction from the feedback process to the organelle is automatically ensured. However, there is no conceptual reason to require the same organelle to do all this. Hence, the three requirements have been separated in the version presented in the text.
95. Stahl, F., *op. cit.* in Note 60.
96. Davis, B.D., *op. cit.* in Note 89.
97. It should perhaps be emphasized that probably none of the proponents of these mechanisms, especially Davis, B.D., *op. cit.* in Note 89, would choose to call the mechanisms "neo-Lamarckian." Moreover, all that is being suggested here is that they are only "neo-Lmarckian" in the sense that they satisfy the criteria of the explication of the term given in Section 2.
98. Cairns, J. *et al., op. cit.* in Note 2. It is, perhaps, worth emphasis that there is no evidence, as yet, in favor of these mechanisms
99. Davis, B.D., *op. cit.* in Note 89, has argued that such a mechanism, itself, would have adaptive value. This is quite likely and such a mechanism might well have arisen in a thoroughly neo-Darwinian fashion. However, the new mutations that such a mechanism would make possible would still show the characteristics of being neo-Lamarckian. Thus the basic question, whether there is any neo-Lamarckian mutagenesis, remains the same.
100. See Sarkar, S., *op. cit.* in Note 3, for a detailed examination of this issue.
101. For details, see Sarkar, S., *op. cit.* in Note 3, and Wimsatt, W.C. Reductive explanation: a functional account. *Boston Studies in the Philosophy of Science* 32: 671–710, 1976.
102. Cairns, J. *et al., op. cit.* in Note 2, p. 145.
103. Sarkar, S., *op. cit.* in Note 3, and Wimsatt, W. C. Reduction and reductionism. In *Current Research in the Philosophy of Science*. P.D. Asquith and H. Kyburg, editors. Philosophy of Science Association. pp. 352–377, 1978.
104. Sarkar, S., *op. cit.* in Notes 3 and 4; Kauffman, S.A., op. cit. in Note 4; Wimsatt, W.C., *op. cit.* in Note 101.
105. The classic accounts of "theory reduction" are Nagel, E. *The Structure of Science: Problems in the Logic of Scientific Explanation.* New York: Harcourt, Brace & World, 1961, and Schaffner, K. Approaches to reduction. *Philosophy of Science* 34: 137–147, 1967. For arguments that attempt to show that such classic accounts of "theory reduction" cannot capture the flavor of research in molecular biology, see Sarkar, S. *op. cit.* in Note 3, Wimsatt, W.C., *op. cit.* in Note 101, and, especially, Hull, D. Reduction in genetics—biology or philosophy? *Philosophy of Science* 39: 491–499, 1972.
106. Luria, S.E. and M. Delbrück, *op. cit.* in Note 30. The same point can be made about most of the research of the Phage Group which usually tried to infer informations about processes within the virus by looking at interactions of the virus, as a whole, with other entities. Though, the present article argues for reductionist research in the context of the problem being considered here, examples such as this clearly show the value of non-reductionist research in biology in other contexts. Of course it is almost trivial to assert the same value for non-reductionist research in most of evolutionary biology.

107. Cairns, J. *et al., op. cit.* in Note 2, p. 145.
108. Delbrück, M. A physicist looks at biology. *Transactions of the Connecticut Academy of Sciences* 38: 173–190, 1949; Stent, G.S. That was the molecular biology that was. *Science* 160: 390–395, 1968; Bohr, N. Light and life. *Nature* 131: 421–423, 457–459, 1933.
109. The important point is that the willingness to use such tests show that there was no prior commitment only to investigate biological problems in a reductionist fashion. It is not being suggested that Luria and Delbrück were deliberately attempting to keep their research strategy non-reductionist. In fact, Delbrck at least partly operated with an assumption that the only way in which his non-reductionist hopes might be realized is by pushing reductionist physical methods to the extreme and finding a paradox that defied such explanation. For details of this history, see Sarkar, S., *op. cit.* in Note 3, and Fischer, E.P. and C. Lipson, *op. cit.* in Note 29.
110. The only class of exceptions to the general reductionist explanations offered by molecular biology, at present, is that of functional explanations. For details of the problems that functional explanations pose for reductionist accounts, see Sarkar, S., *op. cit.* in Note 3, and Sarkar, S. Natural selection, hypercycles and the origin of life. In *PSA 1988*, Vol. 1. A. Fine and J. Leplin, editors. pp. 196–206, 1988.
111. 86. Wimsatt, W.C. Teleology and the logical structure of function statements. *Studies in the History and Philosophy of Science* 3: 1–80, 1972.

EDITORIAL COMMENTS TO SOBER BY SARKAR

It seems plausible to suppose that the Self and the individual arise together, at least insofar as the biological Self is concerned. After all any notion of the Self in biology must surely be more general than the one that philosophers conventionally assume, namely, that it is connected inherently with what humans call "consciousness." But, what is the individual? Even within biology different disciplines might well choose to construe it differently. The biochemist might choose to emphasize reasonably decoupled chemical cycles, that is, those cycles in which the interactions within are much stronger than those of the cycle as a whole. The molecular biologist might well choose other (generally simpler) mechanisms. Are they all individuals? From some philosophical points of view — those of Carnap and Quine, for example— there is no factual issue to be settled here: it is all a matter of convention, a matter for choice on pragmatic rather than "substantial" criteria. From other points of view, such as those of the "scientific realists," there is an ontological issue of great importance here and one that has to be settled by a careful consideration of the biology of the processes involved.

Sober's position, in this paper, falls within the latter tradition. His concern is the notion of the individual in evolutionary biology. The discipline chosen is, of course, singularly important—as has often been said, nothing in biology makes sense except in the light of evolution. First, Sober turns to the old debate about whether selection acts only on organisms or also on larger units such as groups. Unlike Maynard Smith and others, he views "group selection" to include "kin selection" where members of a group are closely related. Consequently, it becomes easy for him to show that group selection can occur and that it might even be of some importance to evolution. Of particular interest is his argument that explanations using group selection can well be more informative than explanations using only organismic selection. One important consequence of this argument is that it shows that the usual reductionist thesis that "lower-level" explanations are *always* more informative than "higher-level" ones cannot be maintained. Second, Sober addresses the relation between units of selection and the concept of the individual. In particular, he examines the notion that species are individuals, a position recently championed most vigorously by Hull and Ghiselin. He argues that even if species are units of selection they need not be individuals.

Sober's views are comforting to those who would hold that the Self and the individual emerge together. Were all units of selection individuals, groups of organisms or species would be individuals. If the Self emerges with the individual, such groups and species would have selves. Such a conclusion might well be endearing to radical holists, but would undoubtedly be very troubling to those less romantic. Sober's conclusions preclude such a quandary. But, what of the "individuals" of the theories and mechanisms of the biochemists and molecular biologists? Must these now be considered to have selves in order to maintain terminological uniformity? Perhaps, after all, there are different notions of "individual" and the confusion over the notion of Self will need to be resolved in some way independent of the philosophical considerations on individuality.

ELLIOTT SOBER

ORGANISMS, INDIVIDUALS, AND UNITS OF SELECTION

I. WHY ORGANISMS?

We often think of science—and of philosophy as well—as a process by which puzzlement is removed. This suggests that progress in a science is to be measured by the degree to which it eliminates problems rather than creating them. We sometimes say of an idea that "it raises more problems than it solves." The fact that this remark is used to state a criticism perhaps indicates that we think of problems as if they were rashes on the skin of the body scientific. Scientific progress makes rashes go away.

However much we may accept this image of science, we must not neglect a reciprocal aspect of scientific change. The creation of problems—the discovery that some fact needs to be explained, where earlier it was accepted as obvious and commonsensical—is fundamental to the scientific enterprise. In fact, it is arguable that major scientific breakthroughs often involve the discovery of new problems, as well as the formulation of solutions. A new way of posing a question is often the prelude to a new answer.

One of the most basic facts about our everyday understanding of the living world is that it is populated by organisms. Organisms are functionally integrated entities; they have parts of different sorts and these parts interact so as to sustain the organism and allow it to reproduce. The idea is so familiar that it is easy to miss the fact that it is really very puzzling. It is the nature of this puzzle and the biological framework for solving it that I want to discuss.

The idea that the complexity and adaptedness of organisms requires explanation is as old as the sun. Long before Darwin, philosophers and theologians recognized that this observation needs to be explained. The design argument for the existence of God was introduced to fill the void. Organisms exist and have the characteristics they do because God—an intelligent designer — made them that way. Darwin focused on the same phenomenon, but substituted a naturalistic for a theological explanation.

The question I am trying to isolate is *not* the one that asks: "Why are organisms *adapted*?" Rather, I want to focus on the query: "Why are *organisms* adapted?" What's the difference between these two queries? The first asks why organisms are adapted rather than not adapted. The second asks why it is

Alfred I. Tauber (ed.), Organism and the Origins of Self, 275–296.

organisms, rather than objects at some other level of organization, that are adapted. These questions are as different as the two that ask "Why are we having bagels *for dinner*?" and "Why are we having *bagels* for dinner?"

When I ask why it is organisms, rather than other biological objects, that are adapted, this question may sound like it has a false presupposition. Are not species well suited to their continued survival? Cannot the same be said for multi-species communities? And aren't genes, cells, and organs functionally integrated as well? Naively, it may seem that all of nature—nature at every level of organization—is well-adapted.

It cannot be overemphasized that evolutionary theory has rejected the idea that nature is adapted to the ends of survival and reproduction for objects at all levels of organization. R.A. Fisher [1] puts the point succinctly:

> It will be observed that the principle of Natural Selection ... refers only to the variation among individuals (or co-operative communities), and to the progressive modification of structure or function only in so far as variations in these are of advantage to the individual, in respect to his chance of death or reproduction. It thus affords a rational explanation of structures, reactions, and instincts which can be recognized as profitable to their individual possessors. It affords no corresponding explanation for any properties of animals or plants which, without being individually advantageous, are supposed to be of service to the species in which they belong.
>
> This distinction was unknown to the early speculations to which the perfection of adaptive contrivances naturally gave rise. For the interpretation that these were due to the particular intention of the Creator would be equally appropriate whether the profit of the individual or of the species were the objective in view. The phrases and arguments of this pre-Darwinian viewpoint have, however, long outlived the philosophy to which they belong. It would be easy to find among modern writers many parallels to the thought expressed in the following quotation "Of what advantage could it be to any species for the males to struggle for the females and for the females to struggle for the males?"
>
> This sort of question might appropriately be put to an opponent who claimed that the instincts of animals were in each case due to the direct contrivance of the Creator. As a means of progressive change, on the contrary, Natural Selection can only explain these instincts in so far as they are individually beneficial, and leaves entirely open the question as to whether in the aggregate they are a benefit or an injury to the species.

In the paragraph immediately following this passage, Fisher somewhat tempers this picture of natural selection. He says that species-level adaptation, which requires competition among species, is not impossible, but is probably rather unimportant. He then says that the evolution of sexual reproduction might possibly count as an exception to this constrained picture of how natural selection operates.

Fisher had a very definite picture of which level of organization is the level at which adaptations accumulate. Other evolutionists have disagreed with Fisher's conception. My point now is not that Fisher was right or wrong, but that he recognized that any reasonably precise formulation of the theory of natural selection will have to make decisions about whether it is organisms or species or communities, or some definite combination of the above, that are, as we now say, units of selection. As Fisher remarks, this is a simple, but fundamental, point of advance of evolutionary theory over pre-Darwinian thought about adaptation.

Why can't the theory of natural selection be all things to all levels? That is, why can't it view all adaptations as good for the organism, for the group, and for the community of cohabiting species? The reason is that the interests of these different objects can *conflict*. As Fisher says, a trait of an individual can be good for the species but deleterious for the individual possessing it. If the trait becomes common, then it will be a species-level adaptation, not an individual-level adaptation. If the trait is eliminated, the characteristic that supplants it will be an individual adaptation, but not an adaptation that exists for the good of the species. Because what is good for one level can be bad for another, biologists have had to construct their models of natural selection so that these models allow some kinds of adaptations to evolve, but not others.

Fisher was neither the first nor the last evolutionist to discuss the units of selection problem [2]. Why has the problem lingered so? If models of natural selection make different predictions about what sorts of adaptations will exist in nature, why wasn't the problem of the units of selection quickly settled by a few pertinent observations? The answer to this important question has several parts. Two facts about the power of observation bear mentioning here.

First, it often isn't clear from simple observation what the costs and benefits of a given characteristic are to the various levels of organization one wishes to consider. Fisher mentions the evolution of sexuality. It is true to this day that the evolution of sex is deeply puzzling. There are numerous arguments that show different ways in which sex can be individually advantageous. There also are numerous arguments that show how it can be individually deleterious, though good for the species. There are many models, each of them mathematically correct. It is hard to know which of those models actually bears on the evolution of real organisms. One somewhat daunting possibility is that sex evolved for different reasons in different lineages. If so, there will be no single answer to the question of why sex evolved.

A second reason that it has been hard to settle the question of the units of selection by straightforward observation is that characteristics often have benefits that are unconnected with the reason they evolved. In his book *Adaptation and Natural Selection*, G.C. Williams describes a simple example [3]. Consider a population of deer. These deer are pursued by predators, so it is to each deer's advantage to be able to run faster rather than slower. Let us consider this population at three stages. Initially, all the deer are slow. Then a mutant or migrant fast deer appears. Finally, after selection has run its course, all the deer in the population are fast.

Imagine that we come on this population at the end of the process. We see that all the deer are fast. We do not see the historical process that has produced this result. How are we to explain what we see?

One possibility is to say that there has been individual selection for being fast. This casts the organism as the unit of selection. Another possibility is that we might say that herds of fast deer are less likely to go extinct than herds of slow deer. In this case, we would say that the speed of the herd is a group adaptation.

The problem of the units of selection is an *historical* problem. It isn't enough to say that the trait we observe is *presently* good for the individuals or for group. The relevant question is why the trait evolved. This involves figuring out what the costs and benefits were *in the past*. The difficulty is that the present state of the population is not always a crystal-clear guide to what happened historically.

If selection occurred in the context of a single population, with fast deer outsurviving and outreproducing slow ones, then speed is an individual-level adaptation. If selection occurred among an ensemble of populations, with fast herds outsurviving and reproducing slow ones, then the trait might be a group-level adaptation. However, if each of these hypothetical processes would have reached the same end state, it will be impossible to infer which process actually occurred by observing the end state. The evolutionary process, in this case, will have been *information destroying* [4].

Williams uses the contrast between *group adaptation* and *fortuitous group benefit* to make this point. The problem of the units of selection is not settled by going to nature and seeing whether presently existing characteristics benefit the groups in which they occur. A present benefit may be either a group adaptation or a fortuitous group benefit. Indeed, the same ambiguity attaches to questions about other levels of organization. A trait that now benefits organisms may or may not have evolved by a process of organismic selection.

When we observe the group, we see that it is composed entirely of deer whose top running speed is a certain number of miles per hour. To figure out why this trait evolved, we need to know which other traits were present historically. But these alternatives are not something we observe in the present. Natural selection has destroyed the variation on which it operated. Unfortunately, this means that the process has destroyed some of the evidence that would help us figure out how the process proceeded. This general feature about natural selection has helped make the units of selection problem a difficult one to resolve.

The problem, however, has not been limited to the difficulty of mustering telling observations. There also has been a conceptual component — of knowing how to think about multi-level selection processes in an appropriate way. In what follows, I'll try to clarify some of the key concepts involved in this biological problem. This clarification will involve identifying some red herrings. The relative frequency of red herrings among the arguments that have grown up around this problem suggests that the units of selection problem requires conceptual clarification before it can be settled empirically.

II. ALTRUISM

Historically, the problem of the units of selection was centered on the question of whether organisms have characteristics that are good for the group though deleterious for the organisms possessing them. Biologists have used the term "altruism" to label such traits. The biological term has been stripped of the mental content that it possesses in ordinary language. To say that a plant behaves altruistically when it leeches a pesticide into the soil does not involve attributing a mind or motives to the plant. It merely means that the plant helps other plants at some cost to itself. Help and hurt are calculated in the currency of fitness—of effects on survival and reproduction [5].

An altruistic characteristic, thus defined, cannot evolve by natural selection, if selection occurs within the confines of a single population. If some individuals donate fitness benefits to others, whereas others receive those benefits but fail to reciprocate, the donors will be less fit than the free riders. As a result, the "selfish" characteristic of being a free rider will increase in frequency and altruism will decline. In the limit, altruism will be eliminated by the competition between organisms within the population.

On the other hand, if groups compete against other groups, then groups of altruists may do better than groups of selfish individuals. The success of altruistic groups over selfish ones may be sufficient to offset the fact that within each group, selfish individuals do better than altruists. The result may be that altruism evolves and is maintained in the ensemble of populations. I'll discuss in a moment how this is possible.

Altruism is not only interesting in its own right. It is, in addition, a useful characteristic to look for, if one wishes to know whether the adaptations found in nature exist for the good of the organism or for the good of the group. If adaptation is solely for the good of the organism, then there should be no altruism. On the other hand, if adaptations are sometimes for the good of the group, then one might hope to establish this by finding genuine altruism in nature. Contrary conceptions of adaptation make contrary predictions about the existence of altruism [6].

In the previous section, I mentioned that it is often difficult to obtain clear-cut observational evidence about the level at which adaptations exist. But there is, in addition, a prior conceptual difficulty that needs to be overcome. It is embodied in the following simple argument:

1) Altruists are less fit than selfish individuals.
2) In a selection process, less fit characteristics decline in frequency, and more fit characteristics increase.

Altruism declines in frequency under natural selection.

The first premise of this argument seems simply to report how biologists define the concept of altruism. The second premiss simply reflects a fact about how models of natural selection understand the relationship between fitness differences and evolution. The puzzle is this: why has evolutionary theory debated the problem of altruism for so many years, if it can be resolved by so simple an argument? This argument says that altruism cannot evolve; it appeals to no observational fact about specific populations or organisms. Unless the entire history of this problem has been a wild goose chase, there must be a defect in this little deduction.

The problem is with premiss (1). It does not accurately reflect how altruism should be defined. What is true is that altruists are less fit than selfish individuals *who live within the same group*. But it does not follow from this that altruists will be less fit than selfish individuals *in the ensemble of groups*. To understand the problem of altruism, it is important to see why this claim about the ensemble does not follow from what is true for each group in the ensemble.

This may sound absurd and paradoxical. Indeed, *it is entirely natural to think that what is true within each part also must be true within the whole as well.* This "reductionistic" idea about how wholes and parts are related is deeply ingrained in commonsense; nonetheless, it is wrong. The idea of group selection is hard to grasp because it requires that one overcome this commonsense prejudice.

Fig. 1 displays some of the main ingredients in the concept of evolutionary altruism. The fitnesses of the two traits S (for selfishness) and A (for altruism) are each shown as a function of the composition of the group. Notice that no matter whether altruism is common or rare in a group, selfish individuals are on average fitter than altruists. The broken line represents the average fitness

($\overline{w}$) of individuals in the group. This may be thought of as a measure of the group's fitness; the group produces more offspring if the average fitness of the individuals in it is greater. Notice that groups of altruists are on average fitter than groups of selfish individuals. The numbers on the *y* axis are arbitrary fitness units; think of them as expected numbers of offspring if you like.

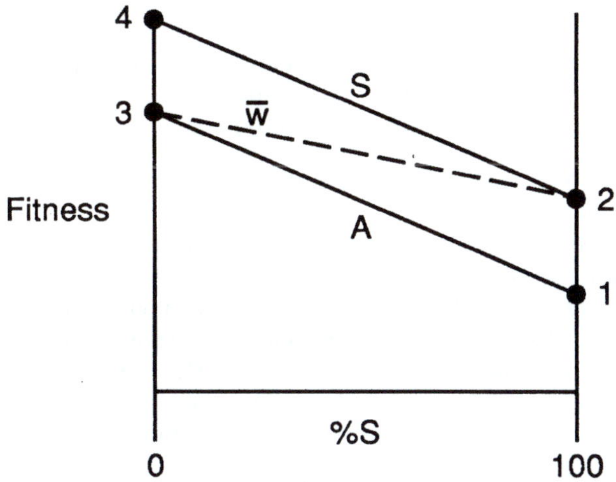

Fig. 1. Within any group, selfish (*S*) individuals are fitter on average than altruistic (*A*) individuals. However, groups of selfish individuals are less fit on average than groups of altruists.

To illustrate the idea about wholes and parts that I just mentioned, let's consider a simple example. Suppose we have two groups, each containing one hundred individuals each. Group 1 is 1% selfish; group 2 is 99% selfish. The approximate fitnesses of the traits, both within each group and averaged across the two groups, can be extracted from Fig. 1:

Group 1	Group 2	Global Average
(1) $w_S = 4$	(99) $w_S = 2$	$w_S = 2$
(99) $w_A = 3$	(1) $w_A = 1$	$w_A = 3$

Notice that what is true within each group is *not* true within the ensemble of groups. Selfishness is fitter than altruism within each group, but the reverse is true within the ensemble.

Statisticians have known about this phenomenon for a long time; it is called *Simpson's Paradox*. Without Simpson's paradox, altruism cannot evolve. A failure to appreciate the significance of Simpson's paradox makes the fallacious argument displayed before look utterly compelling. A simple syllogism makes a biological problem seem as if it can be settled by a definition.

Fitness is a quantity that allows one to compute next generation frequencies of a trait from frequencies in the present. Assuming clonal reproduction and non-overlapping generations, how will the frequencies of *S* and *A* change, both within each group and across the ensemble of groups? The numbers are as follows:

	Group #1	Group #2	Global Average
now:	S=1%	S=99%	S=50%
next generation:	S=1.3%	S=99.1%	S=27%

Selfishness increases in frequency within each group, but declines in frequency in the ensemble.

If common sense enshrines the reductionist idea that what is true within each part must also be true within the whole, even more does it sanction the idea that the *direction of change* in each part must automatically produce the same directional change in the whole. If Democrats are declining in frequency in each state of the Union, doesn't it follow that Democrats are declining in frequency in the United States as a whole? The answer is *no*; it is essential to see this, if one is to be able to grasp even the possibility of the evolution of altruism.

Notice a few vagaries in the example I have discussed. I began the system at 50% altruism, and with like for the most part living with like. Also, I traced the evolution of this system for one generation only. This is, so speak, to look at the system in mid-stream. How did altruism manage to reach the 50% mark? Why is it that similar organisms are living together? And what will happen if the system is allowed to evolve for many more generations?

Let us take the last question first. If the fitness units displayed in Fig. 1 are expected numbers of offspring, then both groups are growing in size. If the two groups hold together indefinitely into the future, each will grow fatter and fatter. Within each group, selfishness is increasing in frequency. In the limit, each group will approach 100% selfishness. In this case, we may safely conclude that the same fate awaits the ensemble of groups. Altruism will be eliminated from the ensemble, if the groups hold together.

To prevent this from happening, it is essential to have the groups reproduce. They must produce offspring colonies. If the groups fragment often enough, and if fitter groups found more offspring colonies than less fit groups, then altruism can be stably maintained in the ensemble.

Alternatively, we could somewhat change our interpretation of the organismic fitness values, and allow groups to go extinct. If groups with many selfish individuals go extinct more often than groups that contain lots of altruists, this may provide a group-level process that helps counteract the individual selection occurring within each group that works against altruism.

Another question mentioned was why similar organisms are living together in the little example I presented. A simple biological device to ensure this is to have relatives live together. Sometimes kin selection is equated with individual selection. My inclination is to see kin selection as a kind of group selection, in which the groups are made of relatives.

The reason I find this apposite is that relatives can compete with each other just as much as nonrelatives. If sibs live together, selfish sibs will do better than altruistic sibs within the group. The fact that they happen to have the same parents does not affect this point. What determines an organism's fitness is its phenotype, not its pedigree [7]. When groups are composed of relatives, it still is true that the within-group process works against altruism. To counteract this, one needs a between-group selection process. This is why kin selection is appropriately construed as a form of group selection.

I hope it is clear from all this that the existence of altruism cannot be settled by *a priori* argument. Rather, what is relevant is the biological properties of the system under study. One needs to have an ensemble of groups that differ in their local frequencies of altruism and selfishness. The groups must go extinct and found colonies sufficiently often to counteract the within-group process that promotes selfishness. It is possible that some natural systems should obey these requirements whereas others may not. The evolution of altruism is ultimately a question about population structure.

III. THREE RED HERRINGS

Having said this, I want to identify three red herrings that have distracted biological discussion from the appropriate empirical questions. Williams [3] argued that group selection hypotheses are "unparsimonious" and so should usually be rejected in favor of lower-level selection hypotheses. Dawkins [2] repeats this argument.

Maybe it is true that the kind of two-tiered process I have just described is "more complicated" than straightforward Darwinian selection within the confines of a single population. This may be right, but I can't see that it matters. The question is which sort of model of selection is more realistic and plausible; parsimony has nothing much to do with this [8].

Perhaps there is some sense to the idea that the requirements for the evolution of altruism are so constraining that they will rarely be satisfied in nature. If this is right, then a population drawn "at random" from nature will probably not contain altruistic characteristics [9]. This argument, it is important to see, rests on a biological premiss, not simply on an *a priori* principle of methodology.

A second red herring is the idea that genes, not groups, must be the units of selection because genes are the units of heredity. This simply doesn't follow, if "unit of selection" means the level at which adaptations accumulate. In the previous discussion, I talked about *A* and *S* as phenotypes of organisms. The discussion would have been no different if I had said that *A* and *S* are genes in a haploid organism. If the *A* gene evolves, this must be because there is sufficient between-group selection to balance the (within group) individual selection that works against it. Defenders of group selection hypotheses do not doubt the correctness of Mendelism.

A third red herring is provided by the following argument: The gene, not the organism or the group, must be the unit of selection, because all selection processes can be represented in terms of allelic fitnesses and frequencies. Again, there is a sense of "represent" according to which you can provide genic descriptions of selection at any level. If gene frequencies change owing to natural selection *at any level whatever*, there will be variation in the average fitness of the genes. However, this has nothing to do with the question of what sort of selection process is occurring.

A variant of this variant, defended by Sterelny and Kitcher [10], argues that it is a matter of convention whether we say that the gene or the group or the organism is the unit of selection, again on the grounds that a genic representation is always available (though the same cannot be said for higher-level representations). I grant that it is a matter of convention how we choose to describe the biology, in the sense that various mathematical formalisms may each have their utility. But what is not conventional is the biological process itself [11]. Whether altruism evolves is not a matter of convention (unless everything is a matter of convention!).

I do not reject the utility of thinking about evolution from the point of view of genes. A genetic model is obviously very important to understanding an evolutionary process. When biologists talk about phenotypes, and model an evolutionary process accordingly, they are basically assuming that the phenotypes are heritable—roughly, that phenotypes and combinations of genes are correlated. This is what I did before when I initially treated *S* and *A* as phenotypes, but then shifted to thinking of them as genes.

IV. HOW HIGHER- AND LOWER-LEVELS OF SELECTION ARE RELATED

In models of group selection that involve altruism, selection occurs *within* groups as well as *between* them. Different groups have different productivities; and organisms in the same group also have different productivities. The interaction of these two sorts of process determines the trajectory of the system as a whole.

Precisely the same thing can happen between and within organisms. Organisms can survive and reproduce to different degrees. And inside an organism, some genes may be more reproductively successful than others. Characteristics evolve because of the interaction of these two sorts of processes.

Before describing the within- and between-organism analog of group selection, I want to reformulate the description I gave in the previous section of how group selection works. I'll continue to talk about traits *S* and *A*, but now, for the sake of generality, I won't always mean selfishness and altruism by those letters.

Suppose that organisms live together in groups of four. This means that there are five possible group compositions, ranging from *SSSS* to *AAAA*. Assume further that a group's productivity is an increasing function of the number of *A* individuals it contains. The fitness functions of the five group types are displayed in the left-hand column of Fig. 2. This describes the between group part of the selection process [12].

We now may ask what happens within each group. We can compute this (assuming a haploid model) by describing the percentage of *A* offspring that are produced within groups of different compositions. Different assumptions about this are represented by the right-hand column of Fig. 2.

In the first row right column, the frequencies of *A* and *S* types in the offspring generation are precisely the same as the frequencies in the parental generation. This means that *S* and *A* are equally fit within each group. So the first row of Fig. 2 represents a case in which there is between group selection favoring *A* and no within group selection at all. The result is that *A* will go to fixation.

In the second and third rows (right column), we see represented the idea of within group selection against *A*. Note that the offspring frequency of *A* is always *less* than the parental frequency, if both *S* and *A* are present in the same group. This means that *S* individuals are producing more offspring than *A*

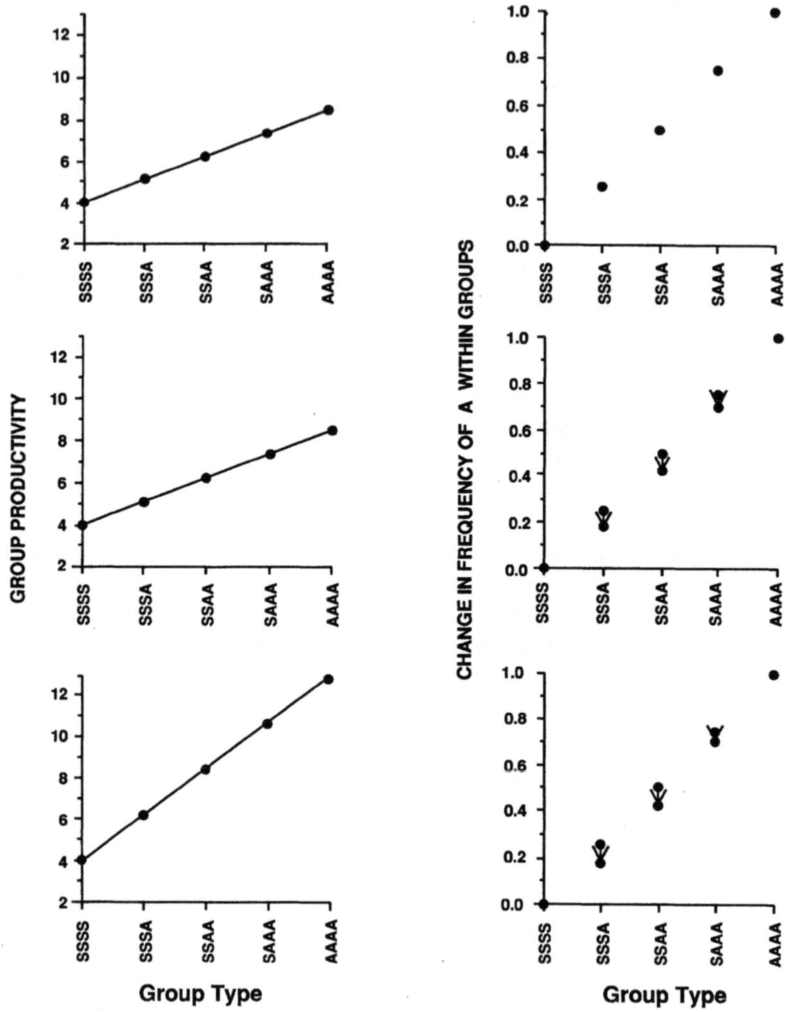

Fig. 2. The left-hand column shows how a group's productivity is influenced by the percentage of *A* individuals it contains. The right-hand column shows whether *A* individuals are less reproductively successful than *S* individuals within the same group. Line 1 depicts group selection favoring *A*, but no (within-group) individual selection, so the result is that *A* goes to fixation. Lines 2 and 3 describe a conflict between group selection favoring *A* and individual selection favoring *S*. The fate of the population is determined by the relative magnitudes of these two forces.

individuals in the same group. So in both the second and the third rows of Fig. 2, there is between group selection favoring *A* and within group selection against *A*.

What will be the result? Since the within group components are the same in these two cases, the result depends on the magnitude of the between group selection. Note that there is more between group selection in row three than there is in row two; the slope of the between-group line is greater in row three than it is in row two. In row two, the within group component of selection is sufficiently powerful to overwhelm the between group process; the result is that *S* goes to fixation. However, in row three, the result is a polymorphism, with both *S* and *A* stably maintained in the ensemble.

In Fig. 3, the comparison of between- and within-group selection is transposed to allow us to compare between- and within-organism selection. The left-hand column describes some standard representations of one locus, two allele models. In row one, *aa* individuals are less fit on average than *Aa* individuals, who in turn are less fit than *AA* individuals. In row two, left column, there is no variation in fitness between individuals. And in rows three and four (left column again), organisms with different genotypes differ in fitness.

The right-hand column of Fig. 3 represents what happens within organisms. We now ask what percentage of *A* gametes each of the three genotypes produces. In the top row (right column), we see the result of a "fair" Mendelian process. The heterozygote produces 50% *A* bearing gametes and the two homozygotes produce 0% and 100%, respectively. This cell describes a case in which there is no within organism selection. Here we are thinking about the genes at a locus (or the gametes or chromosomes in which those alleles occur) as parts of an organism, just as earlier we thought of organisms as parts of a group.

Rows two through four, right column of Fig. 3, describe the process of *meiotic drive*. In this case, the fair Mendelian meiotic ratio is subverted; heterozygotes produce fewer than 50% *A* gametes. A selection process is at work within such organisms; genes (or chromosomes or gametes) are competing with each other. This is precisely analogous to rows two and three (right column) in Fig. 2, in which heterogeneous groups are such that *A* organisms do less well than *S* organisms.

Let's now consider what the result is of the four processes described in Fig. 3. In row one, there is between organism selection but no within organism selection; the prediction of this standard one locus two allele model is that *A* will go to fixation. In row two, there is no selection between organisms, but

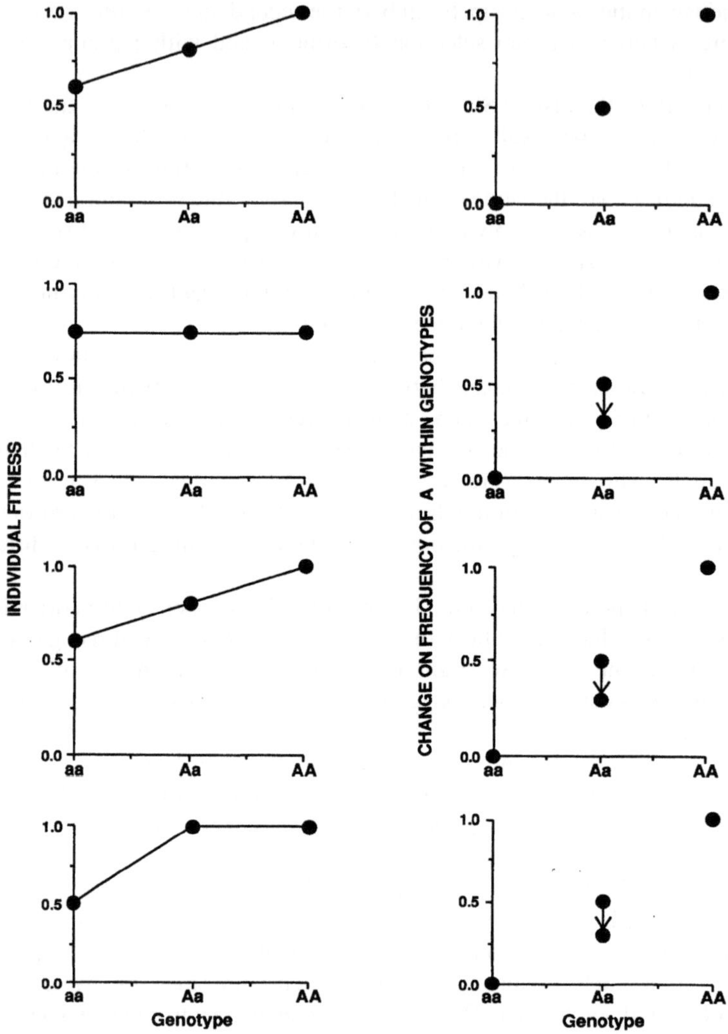

Fig. 3. The left-hand column shows how an individual's fitness is influenced by the percentage of *A* genes it contains. The right-hand column shows whether *A* genes are less reproductively successful than *a* genes within the same individual. Line 1 depicts individual selection favoring *A*, but no meiotic drive, so the result is that *A* goes to fixation. Lines 2 and 3 describe a conflict between individual selection favoring *A* and meiotic drive favoring *a*. The fate of the population is determined by the relative magnitudes of these two forces.

meiotic drive occurs; this form of within organism selection causes *a* to go to fixation. In rows three and four, there is both between organism selection (favoring *A*) and within organism selection (favoring *a*). The magnitude of the segregator distorter effect is the same in both cases; what distinguishes the two rows is the magnitude of the between organism effect. In row three, *a* goes to fixation; meiotic drive is "stronger" than the opposing between organism selection. In row four, a polymorphism evolves; the two opposing forces are more nearly in balance.

Some useful lessons flow from placing Figs 2 and 3 side by side. First, no absolute claim can be advanced about what will happen when a higher-level selection process is opposed by a selection process at a lower level. When between group and within group selection oppose each other (lines two and three of Fig. 2), the result can be a polymorphism or fixation of the trait favored by the lower-level process. When between organism and within organism selection oppose each other (lines three and four of Fig. 3), the same two options are available. It is an oversimplified reductionism that claims that the result of such processes is always determined by the lower-level component.

If we find ourselves too much at home in the Darwinian framework of adaptations that exist for the good of the organism possessing them, the parallelism between these two figures may provide a useful corrective. If it strikes us as natural and inevitable that group adaptations rarely if ever evolve, we also need to ask why organismic adaptations are so commonplace. For just as what is good for the group can be subverted by what is good for the organism, so what is good for the organism can be (but need not be) subverted by what is good for the gene. If organisms are functionally integrated, with parts working for the good of the whole, this is a fundamental fact about the living world that demands explanation. To see it as natural and inevitable is to miss the fact that this is a contingent truth that requires explanation.

A segregator distorter gene that has a strong distortion effect will go to fixation, even though its average effect on organisms is deleterious, if the effect on organisms is, so to speak, less severe. In this case, what evolves will be bad for organisms, but good for the truly selfish genes themselves. The more loci for which this is true, the less apt it will be to regard the organism as the preeminent level at which adaptations have accumulated. Rather, traits that happen to be good for the organisms possessing them merely constitute fortuitous organismic benefits; they will not be organismic adaptations.

We may put the point the other way around. What sort of circumstance is most conducive for the evolution of higher-level adaptations? "Higher level" is here used as a relative notion. We are asking simultaneously how to get organismic adaptations rather than genic adaptations, and how to get group adaptations rather than organismic adaptations. One thing that helps this happen is the suppression of selection at the lower level. If there is no variation in fitness at lower levels, then only higher-level biases can direct the process. In the within- and between-organism case, this means the suppression of meiotic drive. Organisms may evolve modifier genes that reduce segregation distortion. Or, alternatively, segregation distortion may be allowed to do its work, after which point new allelic variation is introduced that is not vulnerable to the driving genes.

The other way to allow higher-level adaptations to evolve is to increase the strength of the higher-level selection process. An altruistic characteristic is better able to evolve if it dramatically increases the fitness of the group in which it occurs, rather than just modestly augmenting it.

Allee and Emerson *et. al.* [13] and Wynne-Edwards [14] thought that group adaptations were to be found in long-lived groups. These groups were able to persist for so long precisely because the organisms in them are altruistic. More recent quantitative modelling has shown that long-lived groups are not at all the best places to look for group adaptations. Groups that hold together for a long time allow selfish individuals to displace altruists. What helps altruism to evolve is groups that are *short*-lived.

We often think of traits that extend the lifespan of an organism as organismic adaptations. This may be true, but the present point is that such traits can provide a less than optimal setting for other organismic adaptations to evolve. For a long-lived organism is a being in which selfish genes have more time to displace genes that benefit the organism at their own expense. Organisms are remarkable kinds of things. There is no *a priori* reason why so much of the earth's adaptations should be devoted to their needs. To the degree that the organism is the preeminent (if not the only) unit of selection, this fact is a very important one to explain.

V. INDIVIDUALS AND UNITS OF SELECTION

The idea of individuality has recently had a large impact on some questions in systematics. In this section, I want to describe how this concept is related to the units of selection problem. It is sometimes thought that the idea that *species* are individuals entails that species are units of selection. Conversely, Hull [15] has argued that a unit of selection, at whatever level it exists, must be an individual.

Ghiselin [16] and Hull [17] advanced the idea that species are not natural kinds, but are spatio-temporal individuals. This is a claim with two separate parts, and so two different arguments are needed to establish it.

A natural kind is definable by some property that all and only the members of that kind must possess. More specifically, that property must leave open where and when the instances of the kind can be found. So, for example, *gold* is a natural kind, defined by its atomic number. The lumps of gold that populate the universe may be scattered anywhere; and as far as the definition is concerned, they may exist at any time.

A biological species, on the other hand, does not possess a characteristic that is both necessary and sufficient for membership in it. This part of the individuality thesis continues Hull's [18] war against essentialism. In addition, membership in a species requires a physical connection with other organisms who also are in the species. If there were organisms in some distant galaxy that are physically just like tigers, they would not *be* tigers, according to Ghiselin and Hull, if they have no historical connection with terrestrial tigers. This is why species are not natural kinds.

A number of commentators have realized that it takes more to show that species *are* individuals than showing that they are *not* natural kinds [19]. Consider the purple objects that are within a hundred miles of Paris and whose entire existence falls between the years 1950 and 1990. Call any such object a PURPLE, for short. Each PURPLE is an individual, but there does not seem to exist an individual of which they (and only they) are the parts. In general, an arbitrary assemblage of individuals does not comprise an individual. Yet, the PURPLES bear relationships to each other; and the definition of the PURPLES imposes a spatio-temporal restriction. What, then, prevents this "whole" from being an individual?

The natural answer is that the parts of an individual must causally interact with each other in ways that they do not interact with things outside the individual [20]. Organisms are paradigm individuals; organ systems and cells causally interact with each other in ways that they do not interact with things outside the organism. This figure/ground contrast between "self" and "nonself" is important. It isn't enough to require that the parts of an organism causally interact with each other. That is too weak, because every physical object exerts a gravitational force on every other. This is true for all the PURPLES. The reason the PURPLES don't constitute an individual is that they are not doing anything with each other that they aren't also doing with things outside.

Typically, the parts that comprise an individual will be physically contiguous; again, familiar organisms provide our paradigm. But this is not strictly necessary. As Ghiselin has noted, the United States of America is an individual nation state, even though its constituent states are not physically contiguous. The key point is that the states interact with each other in ways that they don't interact with things outside.

The United States is a *political* entity. This tells us, roughly, what sorts of interactions we are to expect to occur within nations but not between them. Now let us consider *species*. To say that they are individuals requires more than simply that their parts be genealogically connected. After all, the entire tree of life is genealogically connected, but it doesn't constitute one big species. So what physical interactions carve out the species?

Hull and Ghiselin have opted for the Mayrian [21] idea that reproduction is the cohesive force that binds together the parts of a species. Organisms (or populations) in a species exchange genetic material to produce offspring. Organisms in the species do not do this with organisms that are outside [22].

An immediate consequence of this idea is that there are no asexual species. The organisms in an asexual species often have no more to do with each other than do organisms in different species that happen to belong to the same genus or family. They have a common ancestor, but they go their own ways. Asexual species and higher taxa might perhaps be called "historical entities" to distinguish them from natural kinds [19]; but they aren't individuals.

Another consequence of this proposal is that it appears that many so-called sexual species are not species at all. Species are often made of semi-isolated local populations. If migration between local populations binds the populations to each other genetically, then they will form good individuals. But if there is no gene flow, then these parts are not causally connected with each other in the requisite way [19]. Of course, it is an empirical question how much gene flow there is.

Ghiselin and Hull have accepted these consequences of their proposal. They are prepared to deny the existence of asexual species and to withhold the species appellation from sexual meta-populations in which there is no gene flow between subpopulations. I won't discuss the plausibility of these conclusions; rather, I want to identify the connection of the individuality thesis with the units of selection problem.

Species selection is an idea that was introduced in connection with the idea of punctuated equilibria. But here again we must be careful. The idea of punctuated equilibrium is a claim about when change occurs in the lifetime of a species; it says that most of the change occurs early on, with the species remaining relatively static thereafter. This claim says nothing about the causes of speciation, or about why some clades contain more species than others. It is to the latter idea that the hypothesis of species selection pertains. In brief, punctuated equilibrium is a claim about *pattern*, whereas species selection is a claim about *process*

The rough idea behind species selection is that there is a nonrandom process by which some species give birth to daughter species more often than others do. Species reproduce differentially, in a way that cannot be accounted for as the upshot of purely organism-level selection [23].

One question about the relationship of the individuality concept to the idea of species selection is easy to answer. A good Mayrian species will be an individual. But this does not require that any sort of species selection should have taken place. Mayrian species are perfectly at home in the framework of strict Darwinian organismic selection. The individuality of species does not entail that they have been units of selection.

The relationship in the other direction is a bit more subtle. Does the idea of species selection require that species be individuals? I think the answer to this question is *no* as well. Consider the case of a so-called sexual species whose local populations do not exchange genetic material with each other. This "species" is not an individual in the sense required by Ghiselin and Hull. However, there is no reason why a set of such "species" should not have different propensities to speciate. This would be species selection, but without the species being individuals.

It is important to recognize that the idea of individuality comes in degrees. Organisms may vary with respect to how mutually dependent their parts are on each other. Tigers are better individuals than a field of dandelions connected by underground runners. The parts of a tiger show a great deal more functional integration than do the parts of this dandelion network. Kill a part of the tiger and the rest of the tiger may die as well; this is not nearly so true for the dandelion.

Although organisms may differ in the degree to which they are "individuals," it is perfectly possible that different degrees of individuality should have evolved by straightforward Darwinian organismic selection. The high degree of individuality found in tigers may be the result of individual selection; the same may be true of the lower degree of individuality found in dandelions. Functional interdependence of parts may be advantageous for some organisms, but disadvantageous for others. Varying degrees of individuality may be varying responses to the pressures of individual selection.

Some groups are far more functionally integrated than others. And two groups may both display high degrees of functional integration, but for historically quite different reasons. A Mayrian species shows a great deal of interdependence between the two sexes. A eusocial colony of insects may also display a high degree of functional integration. However, the causal explanation may be quite different in the two cases. If sexuality has an explanation in terms of organismic selection, then the functional integration of sexual populations will be an artifact of lower-level selection, not a species-level adaptation. On the other hand, the functional integration of some haplo-diploid species may be due to competition between groups.

In short, individuality is a way of characterizing the present configuration of an object, whether that object happens to be an organism, a group, or a species. Individuality comes in degrees, depending on how much the parts of the object depend on each other. In contrast, the idea of a unit of selection describes the history of an object, not its present configuration. The individuality of an object may have different possible explanations; and two objects may have evolved different degrees of individuality as the result of selection at the same level. Selection at a given level of organization does not have to produce objects at that level that are highly individualistic. And an object at a given level may be highly individualistic even though it is not the result of selection at that level.

VI. CONCLUDING REMARKS

Organisms are marvelously well-adapted entities. Darwinian selection between organisms is an attractive, if not exclusive, mechanism for explaining why organisms are the way they are. But to simply say that a trait evolved because it was advantageous for the organism possessing it is to gloss over a complication that any adequate theory of adaptation must address.

Organisms are made of parts. If evolution largely suppressed competition between parts and produced organisms whose parts behave in a highly cooperative manner, why did this happen?

Organisms are themselves parts of larger wholes. If evolution has allowed organisms to compete with each other, instead of forcing them to cooperate in the interests of the wholes to which they belong, why did this happen?

Since the adaptive interests of objects at different levels of organization can come into conflict, an adequate adaptationism must say whether and why different levels of organization managed to accumulate characteristics that furthered their interests in surviving and reproducing.

Individuality—the tight integration and interdependence of the parts of a whole—is one of the major results of evolution. We find its preeminent examples at the level of organisms. This does not mean that all organisms are individuals to the same degree; nor does it mean that objects at other levels are always wholly lacking in individuality. A major project for evolutionary theory is to explain the variation in degree of individuality that we find within and between levels.

University of Wisconsin

REFERENCES

1. Fisher, R.A. [1930]: *The Genetical Theory of Natural Selection*. New York: Dover, 1958, pp. 49–50.
2. The dispute about the units of selection has waxed and waned throughout the history of evolutionary theory. Darwin and Wallace disagreed about it. During the forging of the Modern Synthesis, Fisher *op cit*., J.B.S. Haldane, (*The Causes of Evolution*, New York: Cornell University Press, 1932), and S. Wright, ("*Tempo and Mode in Evolution*: A Critical Review." *Ecology*, 1945, 26: 415–419) were each rather skeptical that altruism could evolve, whereas W. Allee, A. Emerson, L. Park, T. Park and K. Schmidt (*Principles of Animal Ecology*. Philadelphia: W.B. Saunders, 1949) and Wynne-Edwards (*Animal Dispersal in Relation to Social Behavior*. Edinburgh: Oliver and Boyd, 1962) took the opposite view. G.C. Williams (*Adaptation and Natural Selection*. Princeton: Princeton University Press, 1966) opposed group selection and Dawkins (*The Selfish Gene*. New York: Oxford University Press, 1975) popularized Williams' position and arguments. Maynard Smith ("Group Selection." *Quarterly Review of Biology* 1976, 51: 277–83) has lined up against group selection; David Wilson (*The Natural Selection of Populations and Communities*. Menlo Park: Benjamin/Cummings, 1980) and Michael Wade ("A Critical Review of the Models of Group Selection." *Quarterly Review of Biology* 1978, 53: 101–14), among others, have been less dismissive of the idea.
3. Williams, G.C. [1966]: *Adaptation and Natural Selection*. Princeton: Princeton University Press.
4. Sober, E. [1988a]: *Reconstructing the Past: Parsimony, Evolution, and Inference*. Cambridge: MIT Press.
5. See Sober ("What is Evolutionary Altruism?" B. Linsky and M. Matthen (eds.), *New Essays on Philosophy and Biology*, Canadian Journal of Philosophy Supplementary Volume, 1988, 14: 75–99) for further discussion of the contrasts and connections between the evolutionary concept of altruism and the ordinary language concept.
6. The fact that altruism is an important test case does not mean that individual and group selection must oppose each other. As I'll explain shortly, group and individual selection can work in the same direction. And it also is possible to have group selection without any individual selection at all.
7. I also should mention that some models of group selection allow altruism to evolve even when groups are randomly assembled. See Wilson ("Weak Altruism, Strong Group Selection," *Oikos* 59 (1990) 135–40) for discussion.
8. Sober, E. "Let's Razor Ockham's Razor." D. Knowles (ed.) *Explanation and Its Limits*. Royal Institute of Philosophy Conference. Cambridge University Press. 1990, 73–94.

9. Wade, *op. cit.*, reviews a number of quantitative models of group selection that claim to establish this sort of thesis; Wade argues that the models in various ways *a priori* bias the case against group selection.
10. Sterelny, K. and Kitcher, P. [1988]: "The Return of the Gene." *Journal of Philosophy* 85: 338–61.
11. Compare: (i) It is a matter of convention whether we use the words "dogs have four legs" to express a truth rather than a falsehood; (ii) it is a matter of convention whether dogs have four legs. I accept (i) but reject (ii). (i) embodies the idea of "trivial semantic conventionalism."
12. Figures 2 and 3, and the idea of using them to illustrate the parallelism of between/within group selection and between/within organism selection, are due to D. Wilson, "Weak Altruism, Strong Group Selection," *op. cit.*
13. Allee, W., Emerson, A., Park, L., Park, T., and Schmidt, K. [1949]: *Principles of Animal Ecology*. Philadelphia: W.B. Saunders.
14. Wynne-Edwards, V. [1962]: *Animal Dispersal in Relation to Social Behavior*. Edinburgh: Oliver and Boyd.
15. Hull, D. [1980]: "Individuality and Selection." *Annual Review of Ecology and Systematics* 11: 311–32.
16. Ghiselin, M. [1974]: "A Radical Solution to the Species Problem." *Systematic Zoology* 23: 536–44.
17. Hull, D. [1978]: "A Matter of Individuality." *Philosophy of Science* 45: 335–60.
18. Hull, D. [1965]: "The Effect of Essentialism on Taxonomy: Two Thousand Years of Stasis." *British Journal for the Philosophy of Science* 15: 314–26, 6: 1–18.
19. Guyot, K. [1987]: "Specious Individuals." *Philosophica* 37: 101–26; Ereshefsky, M. [1988a]: "Individuality and Macroevolutionary Theory." *PSA 1988*, vol. 1. East Lansing, Michigan: Philosophy of Science Association, pp. 216–22; Ereshefsky, M. [1988b]: *The Ontological Status of Species: A Study of Individuality and Its Role in Evolutionary Theory*. University of Wisconsin: Ph.D. Thesis; Wiley, E. [1981]: *Phylogenetics: the Theory and Practice of Phylogenetic Systematics*. New York: Wiley and Sons.
20. Ereshefsky, M. [forthcoming]: "Species, Higher Taxa, and the Units of Evolution." *Philosophy of Science*.
21. Mayr, E. [1963]: *Animal Species and Evolution*. Cambridge: Harvard University Press.
22. Of course, the appropriate formulation of this idea is not that (viable and fertile) interspecific hybrids never occur, but just that they are much rarer than viable and fertile offspring of within-species reproduction.
23. Sober, E. [1984]: *The Nature of Selection*. Cambridge, Mass.: MIT Press.

EDITOR'S COMMENTS TO WILLIAMSON

This paper accurately presents Don Williamson's lectures at Boston University and later given at the Woods Hole Marine Biological Laboratory. To experts who were present, Don's hypothesis was far-fetched, if not simply outrageous. This incredulous reception was well-expected, since the original theoretical paper published in *Progress in Oceanography* [19: 87–116, 1988] had been rejected by eight other journals. The editor's comments are of particular interest:

> It is sometimes stated that papers involving ideas are more difficult to get published than those that follow the conventional wisdom of the time. Iconoclasts are rarely popular with their peers. Here is a paper that I am including in Progress in Oceanography as something of an editorial indulgence, but which I am confident that in time the decision will be vindicated.
>
> ... Darwin would have probably had less trouble submitting a draft of the *Origin of Species* to the Bishop of Oxford. From the content of some of the reviewers' comments, I am convinced that prejudice has blinded them to the logic of the arguments presented by Dr. Williamson.
>
> ... In this paper Williamson takes a fresh look at what is known about several of the larval forms familiar to planktonologists. With a simple, but to some unacceptable, conceptual leap, he presents an explanation for the evolution of certain of the recent larval forms which seems both rational and feasible. Although conclusive proof will never be forthcoming, many of the elements of Williamson's theory are testable. Biochemical techniques can be used to demonstrate just how close some of the proposed relatonships really are. Similarly, the possibility that the fusion of disparate genetic material can occur and hence might occur under natural conditions, can be further explored. Already for many cell biologists the theory will not appear so very far-fetched, because the origins of organelles which are cytoplasmically interited, such as mitochondria and chloroplasts, are thought to lie in the fusion of two different taxa to form a single entity. The original association is thought to have come about through a symbiotic or parasitic relationship. There are many examples of associations which may be analogous to the intermediate steps.
>
> ... The fact that genetic engineering techniques are based on over-riding the [immune] responses, shows that even the more elaborate systems of higher Metazoa can be overcome ... There will still be those who believe that this paper "pollutes the scientific literature", but I remain unapologetic. The essence of science is the development and testing of ideas and theories, and this paper will certainly stimulate debate and experiment.

When Don Williamson attempted to publish the first studies where he believed that he had in fact observed cross fertilization that supported his chimera theory, the rejection comments from leading journals included those questioning his discriminatory and observational powers, aside from more objective scientific queries. The most reasonable issues may be summarized as:

1. A swimming ciliated blastula inside an ascidian chorion or hatching from it must be shown.
2. The rate of self-fertilization must be ascertained. (Ascidians, which are hermaphrodites, may be self-fertilized in the process of removing their eggs.)
3. Related to No. 2, the extent of hermaphroditism must be measured.
4. Chromosome analysis, including whether the hybrids have ascidian mitochondrial genes are also of importance.

Don Williamson has functioned as a solitary, 19th century invertebrate biologist. He has had no access to modern methodology, and as a retired Lecturer at the University of Liverpool, no resources to pursue the scientific problem with current techniques. His theory is based solely on observation, intuition, a broad grasp of phylogenetic relationships, and the tentalyzing first experiments that appear to support his argument. There is of course a poignancy in his pleas for colleagues to apply modern techniques to study the chimera theory. Irrespective that he has not as yet refuted his critics (and this in part explains the tone of his argument), this paper is included here as an exciting new hypothesis that deserves attention, experimentation, and thorough evaluation. The importance of considering it as a new concept of the biological Self, and its implications for evolutionary theory, are self-evident.

DONALD I. WILLIAMSON

SEQUENTIAL CHIMERAS

I. THE HYPOTHESIS

The name "chimera", or its ancient Greek equivalent, was originally applied to a fabulous, fire-spouting monster, with a lion's head, a she-goat's body and a serpent's tail. Most of our information on it comes from Homer, but I have not consulted the original. He seems to have been quite definite that the goat-portion was female, but I do not know whether he specified the sex of the head and tail portions. Whatever the answer to these intriguing problems, this seems to be the first use of the term, so I shall regard it as "definition 1". The same term, chimera, has been applied in more recent years to experimentally produced organisms made up of tissues of two species or of two genetically distinct strains of the same species, by techniques ranging from simple grafting to various forms of genetic engineering. This gives us "definition 2" of chimera. Modern experimental biologists have produced nothing like the creature described by Homer, but their chimeras, like his, simultaneously show identifiable features of different origin. I should now like my readers to consider the possibility, and indeed the strong probability, that organisms with identifiable features of different origin existed long before there were genetic engineers or ancient Greeks, for I maintain that such organisms have been produced naturally, at intervals, throughout much of evolutionary time and that a great many natural animals are really chimeras. (It is possible that some natural plants also are chimeras, but I shall restrict myself to the animal kingdom.) The chimeras I have in mind, however, differ from both the mythical monsters and the modern products of grafting and genetic engineering in that their features of different origin are not usually visible simultaneously but appear consecutively during the life-history of the animal. They are, in fact, "sequential chimeras", each of which starts its development as a member of one group of animals then metamorphoses to become a member of an entirely different group, frequently of a different phylum.

The concept of "self" is quite sufficiently complicated in animals like ourselves which do not metamorphose from one group to another during development, but it seems important in the context of the present discussion to consider forms which, apparently, do undergo such changes and which add another dimension to the problem of identity.

Before giving examples of what I consider to be sequential chimeras, it

Alfred I. Tauber (ed.), Organism and the Origins of Self, 299–336.

should be noted that, according to most dictionaries, chimera has another definition, one often regarded as the most common modern usage. The term is applied to "a mere wild fancy, an unfounded conception, an absurd creation of the imagination", and this is "definition 3". Several of my critics have made it clear that, in their opinion, my concept of a sequential chimera really falls within this definition. My first task is to try to convince my readers that it does not. This will, I fear, take up much of my contribution, but it is important to get others to share my conviction that sequential chimeras are real animals.

I.2. *Astropecten Auranciacus as a Sequential Chimera*

Echinoderms with larvae were some of the first animals which I came to regard as sequential chimeras [1], and as an example for present purposes I have chosen *Astropecten auranciacus* (Fig. 1), a fairly typical sea-star whose development is well described. The adult (Fig. 1a) is a five-armed star, clearly pentaradial, with an oral surface, on which it crawls, and an aboral surface, but there are no anterior and posterior ends, no left and right sides. Its larvae, on the other hand, assume bilateral symmetry soon after they leave the egg (Fig. 1b). They have anterior and posterior ends, and left and right sides, and continue to do so throughout the larval life of the species (Fig. 1c-f), but we should ask why? Why should *Astropecten* and thousands of other echinoderms, all of which are radially symmetrical as adults, spend the early part of their lives as bilaterally symmetrical larvae? Animals of no other group change from bilateral symmetry to radial symmetry during development, or from radial symmetry to bilateral symmetry for that matter (except that most eggs, blastulae and early gastrulae are radially symmetrical, whatever the group). Of course, if one regards *Astropecten* as a sequential chimera which changes phylum during development, then the change in type of symmetry merely means that it changes from a phylum of bilateral animals to one of radial animals. If this were the only evidence in favour of my theory, readers would be quite entitled to dismiss it as an example of "definition 3". There is a great deal of other evidence, but, before considering it, one may ask whether the alternative hypothesis is any more tenable.

I.3. *The Alternative Hypothesis*

The only other explanation of bilateral echinoderm larvae which has ever been put forward assumes that, at one time, the phylum was made up of animals which were bilaterally symmetrical both as larvae and as adults; the larvae

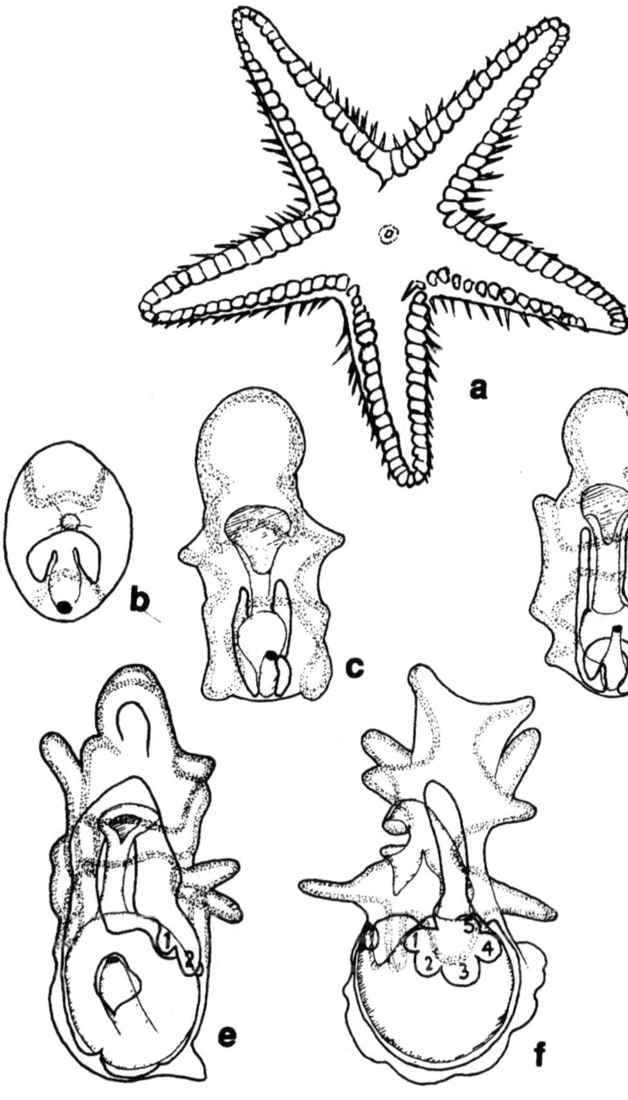

Fig. 1. The sea-star *Astropecten auranciacus* and its development. (a) adult; (b-d) larvae aged 3, 10 and 14 days, ventral views; (e, f) 70 day old larva in ventral view and from left side. 1–5, lobes of coelomic sac. (Redrawn after Hyman [2])

retained their bilateral symmetry while some of the adults evolved radial symmetry and the others became extinct. Now the fossil record of the echinoderms is probably at least as good as that of any other phylum, and quite a lot of early Cambrian forms are known. Some of these were triradial and some pentaradial, but none were bilateral. The triradial *Tribrachidium* occurs in Precambrian strata, but not enough is known about it to say for certain that it was an echinoderm, although several of the experts think it was. There were also some asymmetrical animals with calcareous exoskeletons in Cambrian times, but they should give no comfort to those who think that all echinoderms were once bilaterally symmetrical throughout life. It is very questionable whether these asymmetrical forms were echinoderms, and even more questionable whether they predated radial echinoderms, but even if we assume that radial echinoderms had asymmetrical ancestors, this is very different from assuming that they had bilateral ancestors: we are trying to explain bilaterally symmetrical larvae, not asymmetrical ones. I think it most likely that the earliest echinoderms had no larvae and were radially symmetrical throughout life, but if it should be shown that these radial forms had asymmetrical ancestors, it would not affect my theory. The first echinoderm larvae were, I suggest, acquired in the Carboniferous, and there is no dispute that, by that time, all adult echinoderms were radially symmetrical.

There is no fossil evidence whatsoever to support the theory that bilaterally symmetrical adult echinoderms ever existed. For the sake of argument, however, and in spite of the lack of evidence, let us assume that they did exist. The question then arises, why did they become extinct while the radial forms which are supposed to have evolved from them survived, multiplied and diversified? Generally speaking, bilateral symmetry is assumed to have advantages over radial symmetry, but it is often claimed that the early echinoderms went through an evolutionary phase when all their representatives were fixed forms and that radial symmetry has some advantages for fixed animals. Again the fossil record gives no support to this view. Some of the earliest known echinoderms were certainly a stalked or otherwise anchored, but some of their contemporary relatives were, equally certainly, motile animals. And even if some hypothetical, bilaterally symmetrical echinoderms anchored themselves to the substratum and gave rise to numerous generations of sessile offspring, why should this lead to radial symmetry? Adult bryozoans and brachiopods have been fixed forms at least since the Lower Cambrian and remain so today, but the animals of both these phyla are clearly bilaterally symmetrical and show no tendency to become radial.

Even those who refuse to consider the possibility that echinoderms with larvae might be sequential chimeras should realize that the only alternative hypothesis which has ever been proffered is fraught with difficulties. For about a century successive generations of students have been taught that echinoderms were once bilaterally symmetrical and that their larvae have remained so while the adults evolved radial symmetry, but it is a theory which, as we have seen, receives no support from paleontology, and, as we shall see, it receives even less support from the way in which echinoderms develop and metamorphose.

II. DEVELOPMENT AND METAMORPHOSIS

I would agree with most biologists who have written on the subject that the development of echinoderms and other phyla gives vital clues to their evolution and evolutionary relationships, but I disagree with them on the way in which these clues should be interpreted. Until recently it was generally assumed that all phases in the life-history of an animal must always have evolved together. Darwin [3] thought such a view was axiomatic and placed considerable emphasis on it. I certainly subscribed to it myself for many years and taught it to my students, but now I have changed my mind. This means that, from the point of view of the Darwinians, I am a heretic, but, as T.H. Huxley pointed out in relation to Darwin's theories, "every new truth begins as a heresy". Mine is a relatively small evolutionary heresy, for I agree with Darwin that organisms have evolved from other organisms and that natural selection has played an essential part in determining survival or extinction. I also agree with him that evolution is not restricted to adults; embryos and larvae too have evolved and are evolving. I have, however, come to disagree with his view that embryos and larvae must, in all cases, have the same genealogy as the adults which they give rise to.

Consideration of the development and metamorphosis of echinoderms was a major factor in making me change my mind, but delving into the literature on this phylum soon showed me that I was not the first to reject the conclusions reached by interpreting the clues from echinoderm development along traditional lines. H. Barraclough Fell, the eminent New Zealand specialist on the group, did this as long ago as 1948 [4], and developed his views in a number of later papers. He held that the evidence from larval morphology was quite incompatible with that from adult morphology, and he advocated disregarding the evidence from the larvae in deducing the way in which the phylum had originated and evolved. He insisted that the group was full of larval incongruities but he offered no convincing explanation of them. I agree with Fell that echinoderm larvae seem to have evolved quite independently of the adults over much of the history of the phylum, but while Fell was unable to explain this phenomenon, I think I can.

I postulate that the first echinoderm larva originated in another phylum, that its acquisition by an echinoderm did not occur until after all the existing classes were well established, and that it gradually spread from individual to individual, species to species, and group to group within the phylum, evolving as it spread.

The way in which the larval form was first acquired and later spread is, of course, of great importance. A method is suggested later, and some recent experimental evidence for it is also presented, but let us look first at the facts of echinoderm development and metamorphosis and see how they fit the hypothesis. The fertilized egg of *Astropecten*, like that of any other metazoan, divides repeatedly to first produce a more or less hollow ball of cells, the blastula. The hollow space inside is called the blastocoel. The blastula then develops an in-growing closed tube, the archenteron, whose opening to the exterior is termed the blastopore, and the growing animal is then known as a gastrula. In *Astropecten* and most echinoderms with larvae, the blastula and gastrula are ciliated, free-swimming larvae, but in some groups they are embryos, still enclosed within the egg membrane. The late gastrula of *Astropecten* is the first state to show signs of bilateral symmetry, for pouches form on either side of the archenteron and project into the blastocoel. These pouches soon form sacs which detach from the archenteron to provide the rudiments of the larval coelom, and a mouth develops as an invagination which links up with the inner end of the archenteron. The mouth marks the ventral side, and the blastopore, which has now become the anus, migrates ventrally (Fig. 1b). This type of mouth, which is an opening quite distinct from the blastopore, is a deuterostome, a term also applied to an animal with such a mouth, and this type of coelom, which buds off from the archenteron, is an enterocoel. All echinoderm larvae are enterocoelous deuterostomes, and so are the larvae of the hemichordates, which we shall consider in a moment. In the great majority of invertebrate larvae, however, the mouth does develop from the blastopore (or part of it), making them protostomes, and the coelom forms from splits between the mesenchyme cells which comes to fill the blastocoel, so the coelom is a schizocoel and the larvae are schizocoelous.

In the larva of a sea-star, the surface cilia become concentrated in two bands which develop considerable convolutions and projections as the larva grows (Fig. 1 b-f). This larval form is known as a bipinnaria. The coelomic sacs enlarge considerably, and the sac on the left side of the stomach eventually

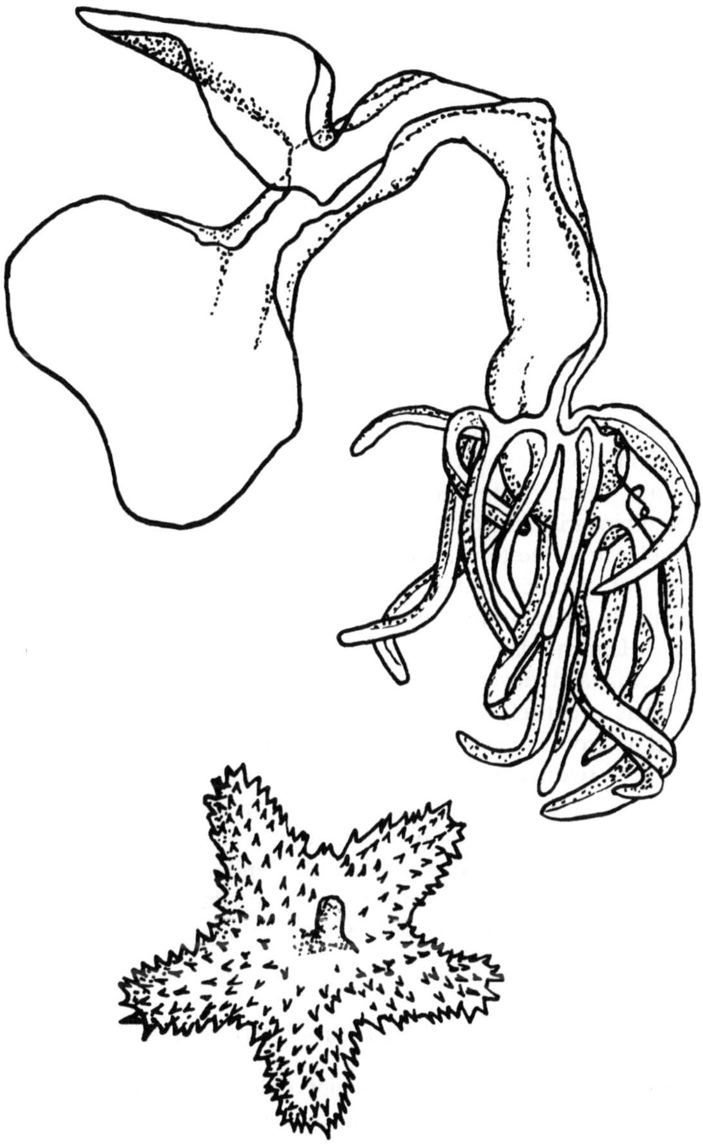

Fig. 2. Swimming bipinnaria larva and settled juvenile of the sea-star *Luidia sarsi*, just after their separation. (Redrawn after Tattersall and Sheppard [5]).

puts out five lobes (Fig. 1e,f). These become the coelom of the radially symmetrical, five-armed juvenile sea-star, which grows within the larva almost like a parasite. It eventually comes to protrude from the larva and then separates from it. The five arms of the juvenile are in no way derived from the arms and projections of the larva, and the independence of the two developmental phases is remarkable.

As *Astropecten* metamorphoses, the larval body degenerates and much of it is absorbed to provide nutriment for the developing juvenile. Most echinoderm larvae undergo similar degeneration at metamorphosis, but some do not. In the sea-star *Luidia sarsi* (Fig. 2) the juvenile detaches from an almost intact larva and crawls away, while the larva goes on swimming and regenerates any tissues which have been damaged by the departing juvenile. In one recorded case in which the metamorphosis took place in captivity, the larva went on swimming for a further three months after the quasiparasitic juvenile had left [5]. This emphasizes the independence of the two phases and the fact that the larva does not "develop into" the juvenile.

Now the bipinnaria larva of *Astropecten* and other sea-stars shows considerable resemblance to the tornaria larva of *Dolichoglossus* and other enteropneusts (Fig. 3c), so much so that tornarias were thought to be the larvae of an unknown family of sea-stars before they were found to belong to another phylum, the Hemichordata. The tornaria, like a bipinnaria, has two bands of propulsive cilia and develops as an enterocoelous deuterostome, but its metamorphosis is entirely different from that of a bipinnaria. The coelomic sacs budded off from the gut of the developing tornaria form a single anterior sac followed by two paired sacs, and, as the larva elongates and eventually settles, these become the unpaired coelom of the proboscis and the paired coeloms of the collar and trunk of the burrowing enteropneust, known as an acorn-worm. Here there is no quasiparasitic juvenile with a different orientation and different form of symmetry from that of the larva, for the tornaria really does "develop into" the juvenile worm by a simple process of differential growth (Fig. 3c-f).

III. ADULT HEMICHORDATES AND ECHINODERMS

The phylum Hemichordata comprises not only the enteropneusts, which develop from tornaria larvae into adult acorn-worms (Fig. 3a), but also the pterobranchs, which develop without a larval stage into forms with a horseshoe-shaped crown of tentacles (Fig. 3b), and some giant larvae whose adult form is unknown. Pterobranchs resemble acorn-worms in having a trimerous

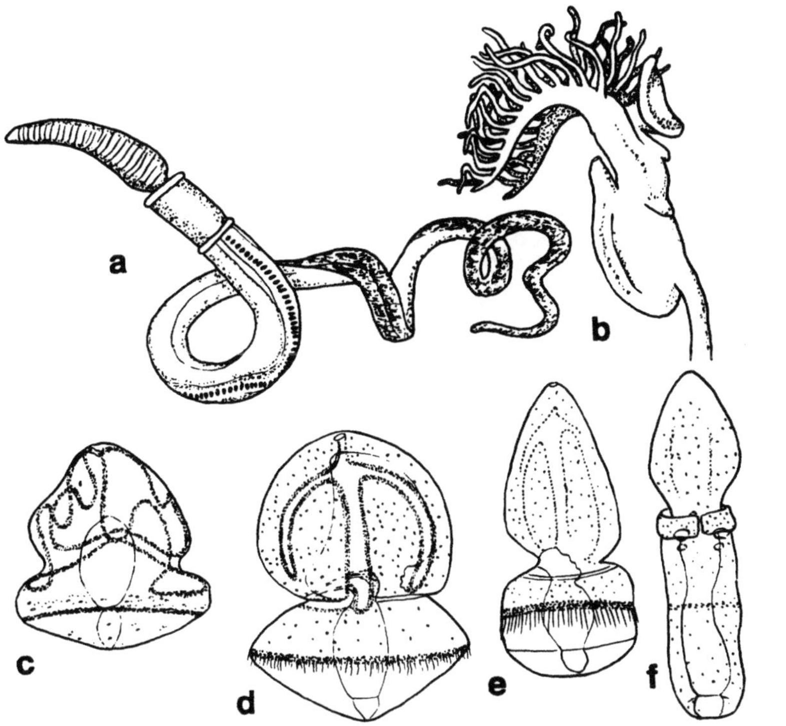

Fig. 3. Adult and larval hemichordates. (a) adult *Dolichoglossus* (Enteropneusta); (b) adult *Rhabdopleura* (Pterobranchia); (c-f) metamorphosis of an enteropneust: (c) tornaria larva; (d,e) metamorphosing larvae; (f) settled juvenile.

body made up of proboscis, collar and trunk, each with its own coelom, but the pterobranch proboscis is much smaller than that of an enteropneust, and the crown of tentacles is an outgrowth of the collar. Adult hemichordates have virtually nothing in common with adult echinoderms except that they are coelomate metazoans, and even the coeloms of the two groups are so differently arranged that this can hardly be claimed as a point of affinity. Some biologists have tried to draw an analogy between the triradial condition of some early echinoderms and the trimerous make-up of hemichordates. It is true that in each case the body may be said to consist of three parts, but surely there is a fundamental distinction between three almost identical parts arranged radially and three very dissimilar regions arranged longitudinally. The trimerous condition of enteropneusts has more in common with that of

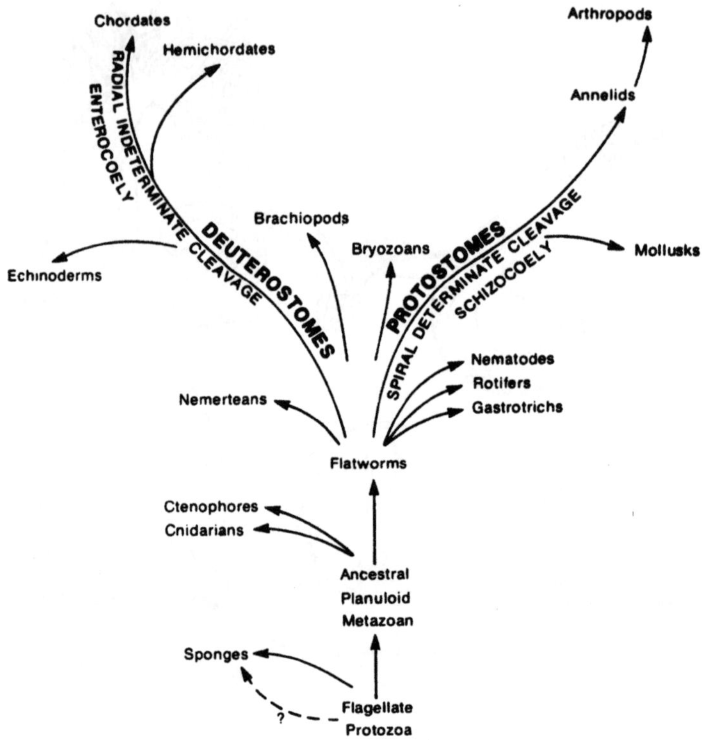

Fig. 4. A phylogenetic tree of the animal kingdom. (From Barnes [6]).

insects than that of echinoderms. It has also been suggested that the tentacles of pterobranchs are comparable with the tube-feet of echinoderms, but really they have much more in common with the lophophores of brachiopods and bryozoans. If it were not for the undisputed similarities between tornaria and bipinnaria larvae and the way they develop from the egg, no one would ever have thought of suggesting that the hemichordates and the echinoderms are offshoots of the same major branch of the phylogenetic tree of the animal kingdom (e.g., Fig. 4), yet such a grouping has been generally accepted since the beginning of the present century. I claim that it is time we chopped down this phylogenetic tree, for the clues from embryos and larvae have been just as badly misinterpreted in the other main branch. For the moment, however, let us continue to concentrate on the echinoderms and their affinities.

IV. ALL ECHINODERMS WITH LARVAE ARE SEQUENTIAL CHIMERAS

There is a considerable range of larval form within the echinoderms, but all the larvae of the phylum are enterocoelous deuterostomes whose evolution from a form resembling a tornaria is not difficult to envisage. What is required is a theory which reconciles not only the similar larvae and extremely dissimilar adults of echinoderms and enteropneusts, but also the totally different form of metamorphosis found in the two groups. The sequential chimera theory does just this.

In elaborating the theory that every echinoderm with a larva is really a sequential chimera, I shall avoid the tedious repetition of "it is postulated that" and ask my readers to take it as read. This is a hypothesis, and for that I make no apology, but it is a hypothesis with an ever growing body of supporting evidence and no known contradictions. Originally (it is postulated) the echinoderms had no larvae and were radially symmetrical throughout life. The first echinoderm to acquire a larva did so from an ancestor of the modern hemichordates, and the larva concerned was similar to a tornaria. I shall discuss the method later, but it must obviously have included the addition of the genetic recipe for a tornaria-like larva to that for an adult echinoderm. When the development of the larva was complete, the genes specifying the form of the juvenile and adult echinoderm would have taken over, but, although these genes had been responsible for the production of such juveniles and adults for countless previous generations, they now had a different starting point. They had to devise an entirely new method of metamorphosis, totally different from that which converted a tornaria into a juvenile hemichordate, but consideration of how development probably proceeded in echinoderms without larvae gives some clues as to how this new metamorphosis was devised.

All echinoderms are radially symmetrical until they reach the gastrula stage, and those which are going to produce larvae then develop lateral coelomic pouches and a ventral mouth. The early echinoderms did not develop these lateral and ventral features and they remained radially symmetrical after the gastrula stage. We can, therefore, regard the gastrula as the starting point for the production of the juvenile echinoderm in these forms. The configuration of the cells of this gastrula which was about to develop into a juvenile echinoderm is probably of great significance. They would form a sort of sac enclosing either a space (the blastocoel) or a mass of mesenchyme cells. Now consider the developing chimera, faced with the task of metamorphosing from a tornaria to an echinoderm. The echinoderm genes had previously produced the juvenile echinoderm from the cells of a gastrula. The only cells in the tornaria on which these genes could act were, I suggest, those with a similar sac-like configuration, those surrounding part of the coelom. They therefore acted on one branch of the coelom, making it grow into radial lobes, then adding other radial structures until the juvenile was complete. The descendents of this animal would develop in much the same way, passing through a tornaria-like stage, and the cells surrounding part of the coelom of this larva would then develop into an echinoderm. There are, of course, lots of questions unanswered, but this gives at least the basis of the explanation of how, today, a radial juvenile echinoderm can develop within a bilateral larva as a quasiparasite.

This first echinoderm to acquire a larva had now added a planktonic dispersal phase to its life-history, which would have given it an obvious advantage over its benthic relatives with no larvae in times of climatic change. Animals with larvae are likely to be much more widespread than those without, and so much less likely to wiped out by a relatively local catastrophe. It may be pointed out, however, that the postulated original echinoderms with no larvae, which were radially symmetrical throughout their lives, they too seem to be merely hypothetical. They occupy a key position in my theory, but is there any evidence that they ever existed? Are they mere wild fancies, unfounded conceptions, absurd creations of the imagination? There are many echinoderms today with no larvae but with relatively large eggs. The embryos within these eggs do not develop the shapes of planktonic larvae, but in most recorded cases development of the juvenile within the embryo follows much the same course as in a free larva. The gastrula does not develop directly into the juvenile. Instead it develops enterocoelic sacs, one of these becomes produced into five lobes, and these five lobes form the radial basis for the development of the juvenile echinoderm. It appears, then, that these forms without larvae have evolved from others with larvae, and they certainly give no support to my theory that once there were echinoderms which developed directly from the gastrula. I was originally inclined to believe that echinoderms with this latter type of development had all become extinct many million years ago, a theory difficult to prove or disprove, but I am indebted to an early paper by H.B. Fell showing that all such forms are not extinct. I have been very critical of Fell's uncritical invocation of convergent and clandestine evolution as explanations of all the incongruities of larval echinoderms, but I am most grateful to him for cataloguing these incongruities and for so many observations which, I claim, support my thesis. His valuable contribution to the present issue was the description of the development of a certain New Zealand brittle-star in 1941 [7].

Brittle-stars resemble sea-stars in usually having five arms, but their arms are slender, flexible, jointed and, as their name implies, brittle. One of the ways in which they differ from sea-stars is in having no intestine and no anus, although there is a mouth and a stomach. The larvae of the group are clearly different from those of sea-stars, but we shall be considering them later. For the moment, we are concerned with the development of a species with no larva. In 1916, Professor H.B. Kirk of Wellington, New Zealand, had published a preliminary account of the direct development of a brittle-star which he thought was *Ophionereis schayeri*, but there was some doubt about the identification. In 1941, H.B. Fell [7] described the development of the same species in much more detail but did not clear up the doubt about its identity. I shall follow Fell in calling its Kirk's brittle-star.

Those who are interested in the details of the development of Kirk's brittle-star should refer to Fell's very full account [7] or to my summary of it [1]. The radially symmetrical juvenile, however, does develop directly from the radially symmetrical gastrula. It shows that an echinoderm which develops as a radially symmetrical animal throughout its life is not just a figment of my imagination, a chimera of the third sort. Such animals exist. There are a few other species of brittle-stars which develop in the same way but their development has not been described in the same detail. Of course, the fact that such animals exist today does not prove that they existed in the remote past and accounted for all echinoderms until sometime in the Carboniferous, but it makes my argument more plausible. At present, we have evidence of how early Paleozoic echinoderms developed in only a few instances, but there is the hope that, in time, we might get such evidence for many more.

Some echinoderms which have direct development brood the eggs within the body of the female, and, in some cases, brood chambers, with or without developing young, have been identified in fossils. In Kirk's species, the eggs are laid on stones, but they do not hatch until the juvenile is quite well developed. Such eggs must carry a store of yolk for the developing embryo and they are consequently considerably bigger than those which hatch as planktonic blastulae: Kirk's species has eggs of diameter 0.5 mm, while eggs of echinoderms which develop as planktonic larvae are usually 0.1 or 0.2 mm in diameter. As pointed out by Jablonski and Lutz [8], there is a correlation between the diameter of the egg and that of the genital aperture of the female in extant echinoderms which seems to be equally applicable to fossils. The available evidence from brood chambers and genital apertures is quite consistent with my theory, but it is far too fragmentary to be conclusive.

The significance of Kirk's brittle-star to the present debate is not limited to the fact that it is radially symmetrical throughout its development. It is perhaps equally important that its coelom develops from splits in the mesenchyme and its mouth develops directly from the blastopore. It is a schizocoelous protostome, while all known echinoderm larvae are enterocoelous deuterostomes, as are those species of the phylum which have secondarily adopted direct development. There are, therefore, two types of early development within the echinoderms: the majority which have been investigated are enterocoelous deuterostomes, and a small minority are schizocoelous protostomes. The ancestral echinoderms must have been one or the other, or, less likely, a mixture of both, so, to arrive at the present position, either the majority or the minority (or both) must have changed. A process must have operated to change the form of early development of at least some echinoderms, and this reveals another phenomenon which is perfectly consistent with the sequential chimera theory but not with the assumption of Darwin and others that embryos and larvae must always have the same origins as the adults with which they are associated. I have already postulated that the earliest echinoderms developed directly, without a larval phase, and were radially symmetrical throughout life, and that later in the evolution of the phylum additional genetic material was acquired which introduced bilaterally symmetrical larvae into the life-history. I now expand this to suggest that the earliest echinoderms developed in much the same way as Kirk's brittle-star, as schizocoelous protostomes, and that the additional genetic material which prescribed bilateral larvae also prescribed enterocoely and deuterostomy. Whether or not one accepts my postulations, it seems inescapable that the form of early development has been changed in some echinoderms, and if a new form of development for both the coelom and the mouth can be introduced into a life-history, why not a new form of larval development?

Those who reject my postulation that Kirk's brittle-star has retained the ancestral form of echinoderm development will probably suggest that the ancestors of this species, by a series of mutations, lost all traces of larval development, then changed from enterocoelous deuterostomes to shizocoelous protostomes. If they are traditional neo-Darwinians they will insist that the changes must have come about by a series of mutations, because this is the only form of evolutionary change they recognize. They will then, however, find themselves in conflict with another tradition: that which holds that schizocoely and protostomy mark a fundamentally different form of development from enterocoely and deuterostomy. It seems undeniable that, occasionally, one of these forms of development has changed to the other, but there are no known examples of intermediates, and it is difficult to imagine how intermediates could function. Such a change seems to be an example of a saltation, a sudden jump involving a great many genes, rather than the gradual accumulation of a series of small mutations. From the evolutionary point of view, this is the essence of the debate: are saltations possible? I maintain that they are, at least as far as the forms of embryos and larvae, and methods of development, are concerned.

V. SEQUENTIAL CHIMERAS AT ALL TAXONOMIC LEVELS AND IN MANY PHYLA

So far I have put forward the sequential chimera theory to explain why radially symmetrical adult echinoderms have bilaterally symmetrical larvae, and why these larvae show marked affinities with the tornaria larva of hemichordates, in spite of the marked differences between adult echinoderms and adult hemichordates. The same theory can explain why, within the phylum Echinodermata, the affinities of the larvae are frequently in conflict with those of their corresponding adults. Let us start with the relationships between the classes.

There were 19 classes of echinoderms at the end of the Ordovician period [9]; today there are six, namely the sea-lilies (Crinomorpha) (Fig. 5a), sea-stars (Asteromorpha) (Fig. 5b), brittle-stars (Ophiuromorpha) (Fig. 5c), sea-urchins (Echinomorpha) (Fig. 5d), sea-cucumbers (Holothuromorpha) (Fig. 5e), and sea-daisies (Concenticyclomorpha). The sea-daisies (not illustrated) consist of a single species, *Xyloplax medusiformis*, discovered in 1986 [10]. *Xyloplax* shows a number of primitive features, and it develops directly, without a larval phase. It would be of great interest to know whether it is a schizocoelous protostome, like Kirk's brittle-star, but, so far, there is no detailed account of its development.

There is fairly good agreement among paleontologists as to the way in which the existing classes of echinoderms have evolved. The sea-lilies, which contain the only living stalked echinoderms, were established as a distinct line by the Lower Cambrian. Another group divided into two lines, one of which gave rise to the sea-stars and brittle-stars (and probably also the sea-daisies) while the other gave rise to the sea-urchins and sea-cucumbers. Recent work on ribosomal nucleotides of representative species has confirmed this general picture [11]. When we look at the larvae, however, we are faced with quite another picture. Many sea-lilies have no larvae, but when a larva does occur it

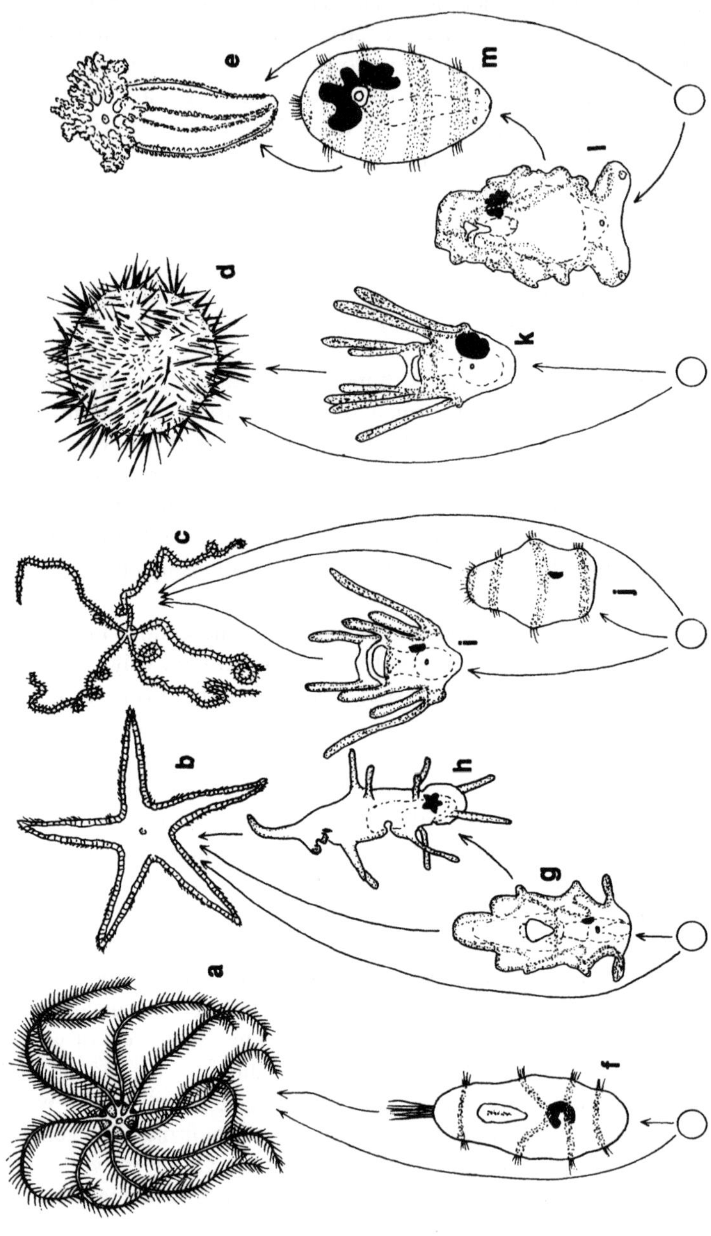

Fig. 5. Adult and larval echinoderms and types of development. (a-e) adults: (a) sea-lily; (b) sea-star; (c) brittle-star; (d) sea-urchin; (e) sea-cucumber. (f-m) larvae (developing juveniles within the larvae are shown black): (f) doliolaria; (g) bipinnaria; (h) brachiolaria; (i) pluteus; (j) doliolaria; (k) pluteus; (l) auricularia; (m) doliolaria. The bottom row represents eggs. Drawn to different scales.

is a barrel-shaped doliolaria (Fig. 5f). This form of larva also occurs as a second larva in the development of sea-cucumbers (Fig. 5m) and as the first larva in the development of a few brittle-stars (Fig. 5j). The first larva of the sea-cucumbers (the form which develops from the gastrula) is an auricularia (Fig. 51), which is not unlike the bipinnaria of a sea-star (Fig. 1b-f, 5g). In some sea-stars, the bipinnaria changes to a brachiolaria (Fig. 5h) which attaches itself to the substratum before the juvenile emerges, but others, including those we considered earlier, have no brachiolaria stage. Another form of larva occurs in all sea-urchins and most brittle-stars which have a larval phase. This is the pluteus (Fig. 5i,k), and it is of generally similar form in the two groups. One may call the sea-urchin larva an echinopluteus and the brittle-star larva an ophiopluteus, but in both the ciliated bands are drawn out into long arms supported by calcareous rods. Echinoderm larvae thus seem to fall into three morphological groups: (i) the bipinnaria of sea-stars and the auricularia of sea-cucumbers, (ii) the pluteus larvae of sea-urchins and brittle-stars, and (iii) doliolaria larvae, which can occur in sea-lilies, sea-cucumbers and brittle-stars. This grouping of the larvae has nothing in common with the evolutionary grouping of the adults.

Traditionalists can explain the unexpected similarity of sea-star and sea-cucumber larvae by postulating that the bipinnaria and the auricularia have evolved little from the original form of echinoderm larva. One of the most important contributions of Willi Hennig to taxonomy was his insistence that the occurrence of a number of primitive characters (which he mistakenly called "plesiomorphic") does not necessarily indicate close relationship. If, however, this is the primitive form of echinoderm larva, the pluteus cannot be, so how is the similarity of echinopluteus and ophiopluteus larvae to be explained? The traditional answer to this anomaly, indeed the only answer that I know of, other than mine, is to invoke convergent evolution. (Convergent evolution, in general, implies that two or more distantly related organisms have evolved similar adaptations to similar environments, resulting in superficial similarity in appearance.) In the case of echinopluteus and ophiopluteus larvae, the postulated similar adaptations are to a planktonic environment, but, while there is no question that both types of larvae are planktonic, I sincerely question whether planktonic life results in convergence to any particular shape of body. No planktologist would have advanced this argument. Anyone who has spent an appreciable amount of time looking at plankton cannot fail to be impressed by the diversity of planktonic shapes. There is not one shape for planktonic animals, or even a few, but thousands, and, of these numerous forms, the only ones to have ciliated arms supported by calcareous rods are the pluteus larvae of sea-urchins and brittle-stars. The larvae of sea-stars, sea-cucumbers and sea-lilies are also planktonic, but they are not plutei. Also, within echinopluteus and ophiopluteus larvae there is considerable variation in the length, shape and number of arms: a situation much more consistent with divergence from an ancestral form rather than convergence to a particular form. And neither convergent evolution nor any other form of neo-Darwinian evolution seems capable of explaining why a minority of brittle-stars do not have pluteus larvae but have doliolarias, rather like those of sea-lilies.

I do not reject neo-Darwinian evolution as an explanation of a great deal of phylogeny. Genes do mutate and recombine to produce variations in form, and natural selection does operate on the resulting organisms, but such processes do not explain the distribution of larval forms in the different classes of echinoderms any better than they explain how echinoderm adults and larvae have different forms of symmetry, or how the echinoderms evolved such a bizarre and wasteful way of metamorphosing from bilateral to radial symmetry during development. If however, one regards all echinoderms with larvae as sequential chimeras, all these paradoxes find ready explanation. Everything is consistent with the suggestion that no echinoderms had larvae until all the existing classes of adults had evolved and were well established. The genetic recipe for a bilaterally symmetrical, planktonic larva was then acquired from a hemichordate by a sea-cucumber, and it spread from there not only to other members of the same species but also to several other members of the class and, eventually, to a sea-star and to a sea-lily. There was limited spread of the larval form within each of these classes and there was evolution of larval form by the accumulation of mutations. Within the sea-cucumbers, developmental acceleration (adultation) of the genes originally responsible for radial symmetry in the adults produced barrel-shaped late larvae with a considerable degree of radial symmetry, while larval genes for bilateral symmetry continued to operate after metamorphosis to give the adults some degree of bilateral symmetry. Within the sea-stars the larval arms tended to lengthen, and in one evolutionary branch the adult genes prescribing organs of attachment became operative in the late larvae to produce the first brachiolaria. Another example of heterochrony occurred in the sea-lilies, when, as in the sea-cucumbers, the genes responsible for radial symmetry in the adults became operative in the larval phase to produce barrel-shaped larvae. After many millions of years, a sea-urchin acquired a planktonic larva from a sea-star. It was a sea-urchin larva which first evolved an internal skeleton to produce a pluteus, and similar skeletons, with evolved variations, have been inherited by all existing sea-urchin larvae. The brittle-stars were the last group to acquire larvae, and they might not have started to acquire them until the Tertiary era. One of them acquired pluteus larvae from a sea-urchin, and from there this larval form spread to many other members of the class; another acquired doliolaria larvae from a sea-lily, and this larval form spread to a few other brittle-stars; and a few brittle-stars, like Kirk's species, have not yet acquired larvae. The acquisition of larvae by echinoderms did not begin until about 200 million years after the existing classes had become established, and the spread of larvae from class to class was quite independent of the evolutionary relationships of these classes.

So far, I have suggested that the sequential chimera theory offers a tenable explanation of why radially symmetrical echinoderms develop from bilaterally symmetrical larvae and why they develop as qauasiparasites within the larvae, and also why there seem to be two quite different groupings of the classes of echinoderms, one based on the adults and the other on the larvae. I should now like to draw your attention to a few species of sea-urchins and brittle-stars which suggest that the acquisition of new larval forms and the modification of old ones is an on-going process.

Adult sea-urchins of different species of the genus *Lytechinus* show only minor differences, and larvae have been described for four of these species. The pluteus larvae of *L. variegatus*, *L. amnesus* and *L. panamensis* are all quite similar, not only in general appearance but also in the form of the skeletal rods and their configuration [12] (see Fig. 6a for *L. variegatus*). *L. verruculatus*, however, has a quite different form of pluteus larva [13] (Fig. 6b). Not only does it develop very much more slowly, but the skeletal rods are strikingly different. The main rods of *L. verruculatus* are smooth and fenestrated, as opposed to spiny and non-fenestrated in the other three species, and all the rods are joined to form a basket-like structure, a characteristic feature of the larvae of several other families of sea-urchins but otherwise unknown in the family Toxopneustidae, to which *Lytechinus* belongs. There are a few other comparable examples of incongruous echinopluteus larvae (listed in ref. 1), but, in general, the larvae of sea-urchins can be classified into genera, families, orders and subclasses which match the corresponding groups of adults. In general, then, adult and larval sea-urchins seem to have evolved together, as classical theory tells us they should. It is, however, difficult to escape the conclusion that the larva of *L. verruculatus* has been profoundly modified, or an entirely new larva substituted, after *Lytechinus* became established as a separate genus. I attribute this to the comparatively recent input of sufficient genetic material to prescribe (or profoundly modify) a larval form, but this input has left the form of the adult virtually unaltered.

A rather similar case is known among the brittle-stars. *Ophiura albida* and

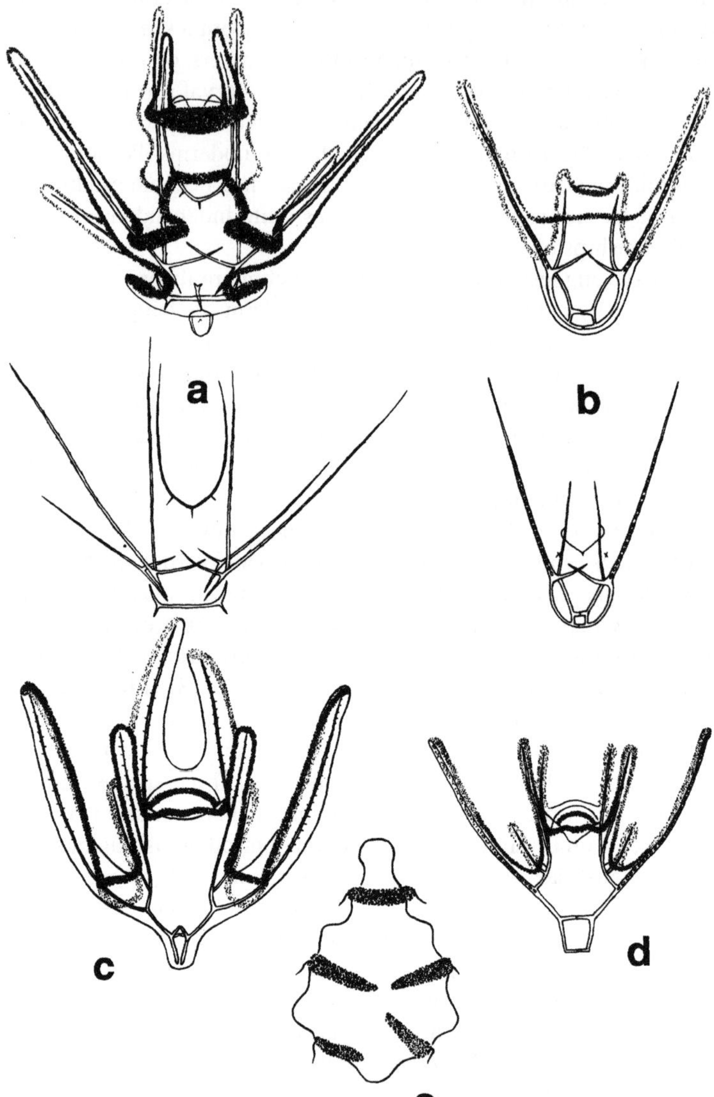

Fig. 6. Dissimilar larvae of similar adults. (a,b) class Echinomorpha, family Toxopneustidae: (a) pluteus of *Lytechinus variegatus* and its skeleton; (b) pluteus of *Lytechinus verruculatus* and its skeleton. (c-e) class Ophiuromorpha, family Ophiolepidae: (c) pluteus of *Ophiura albida*; (d) pluteus of *Ophiura texturata* ; (e) doliolaria of *Ophiolepis cincta*. (Redrawn after Mortensen [12, 13] and Fell [14]).

O. texturata are quite difficult to distinguish as adults, and there seems to be no doubt that they are correctly classified in the same genus. Their pluteus larvae, however, are very obviously different, and again the differences are not confined to general appearance but also affect the form and configuration of the skeletal rods (Fig. 6c,d) [13]. In fact, the pluteus of *O. albida* seems to provide the only known case of fenestrated rods in the larva of a brittle-star, although this condition is usual in sea-urchin larvae of several families.

There are certainly large and unexpected differences between the larvae of *Ophiura texturata* and *O. albida*, but an even more striking case of larval incongruity occurs in a closely related brittle-star. *Ophiolepis cincta* belongs to the same family as *Ophiura*, namely the Ophiolepidae. If its larva were unknown, one might speculate as to whether it would be more like that of *Ophiura texturata* or *O. albida*, but it is known and it is nothing like either. It is a doliolaria larva, with neither arms nor skeletal rods [14] (Fig. 6e). I have already suggested that the brittle-stars were the last group of echinoderms to acquire larvae, and some species have not yet done so. I now suggest that the Ophiolepidae have acquired their larvae quite recently, after the present species became differentiated, and that *Ophiura texturata*, *Ophiura albida* and *Ophiolepis cincta* each acquired its larva from a different source. *Ophiolepis cincta* seems to have acquired its larva from a sea-lily.

So far I have been pursuing the idea that all echinoderms with larvae are sequential chimeras and some of them have become so quite recently. I must point out, however, that apparent sequential chimeras are by no means confined to the echinoderms, for animals in many phyla have incongruous larvae which I regard as later additions to the life-histories in which they now occur. Many such cases are mentioned in an earlier paper [1] and are described in more detail in my forthcoming book [15]. Here I shall list them very briefly.

The larvae considered in the previous sections have all been ciliated, deuterostomatous, enterocoelous forms. Another form of larva is, however, much more widespread among the invertebrate phyla. These larvae, too, are ciliated, but the cilia are arranged differently. There is always a preoral ciliated band, the "prototroch", and there is usually an apical tuft of longer cilia springing from a sensory area. The mouth of these forms develops from part of the blastopore, so it is a protostome, and the coelom, when there is one, is formed from splits in the mesenchyme, so it is a schizocoel. The best known larva of this type is the trochophore (Fig. 7a), in which the prototroch is equatorial and not produced into lobes or lappets. Trochophores are found in at least some members of four different phyla, namely (i) the Annelida (Fig. 7f), which are coelomate, segmented worms, (ii) the Echiura (Fig. 7g) and (iii) the

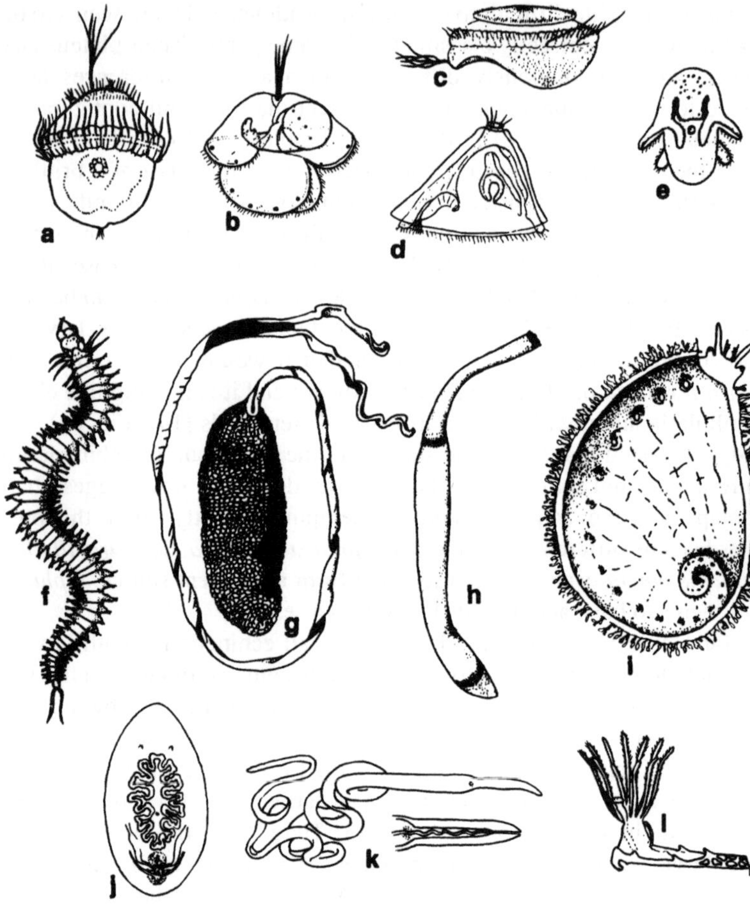

Fig. 7. Trochophores and related larvae and their corresponding adults. (a) trochophore larva: occurs in some Annelida Polychaeta (f), Echiura (g), Sipuncula (h) and Mollusca (i); (b) pilidium larva with developing juvenile: occurs in some Nemertea (k, with enlarged view of proboscis in rhynchocoel); (c,d) larvae of Bryozoa (l); (e) Müller's larva: occurs in Turbellaria Polycladida (j).

Sipuncula (Fig. 7h), two phyla of worm-like animals which are coelomate but not segmented, and (iv) the Mollusca (Fig. 7i), which includes slugs, snails and octopuses, none of which has a true coelum and most of which are unsegmented. Larvae with different names but built on the same general plan as trochophores occur in a number of other phyla. So we have the pilidium larva (Fig. 7b), in which the prototroch is produced into two large lappets and which occurs in the life-history of some nemertine worms (phylum Nemertea) (Fig. 7k). These are acoelomate, unsegmented worms, with a proboscis. Bryozoan larvae may be unshelled (Fig. 7c) or shelled (Fig. 7d), and the corresponding adults, known as moss animalcules, are coelomate, unsegmented, sedentary and usually colonial (Fig. 71). Müller's larva (Fig. 7e) has the prototroch produced into a number of lobes; it develops into a polyclad flatworm (phylum Platyhelminthes, class Turbellaria) (Fig. 7j), which is another type of acoelomate, unsegmented worm.

Of the larvae just mentioned, only Müller's larva "develops into" its corresponding adult by a process of differential growth, modifying but not discarding the larval tissues and organs. All the others discard at least some of the larval tissues and organs at metamorphosis, including all or virtually all of the nervous system. In many polychaetous annelids, nemertine worms and bryozoans, the independence of the juvenile and the larva is reminiscent of that in the echinoderms. All these larvae seem to have evolved from forms similar to the larvae of polyclad flatworms, and in this group they had their origin. The possession of similar larvae by the very diverse assemblage of adult animals which has been listed is not, however, an indication of close relationship or common ancestry, as has usually been supposed. Rather, I suggest, the polyclad larva and its derivatives spread among the various phyla concerned after they had become separate phyla, in much the same way as the tornaria and its derivatives spread through the various classes of echinoderms. Apart from the polyclad flatworms, all animals with trochophore-like larvae are also sequential chimeras. The polychaetous annelids, sipunculans and molluscs all had planktonic larvae before some of them acquired a trochophore stage, but the trochophore, when it was introduced into the life-history, took developmental precedence, and the original larva became the second larva in the ontogeny of the animals concerned.

In all trochophores and trochophore-like larvae, the coelom, when there is one, develops as a shizocoel and the mouth develops as a protostome, but I should now like to draw your attention to a mollusc without a larva which develops as an enterocoelous deuterostome. In all other molluscs, with or without larvae, which have been studied, the mouth develops as a protostome and there is no coelom, but the mesoderm is derived from the ectoderm. The exception to this rule is a freshwater snail called *Viviparus viviparus*, which belongs to the prosobranch gastropods. As an adult, it seems to be a fairly typical member of this group, but in its development the blastopore becomes the anus, making the animal a deuterostome, and the archenteron develops a ventral bulge which closes off to form a small enterocoel. This coelomic sac soon falls apart, but the endodermal cells which surrounded it are the progenitors of the adult mesodern [16]. Either the recent ancestors of *V. viviparus* have undergone a change in the method of early development or *Viviparus* has retained the original method of early development of all molluscs and the others have changed. In this case I favour the former alternative, implying a change from protostomy to deuterostomy, the same change which I believe has taken place in the majority of echinoderms. If, as is now widely held, the original molluscs were acoelomate, then there has also been a change from acoely to brief enterocoely. The important point for the moment, however, is that either a very small minority or the vast majority of molluscs must have undergone a fundamental change in their method of early development with little or no effect on adult form. And if a new form of embryo can be introduced without affecting the adult, why not a new form of larva?

VI. A HYPOTHETICAL MECHANISM

I hope I have now made the case that there are a great many animals whose form of embryonic or larval development, or both, seems to have been added or substituted after the adults had evolved their characteristic forms, and, in many cases, it is possible to suggest where the new forms of embryo or larva came from. Up to now, however, I have considered only the evidence that embryonic and larval forms have been transferred between distinct evolutionary lines without considering a method. Fortunately it is possible to suggest one. The mechanism in question must be capable of transferring the genetic recipe for an embryo and larva, and I admit that, initially, I spent a great deal of time trying to envisage a way in which this part of the total genome of an animal might be separated from the rest. It eventually occurred to me that if the whole genome of one animal were added to that of another, there is probably an existing mechanism which will select suitable recipes for development in the appropriate order to produce a viable life-history. The obvious method of adding the total genome of one animal to that of another is by cross-fertilization, and, as the great majority of the animals in question shed their eggs and sperm into the sea, contact between eggs and foreign sperm must be common. Certainly there are barriers to cross-fertilization, and natural fertilizations are nearly always between eggs and sperm of the same species. These barriers, however, can be overcome fairly easily in the laboratory, and it does not seem unlikely that they have occasionally been overcome in the sea.

In all animals there must be a mechanism which governs the order of development, so that, if a species has a larval phase, the larva will always develop before the juvenile and adult. Now if a sperm from an animal with a larval phase fertilized an egg of an animal with none, this mechanism governing the order of development would usually ensure that the larva (which happens to be paternal) would develop first. The fact that the heterozygote would initially have only a haploid set of chromosomes from each parent is probably of little consequence, for it is known that many larvae will develop parthenogenetically, from a haploid egg. They probably do not remain haploid, but I know of no account of how or at what stage they double their chromosome number. There may be some exceptions to the general rule that a larval form will always take developmental priority over juveniles and adults. I think it likely that the order of development is governed more by simplicity of body form than by whether or not this form previously belonged to a larva. In general, larvae have simpler body forms than juveniles, but there are some metazoans which mature as very simple animals. I think it very unlikely that a tadpole would take developmental priority over a rotifer, even though the former is a form of larva, the latter a form of adult.

Returning to our hypothetical hybrid which has hatched as a paternal-type larva, now consider what might happen at the end of the larval phase, when further development must involve metamorphosis. The cells of the larva would all have the genetic recipes to make an adult of the maternal species, an adult of the paternal species or a mixture of both. If the parents were distantly related it seems unlikely that a random mixture of their characters would be any more viable than Homer's chimera, but I suggest that the genes to make a maternal and a paternal adult would be unlikely to operate simultaneously. Just as there is a mechanism which, in most cases, gives larvae developmental priority over adults, there is likely to be a similar mechanism which gives some forms of adult body form developmental priority over others. If the maternal adult form took priority over the paternal, the F1 hybrid would develop first as a paternal-type larva then metamorphose into a maternal-type adult, and this new life-history would be heritable. The descendants of such hybrids would have the genetic recipes for the same life-cycle whether they came from crosses between hybrids or between a hybrid and either of the two original species. The parentage of the F2 hybrids would, of course, determine which chromosomes of the zygote were haploid and which diploid, but at present we can only speculate on whether and how this might affect the priority mechanism. If some adult body forms take developmental precedence over others, the paternal form would be expected to take precedence in some cases, and an experimental hybrid which appears to illustrate such a case will be described below. Such hybrids would have paternal body forms both as larvae and as adults, but again we may speculate on the life-histories of crosses between F1 animals and members of the two original species.

Where there is no clear precedence of one body form over the other, another factor might always favour the development of the maternal form. Although the cells of the late larva would have chromosomes derived from each parent, all the other components of each cell would be maternal in origin, and in the non-specialized cells they would have retained their maternal characteristics. There seems to be wide agreement among geneticists that the nuclear membrane and cytoplasm regulate the expression of genes, and this system of regulation is probably responsible for ensuring that larvae develop before adults. If, to put it anthropomorphically, this same mechanism had now to choose between allowing the expression of maternal or paternal genes specifying adult form, it might be expected, in view of its own maternal derivation, to respond to the maternal RNAs and suppress the paternal. Cells specialized to form tissues of the paternal-type larva would probably not behave in this way, but, in most cases which I regard as interphylar sequential chimeras, the origin of the adult can be traced to relatively few, unspecialized larval cells.

It may be assumed that the same recipe which specifies a paternal-type larva would include the specification for a suitable embryo, which may differ from that of the original maternal species. This seems to have happened when the echinoderms acquired hemichordate-like larvae. Also, if the paternal species were without a larva, the paternal type of embryo might still oust the maternal type in some circumstances. This seems to have happened in the case of the gastropod *Viviparus vivaparus*. It may also be assumed that some forms of larvae take a natural priority over others. Thus, when a trochophore stage was introduced into the life-histories of some polychaetous annelids, sipunculans and molluscs, it took developmental priority over the nectochaete, pelagosphaera and veliger larvae which already occurred in the life-histories of these respective groups. It may be suggested that simpler larvae will tend to take precedence over more complex forms and ciliated larvae over non-ciliated forms. In all known existing life-histories, a phase which is propelled by cilia always develops before one which is not.

I started by developing the theory that, during the course of evolutionary time, types of embryos and larvae have been transferred between distantly related animals to produce sequential chimeras, animals which start their development in one group then metamorphose to another. I have now added the theory that these transfers have resulted from heterosperm fertilizations.

Preliminary studies to test this second theory have been initiated, and I shall consider these next.

VII. TESTING THE THEORIES: EXPERIMENTAL HYBRIDS

It is not possible to repeat in the laboratory the actual cross-fertilizations which, it is claimed, led to existing sequential chimeras, but it is possible to try

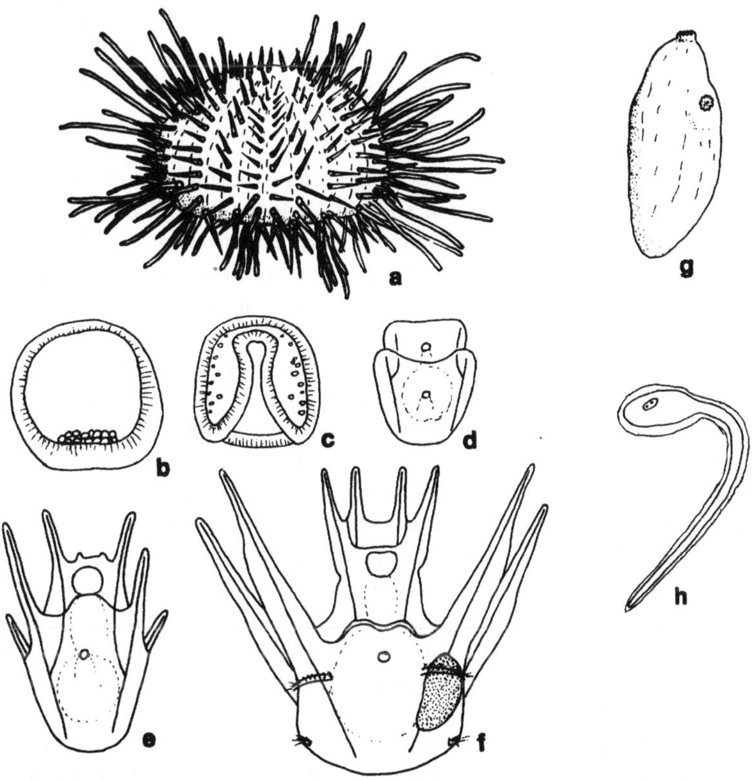

Fig. 8. The species hybridized and their non-hybrid larvae. (a-f) *Echinus esculentus*: (a) adult (about 10 cm diam.); (b) section of newly hatched blastula, 1 day after fertilization; (c) section of gastrula, 2 days after fertilization; (d-f) ventral views of developing pluteus larvae, 3, 12 and 22 days after fertilization (f about 1 mm high, developing juvenile stippled). (g,h) *Ascidia mentula*: (g) adult (about 7 cm high); (h) newly hatched tadpole larva, 1 day after fertilization (about 0.7 mm long).

to produce experimental hybrids between distantly related species. The most obviously relevant crosses would be between one species with a larva and one without, or between two species with very different types of larvae. My few attempts to obtain hybrids between males with larvae and females without have not been successful, but I have had some success in crossing males and females with different forms of larvae. I have used sperm from the sea-urchin *Echinus esculentus* (Fig. 8a) to fertilize eggs of the tunicate *Ascidia mentula* (Fig. 8g) [15]. Ripe males of *Echinus* will usually emit sperm if they are removed from the water and inverted over beakers. Eggs may be obtained from adult *Ascidia* by gently squeezing them. *Echinus* eggs fertilized with *Echinus* sperm hatch as ciliated blastulae which soon develop into ciliated pluteus larvae (Fig. 8 b-f). *Ascidia* eggs fertilized with *Ascida* sperm hatch as non-ciliated tadpoles (Fig. 8h), so the two types of larvae are clearly very different.

In all cases *Ascidia* eggs developed only after exposure to very high concentrations of *Echinus* sperm, which suggests that the natural defenses against cross-fertilization can be overwhelmed [15]. Heterozygotes between these species will have the genetic recipes to develop either as sea-urchins or as ascidians, but, according to my assumptions, the *Echinus* recipe, which prescribes hatching as small, simple, ciliated blastulae, should take precedence over the *Ascidia* recipe, which prescribes hatching as relatively large and complex, non-ciliated tadpoles, and apparently this is just what does happen. On eight occasions, ascidian eggs, after exposure to a milky or creamy suspension of sea-urchin sperm, hatched as ciliated blastulae, and on four occasions some of these blastulae developed into undoubted pluteus larvae (Fig. 9). In each of two experiments I obtained only one pluteus, but in two others I obtained over 200 and over 3,000. The hybrid larvae were morphologically indistinguishable from *Echinus X Echinus* larvae. About 75% of the hybrid plutei were stunted or showed varying degrees of deformity, but similar stunted and deformed larvae occur in non-hybrid cultures and the percentage showing such defects increases in relation to the sperm concentration used.

In all except my most current experiment, none of my larvae showed any sign of metamorphosis, but I now have more than 60 live hybrid larvae with juvenile rudiments and ten settled juveniles. So far, all the juveniles and rudiments have been of *Echinus* form, suggesting that the genetic recipe for juvenile *Echinus* takes precedence over that for juvenile *Ascidia*, at least under the conditions of this experiments. (The chromosomes of the juveniles must be investigated and every effort made to rear specimens to maturity and breed from them; if their development corresponds to that of non-hybrid *Echinus*

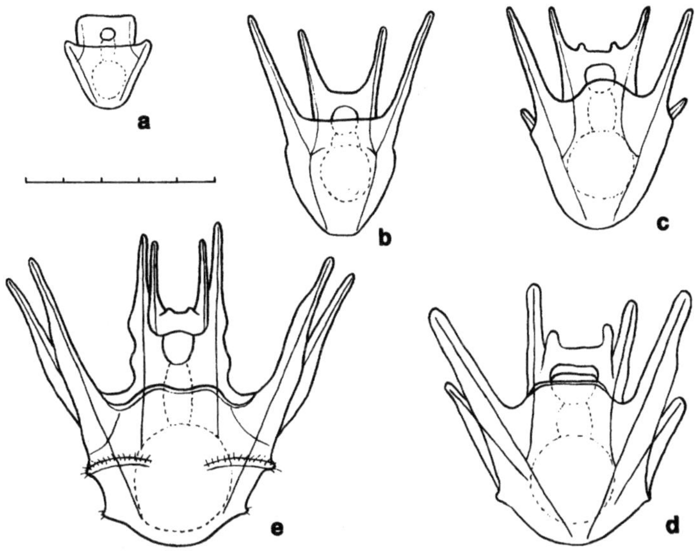

Fig. 9. *Ascidia* × *Echinus* hybrid larvae in ventral view. (a) 3 days from fertilization; (b) 8 days; (c) 15 days; (d) 22 days; (e) 29 days. Scale represents 0.5 mm.

esculentus this may well take at least two years.) The experimental hybrids produced by the fertilization of the eggs of *Ascidia mentula* with sperm of *Echinus esculentus* have not demonstrated metamorphosis from a paternal-type larva to a maternal-type juvenile, but their development is nevertheless quite consistent with the sequential chimera theory. In evolutionary history, the only types of animals which are postulated to have acquired echinoderm larvae by hybridization are echinoderms. Ascidian x sea-urchin hybrids, however, do seem to confirm that a larval phase can be transferred by heterosperm fertilization and that one larval phase can take developmental precedence over another. It will be necessary to confirm the form of the metamorphosed juveniles as they grow, but the development of these hybrids seems to be fully in accord with my assumption that crosses between distantly related animals do not show a phenotypic mixture of the characters of both parents, but are clearly either paternal or maternal in form, both as larvae and as adults.

There are, I know, those who prefer to disbelieve my results rather than accept their implications, but I urge such skeptics to carry out similar experiments themselves. Anyone can do them. I hope I have convinced at least a few people that the experimental hybridization of distantly related species is not a fruitless pastime, and I am confident that further hybridizations will produce more evidence relevant to my theories. If such evidence continues to be fully consistent with the theories, it will further raise the probability that comparable heterosperm fertilizations could have taken place in nature, but it will never prove conclusively that they did. There are, however, other ways of testing whether sequential chimeras really exist.

VIII. TESTING THE THEORIES: GENES AND THEIR PRODUCTS

The existing evidence that many animals are sequential chimeras comes largely from their morphology, as embryos, larvae and adults. (The evidence from their methods of metamorphosis is also very important, but as metamorphosis is the change from one morphology to another, this may be regarded as a branch of the morphological evidence.) The need for a new theory arose because the relationships between groups of animals suggested by their embryonic and larval morphologies often seem to be at variance with those suggested by their adult morphologies, and the previous explanations of these anomalies are unconvincing. The sequential chimera theory seeks to explain the anomalies by taking the observations at their face value: it postulates that the affinities of incongruous embryos and larvae not only appear to be different from those of the adults they give rise to, but they really are different, and these differences denote different evolutionary origins. But we also require assessments of affinity which are independent of morphology and which can be applied separately to larvae and adults, by studying genes which are active during the different phases of development. I shall return to such matters in a moment, but investigations of the total nuclear genome might also be relevant. In some cases, the chromosomes, or the total nuclear DNA, of an animal may give evidence that embryonic or larval forms have been transferred by hybridization. In the original hybrid, the paternal and maternal chromosomes would probably be separate, and in many cases they would be recognizable. These chromosomes of different origin would probably remain separate for many generations but not indefinitely, for chromosomes can break and fuse to change their number and shape. This would not directly affect the total amount of DNA, but microscopic examination of the chromosomes would probably be more rewarding in cases of recent, rather than ancient acquisition of larval forms. Comparisons may be made between the chromosomes of forms with similar adults but different larvae or no larvae, e.g., particularly promising examples for investigation are provided by the brittle-stars. Kirk's brittle-star has been mentioned as a form with no larva either in its own life-history or in those of its ancestors. Its ancestry probably involved no hybridizations with members of other evolutionary lines and is therefore simpler than that of most other brittle-stars. Brittle-stars with larvae would be expected to have more chromosomes and more DNA than Kirk's species, and within the family Ophiolepidae mention has earlier been made of *Ophiura albida, Ophiura texturata* and *Ophiolepis cincta* whose larvae appear to have been acquired from quite different sources (Fig. 6). The hybridizations which introduced the larvae into the life-histories of these species probably took place comparatively recently and they may well be reflected in different karyotypes.

I postulate that in each hybrid with a new life-history, the larva will be paternal in form, the adult maternal. The animal with this new life-history will, however, also have the unused genetic prescription for an adult of paternal form, and an animal with several hybrids in its ancestry could potentially have the unused prescriptions for several diverse adult forms. I suggest that the ophiopluteus larval form of brittle-stars was derived from an echinopluteus of a sea-urchin, which, in turn, was derived from a bipinnaria of a sea-star, and this was derived from and auricularia of a sea-cucumber, which was derived from a tornaria of an acorn-worm, and each larval transfer was the result of a cross-fertilization. A modern brittle-star with a pluteus larva would almost certainly not have the intact genetic prescriptions for the adults of a sea-urchin, a sea-star, a sea-cucumber and an acorn worm unused in its nucleus, for there seems to be a mechanism whereby inactive DNA is changed into heterochromatin, condensed and ultimately lost. Total nuclear DNA is therefore no more likely than the karyotype of an animal to give evidence of hybridizations between its ancestors which took place hundreds of millions of years ago, but both could give indications of recent hybridizations.

It is a corollary of the sequential chimera theory that the similarities between the larvae of, for example, acorn-worms and sea-cucumbers, sea-urchins and brittle-stars or polychaetes and molluscs, are not due to convergence, but reflect close genetic similarities, while the differences between the corresponding adults reflect wide genetic differences. These predicted genetic similarities and differences could be tested either by studies on the genes themselves or on their products, and I am now urging those with the necessary expertise and opportunities to carry out such tests. Investigations of the total genomes of animals would be much less relevant than investigations which distinguish between genes transcribed in larvae and those transcribed in adults, and results have already been published of a study on adult echinoderms which supplies half the information needed to provide a test of the theory. There should be no insuperable difficulties in following this with a comparable study on larvae to complete the information. In refer to the analysis by Raff *et al.* [11] of one form of ribosomal RNA in adult echinoderms from each of five classes. The grouping of classes suggested by their results confirms the usual interpretation of adult morphology and paleontology: the sea-stars and brittle-stars form one group, the sea-urchins and sea-cucumbers another, and the sea-lilies a group of their own. When the ribosomal RNAs of larval echinoderms are eventually investigated they will almost certainly be found to differ significantly from those of the adults, but this would be expected whether the larvae had evolved in parallel with the adults or independently. *The important result, however, from the point of view of the sequential chimera theory, will be the grouping of classes suggested by these larval RNAs. It is predicted that the results will mirror the spread of larvae from class to class* outlined in the previous paragraph. They will show affinities between auricularias (late larvae of sea-cucumbers and the only larvae of some sea-lilies and brittle-stars) and bipinnarias (larvae of sea-stars), and they will also show affinities between the pluteus larvae of sea-urchins and brittle-stars, with the echinopluteus nearer than the oiphiopluteus to the bipinnaria.

It is usually easier and quicker to study the chemical products of genes than the genes themselves, and several of the techniques of chemotaxonomy could provide important evidence on the sequential chimera theory. Systematic serology has its basis in the immune reaction of animals to foreign proteins by the production of antibodies. Comparable proteins from a range of animals will give a range of reactions with such antibodies, and these reactions can be studied by a variety of methods to give an assessment of the relatedness of the animals concerned. For present purposes, of course, we require separate assessments of the relatedness of adults and of larvae. Another approach is provided by examining the enzymes which promote and regulate growth, e.g., by electrophoretic techniques. In general, the greater the number of common enzymes, the closer the relationship. Both serology and enzyme analysis can produce assessments of affinity which may be related to the conventional method of classification into species, genera, families and higher taxa. I should expect them to confirm the implications of morphological studies, showing, for example, that ophiopluteus and echinopluteus larvae have a genetic similarity corresponding to that between animals in different superfamilies while the brittle-stars and sea-urchins they give rise are much less closely related and are correctly placed in different classes; or, at a lower taxonomic level, the larvae of *Ophiura texturata* and *O. albida* seem to belong to different superfamilies while their corresponding adults are closely related species of the same genus.

IX. IDENTITY AND CLASSIFICATION

The purpose of the preceding argument has been to try to convince my readers that sequential chimeras exist or at least that their existence is a strong possibility, strong enough to encourage the collection of more evidence and the undertaking of more tests of the theory, and strong enough to stimulate serious discussion on its implications. The questions it raises are enormous and probably include many which have never occurred to me. Here, I shall concentrate on those relevant to the general theme, and on the questions rather than the answers.

The experimental hybrids, which I have generated between *Ascidia* and *Echinus* show that a larval form can be transferred, without modification, from one animal to a distantly related one. It is postulated that transfers like this, followed by metamorphoses from paternal larvae to maternal juveniles, resulted in echinoderms and other groups acquiring larval phases or additional larval phases; furthermore, if this is an evolutionary mechanism, I expect that the morphology of the adults were unaffected by the hybridizations (in most cases). There are cases scattered throughout the animal kingdon of what were originally larval characters persisting in the adult, or of what were originally adult characters appearing prematurely in the larva. These are phenomena of considerable evolutionary importance, but they seem to have affected only a small minority of animals. Usually the larval and adult characters have remained quite distinct and have shown no tendency to cross the barrier of metamorphosis, and this seems to apply whether or not the larval form was acquired from another evolutionary line. The first echinoderm egg to give rise to a bilateral larva had been fertilized with an enteropneust sperm. The resulting larva was probably morphologically indistinguishable from larvae of the paternal (enteropneust) species, and the adult which it gave rise to was probably morphologically indistinguishable from adults of the maternal (echinoderm) species. The new combination was just as much a chimera as Homer's lion-goat-serpent, although the only time its two components were in physical contact was while the juvenile was developing within the late larva. Each cell of the bilateral larva would, of course, have carried the same total genome as each cell of the radial juvenile which developed within it, so in terms of the total genome the animal with the new life-history was not a chimera. In terms of transcribed genes, however, it was. The larval form was prescribed by enteropneust genes and the juvenile and adult form by echinoderm genes.

Over the hundreds of millions of years that have elapsed since this echinoderm acquired its enteropneust larva, new forms of echinoderm larvae have evolved and new forms of adults, and generally they have evolved quite independently. In most cases, modern echinoderm larvae seem to be prescribed by the descendents of the enteropneust genes of the original hybrid, whether they have been inherited by direct descent or by a combination of descent and further hybridizations. New forms of larvae have arisen as the result of genetic mutations and recombinations, but the recombinations seem seldom to have involved the genes which originally prescribed the form of adult echinoderm. A possible case of genes exerting their influence on both sides of the metamorphic boundary is seen, however, in the case of doliolaria larvae. These show a considerable degree of radial symmetry, and this may be the result of "acceleration" of the transcription of what were originally adult genes for this feature. Nevertheless, the great majority of echinoderm larvae show no trace of radial symmetry or any other feature of adult echinoderms. *Astropecten auranciacus*, which I used as my first example of a sequential chimera, seems to have its life-history as clearly divided as did the original echinoderm-enteropneust hybrid. The larva of *Astropecten* seems to have evolved from an enteropneust larva and the adult from an adult echinoderm, with no interaction between the genes prescribing the two developmental phases. The problems of identity and classification are much the same for *Astropecten auranciacus* and thousands of other species which have inherited acquired larvae as they were for the original "species" which acquired a larva from another group. They are the same sort of problems as those posed by Homer's chimera, but they are posed by real animals.

Is the juvenile *Astropecten* the same individual as the larva in which it develops as a quasiparasite, or is the juvenile *Luidia* crawling over the bottom the same individual as the larva swimming above it, from which it separated up to three months previously? The answer will depend on how one defines an individual, and my dictionary says it is "a single person, animal, plant or thing considered as a separate member of its species". I think, in the present context, we can ignore the person, plant and thing. We cannot ignore the question of the species, but we can postpone it for a moment. In terms of total genome, the larva and juvenile of *Astropecten* are members of the same clone, and the same applies to the larva and juvenile of *Luidia*, but in each case I should expect transcribed genes to support morphology in placing the larva and juvenile in different phyla. It is not merely that sea-stars change shape and way of life as they develop from larva to juvenile or that the two phases overlap in time. A lepidopteran insect changes shape and way of life as it develops from a caterpillar to a butterfly or moth, and a hydroid budding off medusae provides an example of different forms developed from the same egg leading a free life simultaneously, as do the larva and juvenile of *Luidia*. On the other hand, a caterpillar and a butterfly are both insects and a hydroid and a medusa are both cnidarians, and, in both cases, the two phases in development probably have exactly the same ancestry. I hold, however, that a larval and adult sea-star have different ancestries, with the pedigree of the larva stretching back to the tornaria of an acorn-worm while that of the adult is echinoderm all the way. Eventually I hope the immunologists will be able to tell us whether tissues of a larval sea-star react to those of the juvenile developing within it as if they were its own tissues or as if they were those of a parasite. There must, of course, be a considerable range of reactions of animals to their parasites, but I predict that the reaction of the larval sea-star to the developing juvenile will fall within this range. Sequential chimeras therefore raise new problems concerning the concept of self and the individual; they also raise new problems concerning the concept of the species.

Biological species, according to Mayr [17], are "groups of actually or potentially interbreeding natural populations which are reproductively isolated from other such groups", amended later [18] to "reproductive communities of populations (reproductively isolated from others) that occupy specific niches in nature." As Mayr says, "reproductively isolated" are the key words of the biological species definition. There are obvious difficulties in applying this criterion to embryos and larvae, but if these are regarded as no more than phases in the development of the adult then there is no difficulty in regarding the egg, embryo, larva, juvenile and adult of (say) *Astropecten auranciacus* as all belonging to the same species. *A. auranciacus* might be potentially capable of hybridizing with very distantly related animals, but only under exceptional circumstances, so normally it would remain reproductively isolated from other species. There are, however, many definitions of a species, and, in addition to the "biological species", we cannot ignore the "evolutionary species", for which Mayr [18] quotes Simpson's definition of "a lineage (an ancestral-descendant sequence of populations) evolving separately from others and with its own unitary evolutionary role and tendencies". Now the emphasis has shifted from reproductive isolation to evolutionary lineage, and it is an essential feature of any sequential chimera that it is made up of at least two evolutionary lineages and changes from one to another during its ontogeny. The larva of *Astropecten auranciacus* may, then, be regarded as belonging to the same biological species as the adult which bears the same name, but according to my theory it belongs to a different evolutionary species.

At taxonomic levels above the species, it is difficult to avoid consideration of evolutionary lineages. Even a numerical taxonomist has to make a decision whether or not to include embryonic and larval characters, and his decision is likely to be influenced by whether or not he considers the embryo and larva to have the same evolutionary lineage as the adult. Even if he ignores embryos and larvae in all cases, this is likely to be because he suspects that in some cases they may have different evolutionary lineages from the adults. The situation is further complicated because, I claim, sequential chimeras have arisen as the result of hetrosperm fertilizations between animals at several different levels of relatedness: different phyla, different classes, different orders or perhaps more closely related animals. Such heterosperm fertilizations have taken place, infrequently but with profound repercussions, over hundreds of millions of years of the evolutionary history of metazoans, if not of all animals, and some have probably taken place within the last million years, but they have not affected all metazoans or even all the members of any major phylum. There is ever growing evidence that the origin and subsequent elaboration of eukaryotic cells depended on the uniting of prokaryotes from different evolutionary lines [19]. I am now suggesting that mixed ancestries are not only essential features of the early phylogeny of all plants and animals but that they have also played a vital part in the later and current evolutionary history of several groups of animals.

Biologists who embrace the propositions now being put forward will have to evolve new methods of classification and of depicting relationships. The widely accepted evolutionary tree of the animal kingdom (Fig. 4) seems to show the phylogenies of different types of embryos and larvae fairly correctly, but I claim that it gives a highly misleading representation of the way in which adult animals have evolved. There is a hope that redesigning the tree will divert at least some taxonomists from the present seemingly interminable debate on the merits and demerits of cladism. I should like to remind the many biologists who seem to prefer to reject my postulations without serious consideration rather than contemplate their implications that no idea should be rejected merely because it is inconvenient. Conversely, no idea should be accepted merely because it is convenient, and I am thinking particularly of the principle of evolutionary parsimony. This assumes that evolution has always taken place by the most direct route or involving the minimum possible number of events. The principle is very widely applied in the reconstruction of phylogenies, but while it is undoubtedly convenient there seems to be little justification for adopting it as a law of nature. I would not, therefore, claim that evolution must have followed a particular route merely because it is the most parsimonious. I am, however, postulating that a new life-history can be (and frequently has been) acquired in a single generation, and this may be regarded as an extreme example of evolutionary parsimony. It is also, of course, a postulated example of evolutionary saltation, and it provides a method of explaining the occasional leaps which from time to time seem to have punctuated the general equilibrium of slow evolutionary change [20].

The sequential chimera theory is certainly not the last word on evolution nor even on larvae and phylogeny, but I hope it is a step towards a better understanding of evolutionary processes.

University of Liverpool

REFERENCES

1. Williamson, D.I. 1988. Incongruous larvae and the origin of some invertebrate life-histories. Progr. Oceanogr. 19: 87–116.
2. Hyman, L.H. 1955. *The Inertebrates, Vol. IV: Echinodermata.* The Coelomate Bilateria. McGraw-Hill, New York, pp. 763.
3. Darwin, C. 1859. *The Origin of Species by Means of Natural Selection or the Preservation of Favoured Races in the Struggle for Life.* John Murray, London. (Reprinted 1985. Penquin Books. London, pp. 477).
4. Fell, H.B. 1948. Echinoderm embryology and the origin of chordates. Biol. Revs. 23: 81–107.
5. Tattersall, W.M. and E.M. Sheppard. 1934. Observations on the bipinnaria of the asteroid genus Luidia. In *James Johnstone Memorial Volume.* R.J. Daniel, editor. Liverpool University Press, Liverpool, pp. 35–61.
6. Barnes, R.D. 1980. *Invertebrate Zoology.* Fourth Edition. Saunders College, Philadelphia, pp. 1089.
7. Fell, H.B. 1941. The direct development of a New Zealand ophiuroid. Quart. J. Microsc. Sci. 82: 377–441.
8. Jablonski, D. and R.A. Lutz. 1983. Larval ecology of benthic marine invertebrates; paleontological implications. Biol. Revs. 58: 21–89.
9. Paul, C.R.C. 1979. Early echinoderm radiation. In *The Origin of Major Invertebrate Groups.* M.R..House, editor. Systematics Association Special Volume 12. Academic Press, London, pp. 443–481.
10. Baker, A.N., F.W.E. Rowe, and H.E.S. Clark. 1986. A new class of Echinodermata from New Zealand. Nature London 321: 862–863.
11. Raff, R.A., K.G. Field, M.T. Ghiselin, D.J. Lane, G.J. Olsen, N.R. Pace, A.L. Parks, B.A. Parr and E.C. Raff. 1988. Molecular analysis of distant phylogenetic relationships in echinoderms. In *Echinoderm Phylogeny and Evolutionary Biology.* C.R.C. Paul and A.B. Smith, editors. Clarendon Press, Oxford, pp. 29–41.

12. Mortensen, T. 1921. *Studies of the Development and Larval Forms of Echinoderms*. G.E.C. Gad, Copenhagen, pp. 261, 33 pls.
13. Mortensen, T. 1931. Contributions to the study of the development and larval forms of echinoderms. Det Konglige Danske Videnskabernes Selskabs Skrifter, Naturvidenskabelig og Mathematisk Afdeling 9, Raekke IV(I): 1–39, 7 pls.
14. Fell, H.B. 1968. Echinoderm ontogeny. In *Treatise on Invertebrate Paleontology, Part S Echinodermata* 1. R.C. Moore, editor. Geological Society of America and University of Kansas Press, Lawrence, Kansas. S60–S85.
15. Williamson, D.I. *Larvae and Evolution*. Chapman and Hall, New York. In press.
16. Fernando, W. 1931. The origin of the mesoderm in the gastropod genus *Viviparus* (= *Paludina*). Proc. Roy Soc. B 107: 381–390.
17. Mayr, E. 1942. *Systematics and the Origin of Species*. Columbia University Press, New York, pp. 334.
18. Mayr, E. 1982. *The Growth of Biological Thought. Diversity, Evolution and Inheritance*. Harvard University Press, Cambridge, Massachusetts, pp. 974.
19. Margulis, L. 1981. *Symbiosis in Cell Evolution. Life and its Environment on the Early Earth*. W.H. Freeman, San Francisco, California, pp. 419.
20. Eldredge, N. and S.J. Gould. 1972. Punctuated equilibria: and alternative to phyletic gradualism. In *Models in Paleobiology*. T.J.M. Schopf and J.M. Thomas, editors. Freeman Cooper, San Francisco, California, pp. 82–115.

EDITOR'S COMMENTS TO GILBERT

Gilbert traces the developing embryo as Self, first through a historical analysis of our understanding of induction, and then places the issue of embryonic induction in its evolutionary context. "Ontogeny doesn't recapitulate phylogeny, it creates it" (Garstang). How differences in induction generate different "selves" has been made clearer by recent advances in our understanding of regulatory principles in so called positive and negative induction, correlative development, phyletic constraints, and transfer of competence, each of which is clearly elucidated by Gilbert. He then turns to placing the scientific question within the broader conceptual construction that has served as an implicit theme of these essays, i.e, the dialecticism of such processes. Noting that the concept of induction has readily been acknowledged as explicitly dialectical by 20th century investigators, Gilbert well-illustrates Waddington's intellectual debt to Whitehead in this regard. This example offers an excellent case of metaphysical awareness enriching the scientific enterprise, and more saliently, it captures an essential ephemeral component of our understanding of the Self from the vantage of developmental biology. Here we might review earlier antecedents for a clue regarding the elusiveness of our subject.

"Ontogeny" (like "ecology" and "phylogeny") was coined by Ernst Haeckel, the chief apostle of evolution in Germany during the last half of the 19th century. He popularized the biogenetic law: "Ontogeny is the short and rapid recepitulation of phylogeny" (*Generelle Morphologie der Organismen*, 1988); that recapitulaton somehow reflected the transformation of species had a long history, and in the 19th century was championed by both the French transcendental morphologists and the German *Naturphilosophen*. The imperative to find order in the evolutionary tree supported the hope, as opposed to the reality, that "Human organogeny is a transitory comparative anatomy as, in its turn, comparative anatomy is the fixed and permanent state of human organogeny" (Serres, *Précise d'anatomie transcendante appliquée à la physiologie*). There were dissenters, most notably Karl von Baer, who already in the 1820's rejected recapitulaton as both logically and scientifically untenable; von Baer recognized that confirmaton of a natural, unified scheme of development was inconsistent with the facts. Recapitulation was built on the notion of a scale of beings where only a single line of development—progressive or regressive—was allowed. For von Baer, development was a differentiation of the general to the specific, and most saliently, "The developmental history of the individual is the history of the growing individuality in every respect" (*Entwicklungsgeschichte der Thiere: Beobachtung und Reflexion*, 1828).

Recapitulation as a reflection of species transformaton continues to fascinate and humble our attempts to integrate evolutionary and developmental biologies, for it serves as a focus for assigning morphoregulatory rules that are governed by selective processes, but operative at early developmental stages (see Tauber, this volume, pp. 1–39). Or as Leo Buss succinctly states, "A theory of ontogeny has not been framed" (*The Evolution of Individuality*, 1987). This is a most interesting problem, and part of the difficulty must reside in the core concept. The language, as expected, reflects our confusion.

Returning to Haeckel's original use of ontogeny, we note that he viewed developmental processes as a continuum, or a hierarchy, with ontogeny and phylogeny united under a single set of causes. Until Haeckel, "Entwicklungsgeschichte" was used in each case for "developmental history." But he viewed phylogeny and ontogeny as coordinated branches of morphology: "Phylogeny is the developmental history of the abstract, genealogical individual; ontogeny, on the other hand, is the developmental history of the concrete, morphological individual" (*Generelle Morphologie*, p. 60). Precisely in this attempted fusion of individual (development) and species (transformaton) was recapitulation locked into a conceptual morass: development, or change of the embryo, is *not* the evolution of species.

But there is a more fundamental issue lurking within the semantics. How curious that Haeckel chose "ontogeny" for embryologic development. *Gen* (from *geneia*—Greek) meaning "origin," "production," or "birth" is self-evident, but *onto*, meaning "being" is derived from *ont*—the Greek singular of *on*, neutral present participle of *einai*, to be. Haeckel invoked a deeply metaphysical term to describe individual development.

With Darwin, a materialistic basis was established for explaining the development of species, their transformation, their evolved morphology, the basis of organization. Notwithstanding that few truly understood the argument of natural selection, and fewer still accepted it, it was clear that the preceding natural order had been shaken to its foundation. Haeckel dubbed individualization (i.e., embryological development) "ontogeny," the birth of being, and thus revealed his fundamental misunderstanding of the revolution in which he thought he partook.

Haeckel was, according to students of the era (i.e., Nordenskiöld, *The*

History of Biology, 1929), even more influential than Darwin in promoting evolutionism in the last decades of the 19th century. Ironically, while he acclaimed Darwin, Haeckel gave equal credit to Lamarck and Goethe, offering a profoundly anti-Darwinian explanation of evolution: "Phylogenesis ... is a physiological process ... determined with absolute necessity by mechanical causes. These causes are motions of atoms and molecules". (Haeckel, 1866 translated by S.J. Gould, *Ontogeny and Phylogeny*, 1977, p. 79).

Stephen Jay Gould quotes Darwin's angst: "No one, I presume, would doubt about molecular movements of some kind" (Gould 1977, p. 79). Haeckel offered no other mechanism beyond what Darwin's correspondent George Romanes called these "terms of the highest abstraction" (*ibid.*) Haeckel's bizarre reductionism and rationale for the biogenetic law are explicated by E.S. Russell (*Form and Function*, 1916) and Gould (Gould 1977), but perhaps his most sensitive reader was Ernst Cassirer (*The Problem of Knowledge*, 1950): Haeckel saw only a mechanical explanation triumphing over the teleological as testament to Darwin's achievement, but in tracing natural events to their ultimate "physical" causes, he had only inserted

> into matter the living phenomenon ready made, rather than by deriving it from matter. The monism of Haeckel was a naive hylozoism [the doctrine that life is a property of matter] because he believed that all forces, the purely physico-chemical equally with the organic, were ultimately from the primal life force (Cassirer 1950, p. 163).

So much for Haeckel, the Darwinian evolutionist. But returning to his bestowal of "ontogeny"—the birth of being, various commentators at the turn of the century (e.g., Boveri and Radl) recognized that the crucial factor imposed by Darwin was to place historical analysis at center stage of biological thinking:

> one could not possibly conceive of the true nature of an animal by any analysis, be it ever so profound, or by any comparison with other forms, however comprehensive, because there lies hidden in the organism traces of the past that only historical research is able to reveal (attributed to Radl (1905) by Cassirer 1950, p. 171).

Haeckel understood an ahistorical "being", but he lacked the profound historicism required to appreciate Darwinism. In the Newtonian world, the object was given; there was no ontological issue of the subject-object relationship. In the post-Darwinian world, evolution is an on-going dynamic, where the organic object is under constant pressure to change, and in fact is changing, whether by natural selection, neutral evolution, undefined macroevolutionary mechanics, or whatever. There is a parallel intellectual revolution in physics: the object in the world of Einstein and Bohr became relativistic and indeterminant. No longer given, the object as a modern construct must be defined, held momentarily as subject in the present, however fleetingly, and as object in its past (Alfred North Whitehead). The object has become less an entity than a process, dynamically evolving as ever-different. Being has been fundamentally altered in a hercalitean flux, to a becoming.

Ontogeny is fundamentally at the foundation of our biology, for here the emergence of Selfhood is explicitly portrayed. The processes of human individualization are declared from the gene, to the appearance of gill slits, to the fetus sucking her thumb in a wonderous panorama. Had Haeckel called the post-Darwinian embryology "All being is Becoming", he would have seized the significance of individualization in a new world, and appropriated its essential title.

SCOTT F. GILBERT

THE ROLE OF EMBRYONIC INDUCTION IN CREATING SELF

I. THE EMBRYO AS A SELF

Since our general topic is the creation of self, I wish to discuss the role of embryonic induction in the self-creation of the embryo. First, however, I must speak in general terms concerning the embryo as self.

Developmental biology is a strange science because it denies the hegemony of the adult. Usually when one thinks of a dog, one thinks only of the adult form. When developmental biologists think of dogs, however, we see a time line of forms leading from the zygote through various stages of cell differentiation and organogenesis, through puppihood, to the adult hound. No one stage is given more importance than any other, and the *creation* of the adult form is actually considered more important than its *maintenance*. To a developmental biologist, the expression, "Mayflies live but for a day", is completely fallacious. Granted, the adult form only lasts but a day, but the embryonic and larval states of this organism last the remaining 364.

Moreover, at every stage of the way, the embryo is a functioning organism. This is the main difference between a frog and a Chevy. The Chevy never had to function until it was off the assembly line. The frog has to digest, respire, and excrete while it is a zygote, a two-celled embryo, a neurula, a tadpole, and an adult. It has to digest before it had a mouth, respire before it has lungs, and excrete before it has a kidney. Moreover, each of its cells had to survive, just as they would in an adult organism.

To have an embryonic stage means that one has mortality. Unicellular organisms that reproduce by fission are potentially immortal. When an amoeba divides, the resulting cells are siblings, neither is the ancestor, neither the offspring. The amoeba that one sees under the microscope today has no dead ancestors. Death comes to the amoeba only if it is eaten or if it gets dessicated under the microscope slide. When it does, the dead organism leaves no progeny. Amoeba do not undergo a natural death. As David Kirk [1] has noted, death becomes part of life when there is a division of labor between those cells that constitute the body and those cells that go on to produce the next generation. Even in asexually-reproducing volvoxes, after the gonidia have grown as embryos within the parent volvox and then become separate organisms, the

Alfred I. Tauber (ed.), Organism and the Origins of Self, 341–360.

parent volvox dies. There is even a gene whose function is to destroy the parent volvox after the separation of its progeny. If the concept of "self" involves the concept of death, its negation, then embryos are the *sina qua non* of selfhood.

II. INDUCTION AND THE ORIGINS OF SELF

One of the other most important experiments in developmental biology focused almost entirely on the concept of self. These are the experiments that Hans Driesch [2] performed in 1892 when he separated the first four cells of a sea urchin embryo and, to his surprise, each of them became a separate pluteus larva. In other words, a cell which would normally have generated only a portion of the organism had the potential to generate the entire organism. These experiments are repeated regularly in developmental biology classrooms throughout the world. This result has momentous meaning for the concept of the individual self. If each cell has the ability to become an entire organism, why doesn't each cell form the organism? In fact, how can organisms comprised of different cell types exist at all, if each cell could go about its way, fulfilling its potential to create an entire embryo? Somehow, the existence of neighboring cells instructs the cell to restrict its potency to a fraction of what it actually can achieve. "Even though you have the ability to form an entire larva, all you are to produce is oral ectoderm." Driesch called the embryo a "harmonious equipotential system." It was equipotential since every cell had the ability to form any organ and, indeed, the entire embryo. It was harmonious because there was some mechanism that brought these potentially independent parts together to form a single organism.

How is this done? Basically there are two ways. In most invertebrates, early embryonic cells cannot form an entire embryo when isolated. These embryos are not harmonious equipotential systems. Rather, each cell is separately determined by the cytoplasm residing within it. The zygote cytoplasm contains instructions that are given to the dividing cells that take up certain cytoplasmic regions. If a cell contains the cytoplasmic region where the determinants for muscle cells are located, those cells form muscles. If the cell is given cytoplasm which contains the determinants for ciliated ectoderm, the cells stop dividing after a point and start differentiating ciliated ectodermal structures [3, 4].

But what about Driesch's embryos? How can these potentially independent cells form into a single organism? The answer appears to be induction. Most of us who have taken embryology courses know of positive induction: The optic cup induces the overlying ectoderm to form lens instead of skin; the notochord of the frog interacts with the overlying ectoderm to cause that ectoderm to become nerves instead of skin. But there is also negative induction. Jon Henry and colleagues in Rudolph Raff's laboratory [5] have shown that if one isolates pairs of cells from the animal cap of a 16-cell sea urchin embryo, these cells can give rise to both ectodermal and mesodermal components. However, their capacity to form mesoderm was severely restricted if they are aggregated with other animal cap pairs. Thus, the presence of neighbor cells, even of the same kind, restricts the potencies of both partners. Ettensohn and McClay [6] have shown that potency is also restricted when a cell is combined with its neighbors along the animal-vegetal axis. They have shown that when primary mesenchymal cells are removed from the mesenchyme blastula of sea urchin embryos, a group of secondary mesenchyme cells replaces them. These secondary mesenchyme cells come from the vegetal plate adjacent to the region where the primary mesenchyme cells formed. They have shown that the presence of normal primary mesenchyme cells inhibits the surrounding cells from forming primary mesenchyme. If they are removed, these macromeres will form the primary mesenchyme cells. Thus, there is negative induction occurring here (Fig. 1). Just as in normal induction, where one cell type induces its neighbor to produce a new cell type, here we see that one cell can restrict the potency of its neighbor.

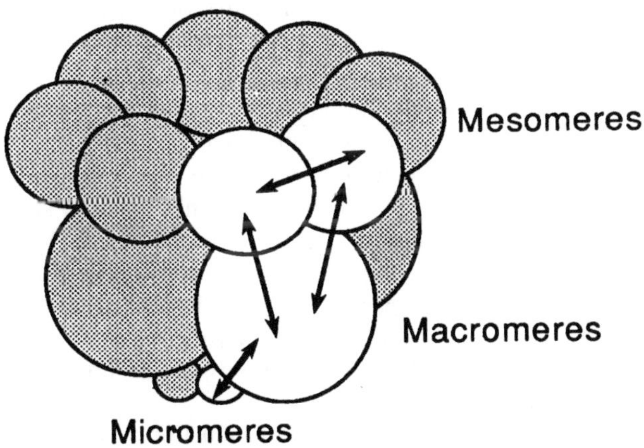

Fig. 1. Negative induction events between sea urchin blastomeres limits the potency of each cell. (From Henry *et al.*, 1989 [5].)

This restriction in potency was followed on the molecular level by Hurley and coworkers [7] who dissociated 16-cell sea urchin embryos into individual blastomeres. The loss of cell contacts between the isolated blastomeres caused dramatic alterations in the synthesis of messenger RNAs from these cells. Some mRNAs, such as that for mesenchymal actin, was expressed by a far greater number of the dissociated cells than normal. Other messages that become common in gastrulating embryos, however, were hardly synthesized at all in the dissociated cells. Thus, the contact between the blastomeres of the cleaving sea urchin blastula restrict each other's potency such that they construct a single embryo.

In this we see the importance of induction in creating the coherent self, a self that is created as an embryo. Perhaps the best examples of this are monozygotic twins and their converse condition, allophenic mice. In human embryos, only one self usually forms from a single fertilization event. The cells form a community wherein the individual cells relinquish some of their potential for the common good. (This community analogy was very popular among biologists of the 1930s and 1940s). Identical twins occur when cells of a single embryo separate before they begin differentiating. This can happen for a long time, since human embryos are very slow to differentiate. If this separation of cells occurs before the trophoblast is formed on day 5, the twins have two separate placentas. If the division occurs between day 5 and day 9 when the amnion is formed, the twins share a common placenta but are enclosed in separate amnions. If separation occurs after day 9, the twins share a common amnion as well. After this period, cell differentiation begins. Any separation after this risks forming conjoined twins.

The converse of this situation is the allophenic mouse [8]. Here, two or more embryos are joined together before differentiation has begun, usually at the four cell stage. What happens if the cells from two embryos are mixed? In mammals where induction is able to regulate the production of embryonic structures, one gets only one self. When mouse embryos from strains having two different coat colors are fused together, the result is a single mouse with regions of different coat colors (Fig. 2). Both embryos play a role in forming the embryo, giving rise to all the parts of the body. Instead of some two-headed or two bodied monster, one has a normal embryo and a normal adult arising from it. As many as three early mouse embryos can be brought together to form a single self [9].

It appears, then, that there is some field of induction which takes whatever cells are there and organizes them into the proper structure. A particularly dramatic example of this is when one fuses an early mouse embryo with cells

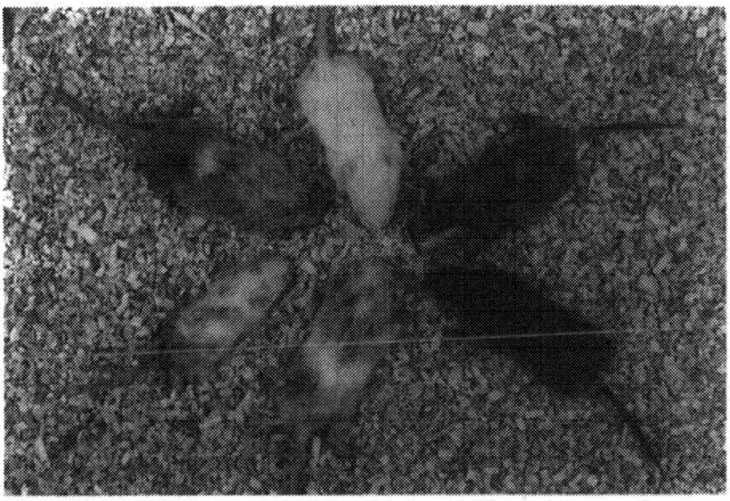

Fig. 2. A ring of allophenic mice. Each mouse has four parents, two white-furred mice and two dark-furred mice. Two embryos, one from two white-furred parents and one from two dark-furred parents, were fused together. (Photograph courtesy of M.L. Oster-Granite.)

from an embryonic tumor [10]. These embryonic tumors, called teratocarcinomas, are malignant cells that have retained the potential to become all the different cell types of the body. However, they are autonomous and the cells do not form a coherent self. Instead of self, one gets a tumor composed of many different cell types. If one of the malignant tumor cells is injected into the early embryo, it will lose its malignant properties and join in to form a portion of the adult organism. Thus, part of the "self" is actually derived from a tumor cell that had been kept in culture. It is obvious that induction is very important in the creation of self.

So far, we have been discussing the embryonic creation of normal selves, that is, cases wherein the embryo regulates so that there is one self per individual. What would happen if the embryo was so manipulated that two selves were generated within the same individual organism? This was precisely the question that confronted Hans Spemann in 1903 and caused him to start upon the series of experiments which led to the concept of the organizer. Spemann [11] relates the following episode as to how he entered upon his research. By *partially* constricting the newt egg without separating the blastomeres, he had caused the formation of two-headed salamander larvae joined at the trunk. Spemann tells us:

> Such animals came to the stage of feeding and it was now most remarkable to see how once the one head and at another time the other caught a small crustacean, how the food moved through the separate foreguts to the joint posterior intestine ... It was probably irrelevant for the well being of the strange creature which head had caught the food; it was for the benefit of the whole. Nevertheless, one head pushed the other away with its forelegs. Hence, two egotisms in the place of one, called forth by the initial separation of the anlagen ... Thus at last—I was then 28 years old —I had found the beginning of my own scientific journey.

The end of this journey was, of course, the formulation of the organizer concept. What does the organizer organize? It organizes a self.

III. INDUCTION AND CONSTRAINTS ON EVOLUTIONARY NOVELTY

If the self can be created through induction, then changes in induction can generate different selves. This was the principle by which Walter Garstang reconnected embryology to evolution. "Ontogeny doesn't recapitulates phylogeny," said Garstang [12], "it creates it." Wilhelm Roux, in setting forth his principles of developmental mechanics, wrote [13] that "an ontogenic and a phylogenetic developmental mechanics are to be perfected." The first sought to discover the mechanisms by which the embryo generated itself; the second sought to explain how changes in that development caused new types of organisms to evolve. We can now see that induction works both as a constraint to evolutionary novelty and as a promoter of such novelty. One of the types of constraints placed upon the evolution of new types of structures are phyletic constraints [14]. Once one has a mechanism of creating a structure by inductive interactions, it is difficult to start over again. The notochord, which is still functional in protochordates, is considered vestigial in birds and mammals. Kleinenberg was probably the first to point out that it is probably retained during development only because it is needed to induce the neural plate. Similarly, Waddington [15] noted that although the pronephric kidney of the chick embryo was considered vestigial since it had no ability to concentrate urine, it is the source of the ureteric bud which induces the formation of a functional kidney during chick development.

This type of phyletic constraint has recently been reviewed by Raff and colleagues [16]. It had been thought until recently that the earliest stages of development are the hardest to change, because altering them would either destroy the embryo or generate a radically new phenotype. But recent work (and reappraising older work) has shown that alterations can be made to early cleavage without upsetting the final form. Modifications of morphogens in mollusc embryos can give rise to new types of larvae that still metamorphose into molluscs, and changes in sea urchin cytoplasmic morphogens can generate sea urchins that develop without larvae but which still become sea urchins. In fact, von Baer was well aware that there was a bottleneck in vertebrate development. There is a stage (called the "pharyngula") where the embryos of all the mammalian classes look very similar. From here, embryonic development diverges. However, birds, reptiles, and fish arrived at this stage after meroblastic cleavages of different sorts, amphibians get to the same stage by way of radial holoblastic cleavage, and mammals reach the same place after constructing a blastocyst, chorion, and amnion. The earliest stages of development, then, appear to be extremely plastic. Similarly, the later stages are very different, as the different phenotypes of mice, sunfish, snakes, and newts amply demonstrate. There is something in the middle of development which appear to be invariant.

Raff argues that the formation of new bauplans is inhibited by the need for global sequences of induction during the neurula stage. Before that period, there are few inductive events. After that period, there are a great many inductive events, but most all of them are local. During early organogenesis, however, there are several inductive events occurring simultaneously that are global in nature. In vertebrates, to use von Baer's example, the earliest stages involve specifying axes and directing gastrulation. Induction has not happened on a large scale. Moreover, as Wray and Raff have shown [17], there is a great deal of regulative ability here such that small changes in morphogen distribution or the position of cleavage planes can be accommodated. After the major body plan is fixed, inductions occur all over the body, but are compartmentalized into discreat organ-forming systems. The lens induces the cornea, and if it fails to do so, only the eye is affected. Similarly, there are inductions in the skin to form feathers, scales, or fur. If they do not occur, the skin or patch of skin may lack these structures. But during early organogenesis, the interactions are more global [18]. Failure to have the heart at a certain place can affect the induction of eyes [19]. Failure to induce the mesoderm in a certain region leads to malformations of the kidney, limbs, and tails. It is this stage that constrains evolution and which typifies the vertebrate phylum.

IV. CORRELATIVE DEVELOPMENT

In addition to constraining evolution within certain bounds, induction also enables evolutionary change to occur without destroying the intricate relationships between the bodily parts. Induction allows the phenomenon of correlative interaction. This concept was drawn from observations that one part of an organ caused the other parts to form, and that these inductions could be reciprocal. This idea that induction leads to the harmonizing of intricate structures such as the eye or the limb excited the founders of developmental mechanics. Curt Herbst [20], one of these founders, celebrated the potentials of induction, and claimed that it might be possible to resolve the entire process of embryogenesis into a series of inductions exerted by one embryonic region upon another. He noted, for instance, that bone structure was influenced by blood vasculature, that taste buds were induced by nerves, and that the lens was induced by the presumptive retina. The harmony of relationships that reciprocal induction generates between organs and between parts of organs was also reviewed by Spemann in 1907 [21].

In many instances, when one part of an organ changes, most all other parts of that organ also have to change. Raissa Berg [22] has called these covarying constellations of characters "correlation pleiades". One of the best examples of such "pleiades" involves the formation of the vertebrate jaw. I am not going to review the evolutionary history leading up to the vertebrate jaw; Stephen Gould has recently done us that service [23]. Rather, I will merely note some changes in jaw structure that have occurred since humans started changing the face of the earth. Humans have artificially selected different facial characteristics in their domesticated animals. In dogs, there are numerous snout shapes from the narrow jaws of German shepherds to the flat faces of chows and bulldogs. Each variation is genetically determined, and it is important to note that each represents a harmonious rearrangement of the different bones with each other and with their muscular attachments. In some cases, such as bull dogs, the breed is selected for a wide face with very little angle between head and jaw. Other breeds, such as the collie, are selected for narrow snouts with a long jaw protruding away from the head. All breeds can move their jaws, shake their heads, and bark, despite the differences in the ways their bones are shaped or positioned. As the skeletal elements were selected, so were the muscles that moved them, the nerves that controlled these movements, and the blood vessels that fed them. This coordination is not quite universal, however. In dogs with greatly shortened faces (such as bulldogs), the skin has not coordinated its development with the bones and therefore hangs in folds from the head [24].

Correlated development has also been shown experimentally. Repeating earlier experiments of Hampe, Gerd Müller [25] inserted a barrier gold foil into the prechondrogenic hindlimb bud of a 3.5 day chick embryo. This barrier separated the region of tibia formation from that of fibula formation. The results of these experiments were two-fold. First, the tibia is shortened and the fibula bows and retains its attachment to the fibulare. Such connections are not usually seen in birds, but they are characteristic of reptiles and the early avian, *Archaeopterix*. Second, the *musculature* of the hindlimb undergoes parallel changes with the bones. Three of the muscles that attach to these bones now show characteristic reptilian patterns of insertion. It seems, therefore, that experimental manipulations that alter the development of one part of the mesodermal limb-forming field also alter the development of other mesodermal components, as well. As with the correlated progression seen in face development, these changes all appear to be due to interactions within a field, in this case, the chick hindlimb field. These are not global effects and can occur independently of the other portions of the body. Thus, induction enables the realignment of embryonic parts such that when one part changes, the other parts can also change.

V. TRANSFER OF INDUCTIVE COMPETENCE

Another way that induction can be used to create developmental changes that may be selected for during evolution is by *transfer of competence*. This idea, popularized by C. H. Waddington, enables the selection of new phenotypes that are favored for survival. In 1936, Waddington, Needham, and Brachet [26] made an unexpected discovery. They had been hoping to find a specific factor from the notochord that would induce neural plates, but instead found that a large variety of natural and artificial compounds were able to cause this induction. This caused them to reinterpret their data and to suggest that the actual neuralizing factor lies dormant in the competent ectoderm, and that a variety of factors were able to release the neuralizing factor from its inhibitor. Waddington then began to focus on the competent cells, rather than on the inducing cells. He thought that many things could act as inducers, but the competent tissue had to have something within it that allowed it to respond to these chemicals. The inducer, he wrote [27], was only the push. It was the competence that was genetically controlled and which was responsible for the details of the development.

Since competence could be achieved independently from an inducer, and since different compounds could induce the same developmental process in this competent tissue, Waddington proposed that a given competent tissue could transfer its ability to respond from one inducing stimulus to another. Moreover, these inducers could be either internal or external. External inducers were already known, and Waddington used the case of sex determination in *Bonnelia* as such an example. He also noted that the ability to form callouses on those areas of skin that abrade the ground was such an example. Here, the skin cells had the ability to form a callus if induced by friction. The genes could respond by causing the proliferation of cells to form the callous structure. While such examples of *environmentally*-induced callus formation are widespread, the ostrich is *born* with them. Waddington hypothesized that since the skin cells were already competent to be induced by friction, they could be induced by other things as well. As ostriches evolved, a mutation appeared that enabled the skin cells to respond to a substance within the embryo. In this way, a trait that had been induced by the environment became part of the genetic heritage of the organism and could be selected. He called this phenomena "genetic assimilation" [28]. This term is unfortunate for many reasons. First, it suggests that changes can only go from externally induced to internally induced. Second, it suggests the very Lamarckian and Lysenkoist notions of evolution that Waddington was fighting against. The term "transfer of competence" is more useful and it can be applied to several areas of developmental biology. I wish to show that there are indeed good reasons for thinking that transfers of competence have occurred and are occurring.

Ant castes. One place where such transfers of competence may be seen is in the evolution of caste separation in ant species. Ant colonies are predominantly female, and the females can be extremely polymorphic. The two major types of females are the worker and the gyne. The gyne is a potential queen. In more specialized castes, a larger worker, the soldier, is also seen. These castes are usually determined by the levels of juvenile hormone seen by the developing larva. If the larva has more juvenile hormone, the larva grows, and this allometric growth forms larger jaws or allows the development of reproductive organs. The determination of the amount of juvenile hormone in the larva can be nutritionally controlled or regulated internally through maternal hormones that act during embryogenesis.

The developmental mechanisms of caste determination have been analysed by Diana Wheeler [29]. In most species, ant larvae are bipotential until near pupation. They are potentially able to become either workers or gynes. At the critical time, differential nutrition switches a larva into one or the other pathway. In *Myrmica rubra*, only larvae that overwinter remain bipotential. After winter, the queen stimulates workers to *under*feed the last instar larvae. This means that as long as there is a queen, no new ones can result. If the larvae are fed, they can becomes gynes. Thus, larvae remain bipotential until late in their last instar.

In *Pheidole pallidula*, however, the queens appear to control caste determination during embryogenesis. There are *no* bipotential larvae in this species; caste determination being decided completely within the embryo. Gynes are produced from those eggs laid just after hibernation has ended, so it is assumed that the eggs laid at different times are biochemically different. Although there appears to be controversy over how the gyne-eggs and the worker-eggs differ, we have here an example of caste determination occurring within the embryo.

Transfer of competence in fish sex determination. If competent tissue can change their developmental triggers from environmental stimuli to internal inducers encoded by the genome, then it would stand to reason that the reverse could also occur. Under some selective conditions, one would expect genetic induction to be replaced by environmental induction. This phenomenon might be the case in certain types of environmental sex determination. Environmental sex determination is found among certain invertebrates (most notoriously, *Bonnelia viridis*) and among vertebrates such as fish, amphibians, and reptiles. In fact, most reptiles have their sex determined rather late in their embryonic development by the temperature. Charnov and Bull [30] have argued that environmental sex determination would be adaptive in certain habitats characterized by their patchiness and by having certain regions wherein it is better to be male and other regions wherein it is better to be female. Conover and Heins [31] provide evidence that in certain fish, females benefit from being larger, since size translates into higher fecundity. It is an advantage to be born early in the breeding season if you are a female *Menidia*, since you would have a longer feeding season and would be grow larger. In the males, size is of no importance. Conover and Heins showed that in the southern range of *Menidia*, the females are indeed born early in the breeding season. Temperature appears to play a major role. However, in the northern reaches of its range, the same species shows no environmental sex determination. Rather, a 1:1 ratio is generated at all temperatures. The authors speculate that this is because the more northern populations have a very short feeding season so that there is no advantage for a female to be born earlier. Thus, this species of fish has environmental sex determination in those regions where it is adaptive, and genotypic sex determination in those regions where it is not. Here again, one sees that the environment can induce the sexual phenotype or it can be a property of the genome as it is with most mammals. Transfer of competence between environmental and genomic inducers is found throughout nature, and it is seen in several mammalian species as well. Delayed implantation of the embryo into the uterus, for instance, occurs in the western variety of spotted skunk (*Spirogale putorius*) but not among its Eastern cousins [32].

VI. TRANSFER OF COMPETENCE IN THE IMMUNE SYSTEM

Let me shift the story of induction from the creation of self to the maintenance of self. In particular, I want to talk about the inductive interactions occurring every second in our immune system. The immune system, first and foremost, is a sensory system that is induced by environmental substances. It has recently come to light that we have several immune systems in our body. Some of them are relatively primitive, some are incredibly sophisticated. In one of our primitive immune systems, common bacterial products induce the differentiation of immunocompetent cells. In other words, we have evolved a system where the bacteria induce their own destroyers.

Bacteria are among the most widespread and dangerous antigens normally encountered by us metazoans. To a bacterium, we are a movable feast, properly warmed to 37°C. As our macrophages develop, they acquire three receptors for the major component of gram-negative bacterial cell walls, bacterial lipopolysaccharide [33]. When bacterial lipopolysaccharide binds to macrophages, these macrophages are induced to differentiate into activated macrophages. They begin releasing interleukin-1, tumor necrosis factor, and numerous arachadonic acid derivatives that can constrict smooth muscles and increase capillary permeability [34]. In other words, macrophages are *competent* to respond to gram-negative bacteria without any specific immunoglobulin or T cell antigen receptor. The ability to bind gram-negative bacteria is also seen in invertebrate coelomocytes [35]. Indeed, the standard assay for bacterial lipopolysaccharide (the endotoxin assay) involves homogenates of horse-shoe crab amebocytes [36, 37]. Thus, the competence to respond non-specifically to gram-negative bacteria is a longstanding ability of macrophage-like cells.

In birds and mammals, macrophages are not the only cells competent to respond non-specifically to gram-negative bacteria. If one wishes to get a population of dividing cells for a karyotype, technicians will take a blood sample and add a drop of bacterial lipopolysaccharide to it. This causes the B lymphocytes to proliferate non-specifically. Moreover, these endotoxin-induced B lymphocytes differentiate and secrete their IgM antibodies! In other words, no matter what the specific immunoglobulin being made by the B cell, the cell will divide and differentiate when exposed to the sloughed product of gram-negative bacteria. B cells, then, are *competent* to being induced non-specifically to this environmental stimulus.

Similarly, a large subset of T lymphocytes are competent to respond to another common antigen, staphylococcus enterotoxin A [38, 39]. No matter what the T cell receptor normally binds, any T lymphocyte bearing particular beta light chains will be stimulated by this microbial protein. So T cells, too, are competent to be induced relatively non-specifically by a common environmental inducer. White and her colleagues have called such inducers "superantigens". I suggest that the actions of these bacterial superantigens represent the non-specific induction of immunocompetent cells by common environmental evocators.

Transfer of competence by transfer of receptors. Now let me return to the notion of competence transfer. The non-specific immune system that I just mentioned is only the innate, the primitive, immune system. In mammals and birds, the immune system has become a great deal more sophisticated. In addition to responding to relatively non-specific environmental cues, we can respond to specific environmental inducers through their cooperation with internal inducers. According to Waddington, the competent cell evolved a genetically controlled (canalyzed) pathway by which it could respond to an inducer. When the competency of the cell was transferred from one inducer to another, the receptor for the signal changes, but the pathway distal to the signal remained the same.

The importance of this principle in the immune system can be seen by looking at the differentiation of the B cell into a plasma cell. This is the adult analogue of an embryonic induction. A B cell is a cell with a huge nucleus and very little secretory capacity. It differentiates into a plasma cell with its relatively small nucleus and its huge web of rough endoplasmic reticulum. The inducing cell is called the helper T lymphocyte. According to the clonal selection hypothesis, the specificity of B cell induction is controlled by its cell surface antigen receptors. Each B cell synthesizes one and only one type of antibody and places it in its cell membrane to bind antigen. Only those B cells that bind antigen are able to divide and differentiate into plasma cells. When the binding of antigens occurs on these cells, the B cell begins to synthesize the receptors for interleukins 4, 5, and 6. These proteins, sometimes called B cell growth and differentiation factors, are inducers secreted by the helper T cells. Thus, there are two inducers of plasma cell differentiation, an external antigenic inducer that controls the specificity of induction, and a set of internal B cell inducers secreted by activated T helper cells. First, the B cell acquires the competence to bind an external inducer (the antigen). This binding then makes it competent to respond to the internal inducers (interleukins).

This provides us an interesting perspective. Once any B cell is triggered to differentiate and divide, it becomes competent to respond to the non-specific interleukins. In other words, the specific competence is provided by the antigen receptor. If a different antigen is presented, a different set of B cells becomes competent to divide and differentiate into antibody- secreting plasm cells. Thus, the binding of antigen to a B cell antigen receptor triggers into action a *preset program* for cell division and differentiation.

We also see transfer of competence when we look at the cell division pathway of the B cell. A G-protein that is bound to the cell surface immunoglobulin activates phospholipase C. Phospholipase C splits phosphatidylinositol bisphosphate PIP_2 into inositol triphosphate which releases calcium from the endoplasmic reticulum and into diacylglycerol which activates protein kinase C with a resultant elevation in intracellular pH. Together, these signals activate the nuclear protooncogenes that initiate cell division [40]. Interestingly enough, these are the same reactions that happen when antigens bind to the T cell receptor. So it appears that lymphocytes, both B and T, use the same pathway for cell division. The antigen binds to the receptor, be it immunoglobulin or T cell receptor, and the signal activates a G protein which activates phospholipase C to split PIP_2 into inositol triphosphate and diacylglycerol; the result being the activation of the nuclear signals to replicate. The reactions are the same. All that changes is the cell surface receptor.

But this should be an already familiar story. Indeed, the lymphocytes are latecomers to this pathway; for this pathway is the series of reactions by which every cell in the body divides. A more general scheme is known in which a growth factor binds to the growth factor receptor. This in turn, activates the G protein which activates phospholipase C which starts the process in motion. But this, too, is not the entire story. The same pathway of division is seen in the reactions that activate the fertilized egg. Here, the activator isn't an antigen or a peptide growth factor; it's a sperm. So here's a pretty old pathway, a super-highway (Fig. 3). If you want a cell to divide in response to something, merely change the receptor. After that, the entire pathway is in place. The pathway exists in cells, and if the cell is going to divide, it merely needs a stimulus to enter that pathway. Moreover, one can see that the pathway is independent of the stimulus that initiates the cell division. Different receptors can be inserted into the membrane, causing the pathway to become activated by new compounds. Indeed, one can experimentally add the gene for a different receptor and get *Xenopus* eggs to initiate cell division when neuro-transmitters or epidermal growth factor are added to their culture [41]. This is also the mechanism by which certain tumors develop, as in the case when the

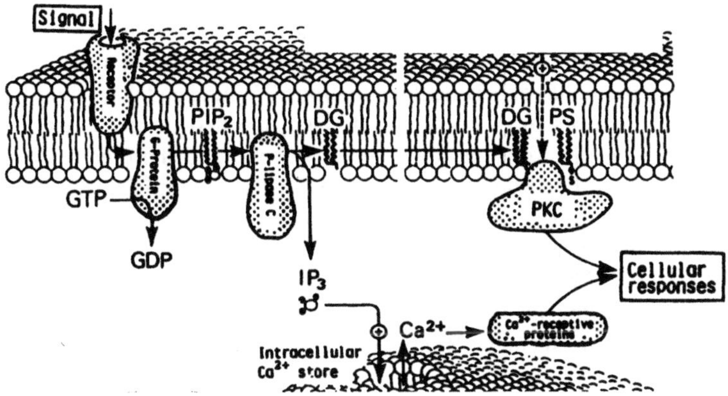

Fig. 3. General pathway for inducing cellular changes from the exterior of the cell.

simian sarcoma virus starts producing platelet-derived growth factor in a cell that already has receptors for it. The ability to transfer its competence from one stimulus to another makes this a cellular analogue of competence transfer.

In the development of deuterostomes, we see that induction is crucial to the generation of self. We saw this in the ability of induction to form an organism out of potentially autonomous and totipotent cells. This induction can even create a single self out of two or more initial embryos. We also saw that in the immune system, induction is necessary for the maintenance of self. The macrophages, T cells, and even in some instances, the B cells, are able to be induced by that which is not self.

INDUCTION AND DIALECTIC

There can be few biological events more obviously dialectical than induction. Here, two tissues interact to produce a novel structure which could not be accounted for without these interactions. A notochord alone generates part of the spinal cartilage. The ectoderm alone would produce only skin. Put them together, and the central nervous system forms. A ureteric bud without a metanephrogenic mesenchyme doesn't branch. A metanephric mesenchyme without a ureteric bud doesn't form a tubular epithelium. Put them together and you get a functional kidney.

The dialectical nature of induction has not been ignored. Workers such as Waddington, Needham, Zavadovsky, and Cooke have celebrated the dialectical aspects of induction. Indeed, the notion of competence came largely from Waddington's notions of dialectic as drawn from Alfred North Whitehead. As Waddington would later remark, "I tried to put the Whiteheadian outlook to actual use in particular experimental situations." Whitehead was a philosopher of interacting processes of becoming. Indeed, the first three elements of his philosophy were the concepts of system, process, and the creative advance into novelty [42]. For Whitehead, no thing existed except in relation to other things, and these *nexus* were always changing, allowing new *nexus* to form. All relationships were in the process of becoming, and all things were linked within systems. These were very useful concepts for biologists who were engaged in studying process, interrelationships, and the emergence of new forms. Many biologists in the Biotheoretical Gathering were deeply influenced by these notions [43].

> As far as scientific practice is concerned, the lessons to be learned from Whitehead were ... from his replacement of "things" by processes which have an individual character which depends upon the "concrescence" into a unity of very many relations with other processes.

This certainly seems to be the case in *Organisers and Genes*, which reads like a Whiteheadian primer on embryology. Throughout this book, Waddington stressed "interrelationships", "causal networks", and "interconnections". (All these terms can be found on the first page!) He sought [44] to identify—not the inducer—but the "causal network underlying this particular process of differentiation", and hoped "to know the whole complex system of actions and interactions which constitute the differentiation."

This tendency to think in terms of process, system, and interaction also distinguished Waddington's approach to induction from those of most other investigators. In 1940 he was not talking so much about the inducer as he was about "The evocator-competence reaction". Similarly, he was not so much concerned with the action of genes as with "the system of tracks and their genetic control". For Waddington, the parts of the embryo were always in dialogue. The evocator was nothing without the competent responding tissue, and the responding tissue was nothing without the evocator. They were linked in a system. Moreover, Waddington interpreted the results of embryology to show that these components interacted to effect development. Waddington looked upon an organism as a developing system, and claimed that natural selection worked upon aspects of development. Evolution was accomplished through hereditable changes in an organism's development. He credited this idea to his Whiteheadian outlook, and called his first paper on evolutionary topics (1941) "the evolution of developing systems."

Waddington was also a Socialist, and he interpreted his data within a frame-work provided by dialectical materialism. In a paper read before a Marxist student organization, he explained his work on competence:

> The analysis of the material which reacts with the organiser is more instructive. The various regions of the egg are only capable of reacting with the organiser during a certain period. During this period, they were, originally, said to have a reaction-potency; and the nature of this was not discussed. Dialectical materialism would, surely, suggest that it could profitably be considered as made up of a pair of contrasting tendencies. Now something very like this is the case. [45].

But he was a critical Socialist. If his data departed from the laws set forth by Hegel and Engels, then dialectical materialism, no matter how correct in its overall analysis, was wrong in the particulars. Waddington told this same audience

> This situation is certainly not one which can be adequately formulated in the aphorism of "the interpenetration of opposites"—and still less in the even more esoteric doctrine of the negation of the negation". At best one could water down "the interpenetration of opposites" to "the interaction of different entities". [45].

Waddington was extremely consistent in seeing biological events as interactive systems. Just as the inducer and the responding cell were partners in an interactive system, so were the nucleus and the cytoplasm. Moreover, whereas most of the investigators of his time (and ours) would say that the nucleus and the inducer were the active agents, putting forth substances that the more passive cytoplasm or responding cells would react to, Waddington empowered both the cytoplasm and the competent cells, seeing them as important partners. In fact, for Waddington, it was the cytoplasm and the competent cells that had the specificity in the partnership. He saw that neither the nucleus nor the inducer is the specific component of their respective systems. The nucleus contained all the genes, while it was the cytoplasm that differed. In his *Principles of Embryology* (1956), Waddington entitles his first chapter on developmental genetics, "The Activation of Genes by the Cytoplasm". He uses four sets of examples [46]. First, he cites the mosaic eggs of the tunicate where the portion of egg cytoplasm contained in each early blastomere determines the fate of the cell. Second, he uses induction as an example of diffusible substances activating a concrescence of genes, a canalized pathway. His third example was that *Drosophila* larval chromosome puffs differed depending on the cytoplasm of the chromosome, and his fourth example dealt with cytoplasmic inheritance in *Paramecia* wherein the presence or absence of a cytoplasmic factor determined whether the nuclear-encoded G antigen was expressed or not. In all cases, there is a dialectical relationship between nucleus and cytoplasm and the specificity is provided by the cytoplasm.

In this 1956 book, Waddington also calls attention to the work being done on adaptive enzymes. Here, Waddington finds evidence for the genome's being controlled by the environment. When Jacob and Monod formulate their operon model, Waddington immediately includes it into his new embryology text, and displays it as a model for embryonic induction. To Waddington, the operon model showed the interaction between cytoplasm and nucleus. Not only did the nucleus make proteins, but these proteins could return to the nucleus and tell the genome which proteins to synthesize. Second, the types of protein made could be determined by substances entering the cells from the outside. In *E. coli*, this would be lactose. In the competent amphibian ectoderm, this could be the inducer. All the inducer would have to do would be to inhibit the inhibitor. If Waddington saw this as the "negation of negations", he didn't say. However, his dialectical view of nucleus and cytoplasm and of inducer and competent cell allowed him to portray the induction of bacterial enzymes as being a model for the induction of vertebrate organs [47]. Thus, by a dialectics philosophy of organism, Waddington attempted to reconcile genetics and embryology. Out of this reciprocal induction, a new science was generated upon what had been embryology, namely, developmental genetics.

Swarthmore College, Pennsylvania

REFERENCES

1. Kirk, D., personal communication.
2. Driesch, H. 1893. The potency of of the first two cleavage cells in echinoderm development. Experimental production of partial and double formations. In *Foundations of Experimental Embryology*. B.H. Willier and J.M. Oppenheimer, editors. Hafner, N.Y.
3. Venuti, J.M. and W.R. Jeffery (1989). Cell lineage and determination of cell fate in ascidian embryos. *Int. J. Dev. Biol.* 33: 197–212.
4. Buss, L.W. (1987). *The Evolution of Individuality*. Princeton University Press, Princeton.
5. Henry, J.J., S. Amemiya, G.A. Wray, and R.A. Raff. 1989. Early inductive interactions are involved in restricting cell fates of mesomeres in sea urchin embryos. *Dev. Biol.* 136: 140 –153.
6. Ettensohn, C.A. and D.R. McClay. 1988. Cell lineage conversion in the sea urchin embryo. *Dev. Biol.* 125: 396–409.
7. Hurley, D.L., L.M. Angerer, and R.C. Angerer. 1989. Altered expression of spatially regulated embryonic genes in the progeny of separated sea urchin blastomeres. *Develop.* 106: 567–579.
8. Mintz, B. 1970. Clonal expression in allophenic mice. *Symp. Int. Soc. Cell Biol.* 9: 15.
9. Markert, C.L. and R.M. Petters. 1978. Manufactured hexaparental mice show that adults are derived from three embryonic cells. *Science* 202: 56–58.
10. Papaioannou, V. 1979. Interactions between mouse embryos and teratocarcinomas. *In Cell*

Lineage, Stem Cells, and Cell Determination. N. LeDouarin, editor. Elsevier North-Holland, New York. 141–155.

11. Spemann, H. 1943. *Forschung und Leben.* F.W. Spemann, editor. J. Engelhorn, Stuttgart. 180 –181.
12. Garstang, W. 1922. The theory of recapitulation: a critical restatement of the biogenetic law. *J. Linn. Soc. Zool.* 35: 81–101.
13. Roux, W. 1894. The problems, methods, and scope of developmental mechanics. *Biological Lectures of the Marine Biology Laboratory, Woods Holl.* Ginn and Company, Boston. 149–190.
14. Gould, S.J. and R.C. Lewontin. 1979. The spandrels of San Marcos and the Panglossian paradigm: A critique of the adaptationist program. *Proc. Roy. Soc. (London) B* 205: 581–598.
15. Waddington, C.H. 1938. The morphogenetic function of a vestigial organ in the chick. *J. Exp. Biol.* 15: 371–376.
16. Raff, R.A. 1990. Personal communication. Santa Fe Institute Symposium on The Theoretical Foundations of Development and Evolution.
17. Wray, G.A. and R.A. Raff. 1989. Evolutionary modification of cell lineage in the direct-developing sea urchin *Heliocidaris erythrogramma. Dev. Biol.* 132: 458–470.
18. Slack, J.M. W. 1983. *From Egg to Embryo: Determinative Events in Early Development.* Cambridge U. Press; p. 64.
19. Jacobson, A.G. 1966. Inductive processes in embryonic development. *Science* 152: 25–34.
20. Saha, M. 1991. Spemann seen through a lens. *In A Conceptual History of Modern Embryology*. S.F. Gilbert, editor, Plenum Press New York. In press.
21. Spemann, H. 1907. Zum Problem der Correlation in der tierischen Entwicklung. *Verhandl. d. deutche zool. Gesell.* 17: 22–49.
22. Berg, R.L. 1960. Evolutionary significance of correlation pleiades. *Evolut.* 14: 171–180.
23. Gould, S.J. 1990. An earful of jaw. *Nat. Hist.* 3/90: 12–23.
24. Stockard, C.R. 1941. The Genetic and Endocrine Basis for Differences in Form and Behaviour as Elucidated by Studies of Contrasted Pure-line Dog Breeds and Their Hybrids. *Amer. Anat. Memoirs* 19. Wistar Institute, Philadelphia.
25. Müller, G.B. 1989. Ancestral patterns in bird limb development: A new look at Hampé's experiment. *J. Evol. Biol.* 1: 31–47.
26. Waddington, C.H., J. Needham, and J. Brachet. 1936. The activation of the evocator. *Proc. Roy. Soc. (London) B* 120: 173–207.
27. Waddington, C.H. 1940. *Organisers and Genes.* Cambridge University Press, Cambridge. For further analysis of Waddington's attempts to integrate development, genetics, and evolution, see Gilbert, S.F. (1991). Induction and the origins of developmental genetics. *In A Conceptual History of Modern Embryology.* S.F. Gilbert, editors. Plenum Press, N.Y.
28. Waddington, C.H. 1942. Canalization of development and the inheritance of acquired characters. *Nature* 150: 563–564.
29. Wheeler, D. 1986. Developmental and physiological determinants of caste in social hymenoptera: Evolutionary implications. *Amer. Nat.* 128: 13–34.
30. Charnov, E.L. and J.J. Bull. 1977. When is sex environmentally determined? *Nature* 266: 828–830.
31. Conover, D.O. and S.W. Heins. 1986. Adaptive variation in environmental and genetic sex determination in a fish. *Nature* 326: 496–498.

32. Renfree, M.B. 1982. Implantation and placentation. In *Reproduction in Mammals 2: Embryonic and Fetal Development.* C.R. Austin and R.V. Short, editors. Cambridge University Press, Cambridge.
33. Wright, S.D. and M.T, C. Jong.1986. Adhesion-promoting receptors on human macrophages recognize *Escherichia coli* by binding to lipopolysaccharide. *J. Exp. Med.* 164: 1876–1888.
34. Morrison, D.C. and R.J. Ulevitch.1978. The effects of bacterial endotoxins on host mediating systems. *Amer. J. Pathol.* 93: 527.
35. Bang, F.B. 1979. Ontogeny and phylogeny of the response to gram-negative enterotoxin among marine invertebrates. In *Biomedical Applications of the Horseshoe Crab (Limulidae).* E. Cohen, F.B. Bang, and J. Levin, editors. A.R. Liss, NY. pp. 109–123.
36. Hodes, D.S., E.J. Bottone, A. Hass, and H.L. Hodes. 1988. Reaction of *Bacillus subtilis* products with amebocyte lysates of the Japanese horseshoe crab, *Tachypleus tridentatus, J. Clin. Microbiol.* 26: 890–892.
37. Uragoh, K., K. Sueishi, T. Nakamura, and S. Iwanaga. 1988. A novel immunohistochemical method for *in vivo* detection of endotoxin using horseshoe crab factor C. *J. Histochem. Cytochem.* 36: 1275–1283.
38. Mollick, J.A., R.G. Cook, and R.R. Rich. 1989. Class II MHC molecules are specific receptors for *Staphylococcus* enterotoxin A. *Science* 244: 817–820.
39. Kappler, J. *et novem al.* 1989. Vβ-specific stimulation of human T cells by staphylococcal toxins. *Science* 244: 811–813.
40. Cambier, J. *et septem al.* 1987. Transmembrane signals and intracellular "second messengers" in the regulation of quiescent B-lymphocyte activation. *Immunol. Rev.* 95: 37–57.
41. Kline, D., L. Simoncini, G. Mandel, R.A. Maue, R.T. Kado, and L.A. Jaffe. 1988. Fertilization events induced by neurotransmitters after injection of mRNA in *Xenopus* eggs. *Science* 241: 464–467.
42. Whitehead, A.N. 1929. *Process and Reality*, Cambridge University Press, Cambridge, p. 151.
43. Waddingtoh, C.H. 1975. The practical consequences of metaphysical beliefs on a biologist's work: An autobiographical note. *The Evolution of an Evolutionist*, Cornell University Press, Ithica, pp. 3 and 5.
44. Waddington, C.H. 1940. *Organisers and Genes, op. cit.*, pp. 3–4.
45. Waddington, C.H. unpublished lecture notes *circa* 1934: Marxism and Biology, University of Edinburgh archives. Courtesy of Dr. P. Abir-Am.
46. Waddington, C.H.1956. *Principles of Embryology*, Macmillan, New York, p. 348.
47. For more details concerning how Waddington used the parallels between microbial and embryological inductions to create a paradigm for developmental genetics, see Gilbert, 1991 [27], and Gilbert (in press), Adaptive enzymes and the entrance of molecular biology into embryology, in *History and Philosophy of Molecular Biology: New Perspectives.* S. Sarkar (ed.) Dordrecht Kluwer Academic Publishers.

DORION SAGAN AND LYNN MARGULIS

EPILOGUE: THE UNCUT SELF

> Speeches and books were assigned real authors, other than mythical or important religious figures, only when the author became subject to punishment and to the extent that his discourse was considered transgressive.
> Foucault [1]

full circle, not based on the rectilinear frame of reference of a painting, mirror, house, or book, and with neither "inside" nor "outside" but according to the single surface of a Moebius strip. This is not the classical Cartesian model of self, with a vital ensouled *res cogitans* surrounded by that predictable world of Newtonian mechanisms of the *res extensa*; it is closer to Maturana and Varela's conception of autopoiesis, a completely self-making, self-referring, tautologically delimited entity at the various levels of cell, organism, and cognition [2]. It would be premature to accuse us therefore of a debilitating biomysticism, of pandering to deconstructive fashion, or, indeed, of fomenting an academic "lunacy" or "criminality" that merits ostracism from scientific society, smoothly sealed by peer review and by the standards of what Fleck calls a "thought collective" [3]. Nor would it be timely to label and dismiss us as antirational or solipsist.

All such locutions stem from the mundane reason, the ethnocentric conception of self which precisely here comes under question. "The philosophy of the subject," writes Jürgen Habermas, "is by no means an absolutely reifying power that imprisons all discursive thought and leaves open nothing but a flight into the immediacy of mystical ecstasy" [4]. On the one hand we position ourselves beyond the 16th-century European Enlightenment, its faith in reason, the arrogance of its secular priests, and the later Darwinian smarm. In this sense we have a post-structuralist, post-modern, non-representational view of self. On the other hand, we dialectically question this position, motionlessly turning it inside out, as it were, and paying heed to the successes of scientific positivism and biochemical reductionism—movements which, philosophically, cannot (at least provisionally) be disentangled from the perva-

Alfred I. Tauber (ed.), Organism and the Origins of Self, 361–374.

sive influence of IndoEuropean grammar, subject-verb-object structures, and the like. In this sense, our view of the organism is less ontological and more biological; the order of metaphysics and physics, the primacy of philosophy over biology, undergoes a reversal more in keeping with the academic notions of self, and the anthological effort to enclose in a coherent, comprehensive, rectilinear manner. Membrane-bounded indeed.

But the membrane is no concrete, literal, self-possessed wall; it is a self-maintained and constantly changing semi-permeable barrier. The idea of the semi-permeable membrane permits us to jump organizational levels, from intraorganismic cell to cellular organism to organismic ecosystem and biosphere. Whether we are discussing the disappearing membranes of endosymbiotic bacteria on their way to becoming organelles, or the breakdown within the global human socius of the Berlin Wall, we must revise this rectilinear notion of the self, of the bounded I. Alan Watts pejoratively referred to it as the "skin-encapsulated ego"; indeed, even though so deeply entrenched, this bounded sense of "self" seems to us to be thoroughly natural—it is neither an historical or cultural universal. For example, the Melanesians of New Caledonia, known in French as the Canaque, are unaware that the body is an element that they themselves possess; the Melanesians cannot see the body as "one of the elements of the individual" [5]. So, too, the Homeric epics never make mention of a body—the flesh-enclosed entity we today take for granted as the definable material self—they speak only of what we would think of as the body's parts, for example, "fleet legs" and "sinewy arms" [6]. "The idea of the 'self in a case...,'" writes Norbert Elias, "is one of the recurrent *leitmotifs* of a modern philosophy, from the thinking subject of Descartes, Leibniz' windowless monads and the Kantian subject of knowledge (who from his aprioristic shell can never quite break through to the 'thing in itself') to the more recent extension of the same basic idea of the entirely self-sufficient individual" [7].

Psychoanalytically, the sense of self on the level of personhood has been construed to be a convenient fiction, an effect of infantile representation that is jubilant but essentially ersatz. (Etymologically, the word "person" means "to sound through"; coming from the Greek *persona*, it refers to a dramatic mask with a speaking hole.) According to Lacan [8], the jubilation that creates the essentially false and paranoid ego in the infant occurs when its gaze confronts the image of a fully contoured and coordinated body at the very time (6–18 months) it is beleaguered by a motor incapacity that renders it more helpless and defenseless than perhaps any other mammal of the same age. The intense motor incapacity and uncoordination, resulting from "prematuration" (or, in evolutionary terms, from neoteny), engulfs the infant in an almost cinematographic world of uncontrollable visions. One of these mystic-like visions is of itself (or the mother) with a coordination and in a place where it does not in fact exist, along the rectilinear mirror plane. This form of mystical identificatory representation with an image or imago Lacan designates as "image-inary." As a fictional form of the I, it is comforting and effects the discrete sense of self from toddler on into adulthood, the sense of self which has been catered to by American ego psychology in contradistinction to the original Freudian insights and painstaking deconstructions of a psyche (psycho-analysis) formerly presumed to be whole. The Lacanian psychoanalytic revamping of the myth of Narcissus suggests that what we perceive to be our body, as the locus of our "self," is in fact plastic, malleable; and indeed, the lability of the imaginary view of self has come to the fore in the first technology-mediated glimpses of a new image of the human body: Earth from space [9]. This rapidly proliferating image, now recognized as our ecological or biospheric home, will, with further population growth, interspecies interdependencies and optimization of global media, begin to be re-cognized as body. Already the shift from biosphere-as-home to biosphere-as-body has become apparent in the scientific work of James Lovelock, whose Gaia hypothesis, with mythical allusions of its own, has inspired a planetary search for "geophysiological" climatological and biogeochemical mechanisms [10]. Biospheric individuality was already recognized by Julian Huxley, who wrote:

> the whole organic world constitutes a single great individual, vague and badly co-ordinated it is true, but none the less a continuing whole with inter-dependent parts: if some accident were to remove all the green plants, or all the bacteria, the rest of life would be unable to exist. This individuality, however, is an extremely imperfect one—the internal harmony and the subordination of the parts to the whole is almost infinitely less than in the body of a metazoan, and is thus very wasteful; instead of one part distributing its surplus among the other parts and living peaceably itself on what is left, the transference of food from one unit to another is usually attended with the total or partial destruction of one of its units [11].

As positivists, materialists, or physical reductionists in the western scientific tradition, we would like to think that the picture of the body as an adequately closed topological surface is necessary and sufficient prima facie self-evidence—for the self. And so it is within a certain rectilinear closure. However, as we—and even this coauthorial "we" must be put in quotation marks as we ponder the self, the subject, person, etc.—intimated, the egotistic I is clear only in the sense of a fundamentally fictional or topologically displaced mirror image; there is nothing behind the mirror. Emphasizing tactility rather than vision, on a sensual level it is easy to imagine a conception of the human environment as beginning with the fingernails, hair, bones and other substances no longer considered to be body parts since they are bereft of sensation. Conversely, technological introjection exemplified by devices such as tele-vision (video, movies, etc.) and tele-portation (automobiles, airplanes, and so forth) suggest a topological extension of the human into what formerly would have been considered the environment. Therefore the body, the material or corporeal basis for "self," has no absolute time-independent skin-encapsulated topological fixity. It is a sociolinguistic psychoanalytic evolutionary construct. Mucus, excrement, urine, spittle, corpses, pornography, and other detachments from and marginal representations of the human body call its essential hegemony, its universal nature, into question. Chastising the Spanish artist for painting unrepresentative cubistic abstractions, a layman withdrew a photograph of his wife from his pocket, and held it up to Picasso with the admonition, "Why can't you paint realistically, like that?" "Is that what your wife really looks like?" Picasso asked. "Yes," replied the man. "Well, she's very small, and quite flat." Our working assumption of what the self is—like the layman's view of what his wife "really looks like"—is based on a model of representation that takes far too much for granted. Representation itself has, in post-modernist philosophy, fallen into disfavor in a manner similar, perhaps, to that in which figurative realistic painting fell into disfavor with the innovation of the camera. This does not mean that the possibilities of representational or propositional truth, of the correspondence theory of reality still so entrenched in science, is necessarily dead; on the other hand, the difficulties posed by the evidence of quantum mechanics, not least of which is the philosophical non-solution of the Copenhagen interpretation of the structure of the atom, suggest that most scientific models of reality may be neither so enlightened nor *au courant* as they assume. Indeed, what is in question is the very possibility of modeling reality at all.

Psychoanalytically, when we broach the topic of castration, amputation, dismemberment, the infant's polymorphic perverse sensations and perceptions of the body being, as in a picture by Hieronymus Bosch, in bits and pieces, is probably close to the true state of nature, if such a state there be. In other words, the infant's primordial pre-socialized experience of the world should not be considered inaccurate but rather, precisely because it precedes sociocultural linguistic norms, less prejudiced and potentially more "realistic." And, apart from parturition, there may be a biological basis for these perceptions which, later in life, are recalled as amputation, castration, dismemberment. Permitting ourselves a wee bit of abstraction here we splice in a couple more comments by Huxley:

> ... certain bits of organic machinery are of such a nature that it is physically impossible for the animal to live at all if they are seriously tampered with. It is just because our blood-circulation is so swift and efficient and our nervous system so splendidly centralized that damage to heart or brain means almost instant death to us, while a brainless frog will live for long, and a heart-less part of a worm not only will live but regenerate. Thus here again sacrifice is at the root ... and only by surrendering its powers of regeneration and reconstitution has life been able to achieve high individualities with the materials allotted her.... We have seen the totality of living things as a continuous slowly-advancing sheet of protoplasm out of which nature has been ceaselessly trying to carve systems complete and harmonious in themselves, isolable from all other things, and independent. But she has never been completely successful: the systems are never quite cut off, for each must take its origin in one or more pieces of previous system; they are never completely harmonious. [11]

Given the abiding prevalence of an image-inary or representational world view in Western science, it is impossible to overestimate the theoretical importance of this relatively abstract, nonrepresentational splicing or grafting which crosses cellular, species, and taxonomic boundaries. Light, no less than matter, cannot be understood simply as a collection of particles but must also be comprehended as a wave: with quantum mechanics the Democritean atomistic Newtonian world view has come to a functional end, although the momentum of scientific discourse has prevented it from reckoning with the consequences of this theoretical shipwreck.

Comparable to the end of the Newtonian age in physics, evidence of the dwindling of an atomistic model of organismic identity in the biological realm is reflected by the debate over the essential unit of selection in Darwinian evolution, whether it is really genetic, the gene—*inside* the organism—or the "individual" competing organism—as Darwin stressed—or group levels such as species or multicellular assemblages. Hierarchy theory entertains species and multicellular assemblages—extended phenotypes *outside* the organism and beyond the traditional confines of the self—to be the crucial units of selection [12]. Certainly the paradoxical notion of group selection seems necessary to explain epochal evolutionary transformations such as those from protoctist colonies to the first plants, animals, and fungi.

The minimal autopoietic, or living, system is the membrane-bounded cell. A cell, or any other autopoietic entity of even more complexity, undergoes continual chemical transformations easily recognizable as "being alive." In the process of this ubiquitous metabolism each living entity is materially contained within at least a membranous boundary of its own making. In addition to the universal plasma membrane of all living cells other boundaries, for example, skin, theca, or cuticles may be self-produced. Such borders include the black smooth skin of humpback whales, the glycocalyx of some amoebae, the hard over-wintering thecal coat of hydra eggs, or the waxy cuticle of a cactus. Minimally the autopoietic unit produces the plasma membrane but often cells and organisms make cellulosic walls, coccoliths, or siliceous spines – complex material extensions found just outside, adjacent or attached to the universally required membrane. All autopoietic entities continually construct, adjust, and reconstruct these dynamic physical structures by which they are bounded.

We recognize autopoietic entities as "individuals," or "individual organisms." A tree, a potted plant, a swimming euglena and a cat are immediately perceived as single living organisms. Minimally, all such autopoietic entities are comprised by at least one genomic system: a DNA-containing genome (i.e., the sum-total of all the genes of the organism) and the RNA-driven protein-synthetic, ribosome-studded internal cellular apparatus associated with that genome.

What is the lowest common denominator of individual life? The minimal autopoietic entity, a single genomic system, is a bacterial cell. Bacteria contain chromonemal DNA, that is, DNA uncoated with histone protein by DNA that codes, via RNA, for an accompanying protein synthetic system itself comprised of RNA and protein. This interacting, metabolizing unit of perhaps some 3000 identifiable genes and proteins bounded by dynamically changing membrane makes and is the bacterial genomic system. Live bacterial cells are single genomic entities in this sense. Whereas single-celled bacteria, uninfected with viruses or plasmids, are comprised of single genomic systems, those so infected have supernumerary genomes—both large (chromonemal) and small replicons (viruses, plasmids). Multicellular bacteria, e.g., *Polyangium, Fischerella, Arthromitus*—there are myriads of them—comprised of many copies of the same genomic system are thus polygenomic. Filamentous, tree-shaped, branched or spherical colonies, such organisms are comprised of homologous genomic systems in direct physical contact with each other. In some cases, like swarms of cyst-forming myxobacteria (e.g., *Chondromyces, Myxococcus*), the component genomes sense each other and fuse forming, a larger structure—no membranes are breached. In others, as when the akinetes of a cyanobacterium float away, the genomic systems disperse. Multicellular bacteria—*Stigmatella, Fischerella,* and the like (Fig. 1)—are polygenomic beings in which each of the comprising genomic systems has very recent common ancestors.

All organisms of greater morphological complexity than bacteria, i.e., nucleated or eukaryotic organisms (whether single or multicellular), are also polygenomic, yet all are comprised of heterologous genomic systems. Since

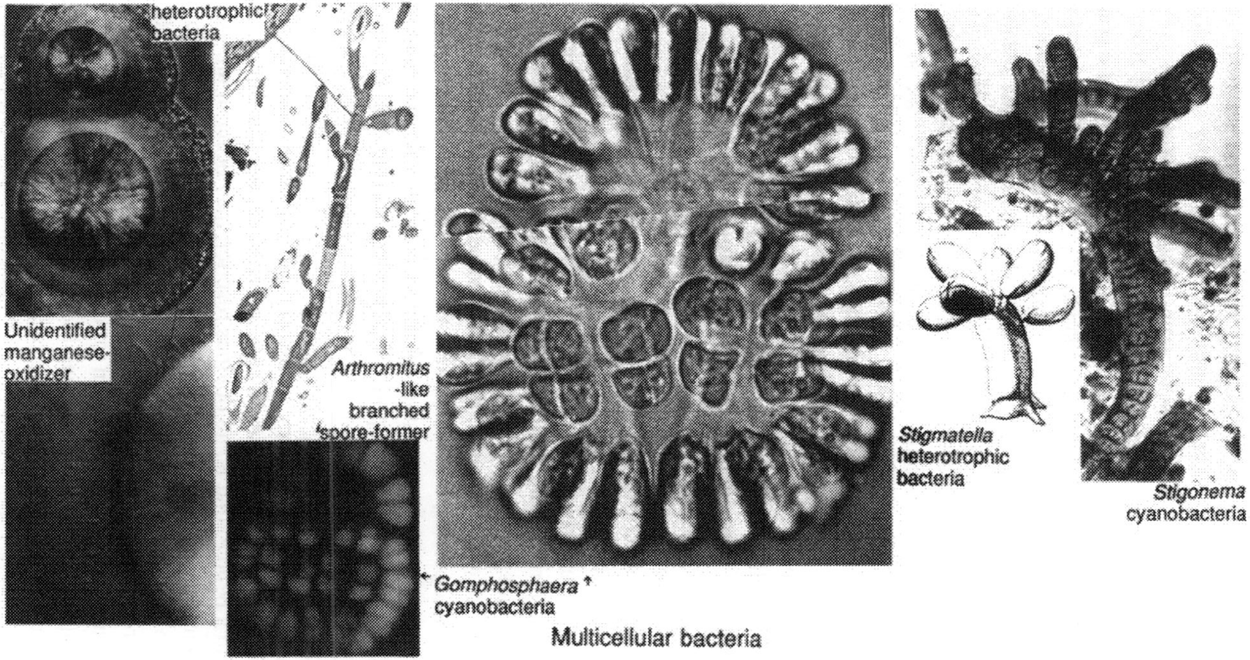

Fig. 1. Multicellular bacteria: examples. Epilithic *stigonema* sp., a cyanobacterium from Norway, grows attached to bare rock. *Stigmatella* sp., a heterotrophic soil myxobacterium, forms this bacteria. The single cells are capable of producing new "tree-like structures". Drawing by C. Lyons. *Gomphosphaera*, a cyanobacterium from a microbial met environment, lacks a single cell stage: it forms these colonies, which reproduce by fragmentation of the entire colony into two smaller, roughly equal colonies (the light micrograph taken with fluorescence microscopy at left indicates the distribution of chlorophyll). *Arthromitus*-like gram-positive spore-forming symbiont from the termite gut shows true branching because single cells are capable of growth at three different sites on the surface. At left are color photos of unidentified heterotrophiccolonial organisms in which the entire colony fragments into two. It is likely that most bacteria in nature are multicellular.

the organelles (nucleocytoplasm, mitochondria, plastids, and so forth) of eukaryotic cells had independent origin among the bacteria, any such cell—any eukaryotic genomic system—must be comprised of heterologous parts. Each component cell is derived as a chimera; ultimately it emerged from a diversity of bacterial ancestors with only remote common ancestry. In a plant cell, for example, the ancestor of the mitochondria is only remotely related to that of the chloroplast—both descend from gram-negative photosynthetic bacteria with complex respiratory pathways. Neither mitochondria nor plastids are very related to the nucleocytoplasm in which they are embedded. The nucleocytoplasm itself is of archaeobacterial ancestry. Such polygenomic eukaryotic systems are intrinsically and unambiguously chimeric, always enclosed within membranes of course, and often within other self-produced structures which lie external to these membranes. In order to qualify as an autopoietic entity, that is, as an individual organism, any such material-metabolizing entities must be bounded by membranes made by their own metabolism. Biologically, any individual is minimally a metabolic system, made of, in some cases, many genomic entities, hetero- or homogenomic, but all are always bounded by a single, continuous covering. The breaching of the boundary signals disintegration or loss of autopoietic status.

We now see a possible correspondence of the "sense-of-self" to "autopoietic entity" or "live individual." All individuals, all living organisms actively self-maintain. From the early Archean Eon (3500 million years ago) and its bacterial inhabitants through the protists of the Proterozoic Eon (3500–2500 million years ago), and the fungi, plants, and animals of the Phanerozoic Eon (570 million years ago to the present), the "sense-of-self" seems synonymous with the nature of autopoiesis; boundaries resist breaching while biochemistry acts to maintain integrity. It is the nature of life to interact with the material world to incessantly integrate its components, rejecting, sorting, and discriminating among potential food, waste, or energy sources in ways that maintain organismal integrity.

What is remarkable is the tendency of autopoietic entities to interact with other recognizable autopoietic entities. These interactions may be neutral, as in an amoeba and a pebble; that is, no obvious reaction may occur at all. Two approaching organisms may be indifferent. Alternatively, two heterologous organisms may be destructive—disintegrative—towards each other. One, for example, may produce extracellular enzymes that destroy the other and, relieving it of its autopoiesis, break it down to component metabolic parts. The resulting chemical breakdown products may then be used as food in a trophic relation whereby the still-intact autopoietic being consumes and incorporates the chemical components of its victim. Though relations between organisms may be disintegrative or neutral, those interactions between autopoietic entities that lead beyond destruction to integrative mergers we find to be the most fascinating. Such mergers (fertilization, partner-integration in symbiosis) lead to autopoietic entities of still greater complexity. For example, the integration of a fungus attacking an alga for nutrients often—perhaps 25,000 times—has lead to a balance between the disintegrative responses of both fungal and algal partner. Eventually a lichen emerges. A lichen is neither a fungus nor an alga—as a "lichen" it is a composite symbiotic complex that itself is an autopoietic entity at a more complex level of organization. The scholars and botanists are not incorrect in naming the lichen a plant—even though, lacking embryos within maternal tissue, one today would not place lichens within the plant kingdom in any classification scheme. In every level of biological organization from beyond bacteria toward the present the "sense-of-self" can be inferred from the integrating and discriminating chemical and motility behavior of the components of what we, after the fact, recognize as the individual organism.

An amoebae, *Paratetramitus jugosus*, with a vacuole is shown in Fig. 2. In the vacuole are two entities. One, interpreted to be a bacterium, is in the process of being broken down, digested, and reutilized as food for the amoeba. Given the terms developed above we can say that the food bacterium, as a disintegrating homologous genomic system, is present in the vacuole. The second structure, a propagule (p), probably a "chromidium," an integrated heterologous genetic system (nucleocytoplasm plus mitochondria) is seen on its way outside the cell. Chromidia are interpreted to be very immature amoebae, that is, stages in the reproduction of these free-living amoebae of the vahlkamfid sort [13, 14, 15]. Thus, at the amoeba level of biological organization, "self" inside the same cell—indeed, inside the same vacuole of the same cell—can already clearly be distinguished from "food." Inspection of the microbiological literature shows, in fact, "sense-of-self" awareness is already present in the virus-infected bacterial world.

Although cell-to-cell mergers are conspicuously lacking in all interacting bacteria, such prokaryotes do accept—take into their membrane-bounded bodies—single genomes in the form of chromonemal DNA: plasmids, viruses, phage. Such DNA is transferred after cell contact directly from a second cell or from the fluid medium. The DNA, from syringe-like bacteriophages—may be forceably ejected through the bacterial membrane. Membranes from more than a single bacterial cell may touch but they never open to accept another live, bacterial being. The only types of bacteria known to be capable of

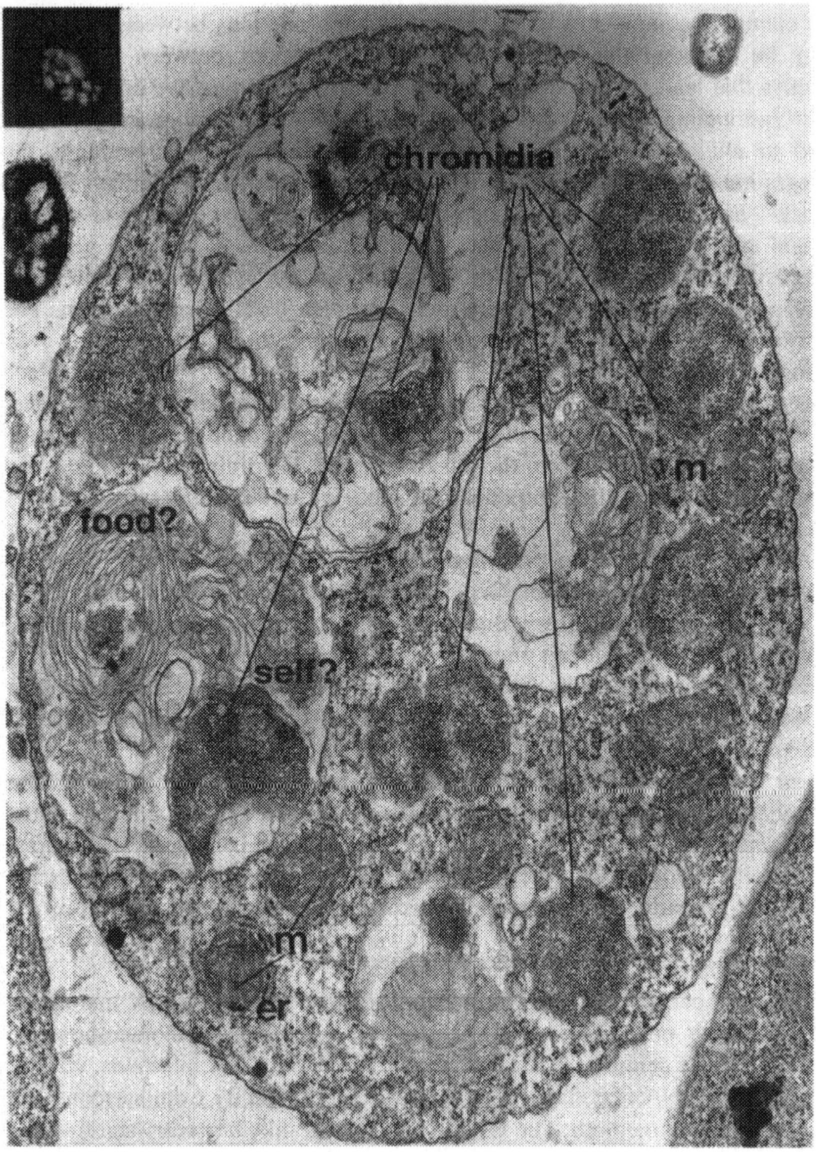
chromidia
m
food?
self?
m
er

penetration of the membrane of a second bacterium prey on and destroy that second bacterium. Predatory behavior involving the breaching of membranes, destruction, and the inevitability of their death is characteristic of *Daptobacter* and *Vampirococcus* attacks on *Chromatium* or *Bdellovibrio* assaults of *Spirillum serpens*, for example [16], in undestructive encounters only naked DNA slips through the membrane of one bacterial cell to another changing its genes, with health and survival of the recombinant as the outcome. Since a virus-infected bacterium becomes immune—it resists superinfection by the same sort of virus—there can be little doubt that an integrating sense-of-self already protects uncontrolled loss of autopoiesis—resistance to death—among the world's smallest creatures. Antigens, parts of proteins, appear on the surface of virus-infected bacteria, signalling to the outside world that these bacteria harbor the viral genome. Although other viruses may attach and even enter the already-infected cell, the humble "immune system" of the bacterium refuses to replicate the new virus which then is lost. Thus signs and signals, self identification, occur already in prokaryotes, of which the human being represents (if we can still use this word) a kind of massive, three-dimensional pointilist elaboration.

With regard to the later-day three-dimensional pointilist elaboration of the arcane immunity of virus-infected bacteria, we are admonished to ponder the connections. The AIDS-infected man differs little—in principle—from the *E. coli* bacterium infected with lysogenic bacteriophage. The "independence" of the nervous system (mind) from the immune system (body) is severely questioned. Candace Pert defiantly speaks only of bodymind or mindbody. Interviewed by her friend Nancy Griffith-Marriott she points to an overemphasis of the blood-brain barrier and the model of the nervous system as a network of penetrating, penile-shaped cells that control the body. Pert emphasizes that monocytes cross that "barrier" within seconds, furthermore these cells of the immune system transform to become the glial cells of the nervous system. (Glial cells are ten times more abundant than neurons in the mature nervous system). Like gut and brain cells, such monocytes bear neuropeptide receptors—surface proteins—sensitive to the endorphin peptides—natural or endogenous drugs inside the individual—of the neuroimmune system that

←

Fig. 2. Food remains of bacteria can be distinguished from structures interpreted to be chromidial propagules in a single vacuole of *Paratetramitus jugosus*, an amoebomastigote taken from a Baja California microbial mat. Perhaps the presence of membrane around the chromidium provides the signal to resist digestion of "self." Electron micrograph in which bar = 1μM; m = mitochondrion, er = endoplasmic reticulum surrounding the mitochondria. At upper left is a comparable photo of the live organism, apparently releasing chromidia. See Margulis and Enzien [15] for details.

bring on feelings of elation and ecstacy. Neuropeptides, small communicative molecules include vasointestinal peptides and endorphins that signal to monocytes. Such protein-like molecules attach to the cell receptors at the surface of gut or brain or monocyte cells at the same place the AIDS virus gets stuck. No, says Pert, there is no mind/body, controller/controlled, male/female, neuron/glial cell dichotomy. Rather there is "mindbody-bodymind," a dynamic system kept informed by devastating news, transforming monocytes, neuropeptide messengers—and hundreds of other integrating mechanisms that mobily confirm the self [17].

Beginning as latter-day evolutions of bounded endosymbiotic bacterial communities we—as densely packed biomineralizing complexes of eukaryotic cells—should not be too sanguine about the longevity of the modern notion of self. Already in the 19th century Samuel Butler clearly and successfully deconstructed personality by parasitizing Charles Darwin's texts. Between the ovum and the octogenarian, held Butler, lie differences greater than those between human beings and other species. What with the vagaries of memory and experience, it is essentially arbitrary to believe that the zygote and the eighty-year old are the same person, whereas the father and the son have different selves. Genotypically we may argue with Butler, but to do so phenotypically would be a far more difficult chore. Butler demonstrates the essential arbitrariness of our definitions of organismic identity, of organic integrity and "individuality" even more strikingly by taking the case of a moth. Here we have a being, Butler says, that undergoes radical bodily change between egg and chrysalis, between pupa and winged insect; and yet the only time we say it dies is after the adult moth form stops moving its wings, despite the other radical phenotypic changes during which the genotype has nonetheless been preserved: we might as easily, Butler reminds us, have chosen to consider the transfer from egg to chrysalis or from chrysalis to moth as "death" —and construed the demobilization of the moth as a sloughing-off similar to the shedding of a skin. Indeed, to seriously consider death at all entails a certain ignorance—a certain disregard for the continuity of the "personality" (let us not be too quick to say germ cells, and invoke the same philosophy of the subject, the self, at a deeper level) despite its radical transformations. So you see that with this figure in which the moth's "self" is held aloft on the tenderhooks of quotation marks "we" have provisionalized identity—not least of all by avoiding the traditional figure of the rectangle which enframes the essay, representing thoughts in an enclosed form that seems to mirror the hegemony of a rigidly structured Platonic body. Topologically the self has no homunucular inner self but comes

TABLE I
Multiple origins of self in evolution[a]

Ancestral bacteria	Extant organelles or organs	Hypothetical minimum number or genomes	"Individual organisms"
Thermoplasma (archaeobacterium)	nucleocytoplasm	2	eukaryotes
Spirochaeta (eubacterium)	kinetosomes, centrioles, microtubule organizing centers	2	most eukaryotes (those with unduli-podial microtubules) [18]
Respiring eubacteria	mitochondria	3	most heterotrophic eukaryotes (with mitochondria)
Cyanobacteria	cyanelle	4	*Cyanophora, Glaucocystis Cyandidium*
Cyanobacteria *(Synechoccocus)*	rhodoplast	4	red algae
Chlorobacteria *(Prochloron)*	chloroplast	4	green algae, plants
Sulfide oxidizing eubacterium *Vibrio fisheri*	thiosome, trophosome	5	Vestiminiferan, tube worms [19]
Carotenoid-producing eubacterim	red nidemental gland	5	*Loligo* (squid)

[a] For technical details of integrations of genomic systems in endosymbiotic origin of eukaryotic cells see ref. [20]; for nontechnical, see ref. [21].

University of Massachusetts at Amherst

REFERENCES

1. Foucault, M., 1977. What is an author? In *Language, Counter-Memory and Practise: Selected Essays and Interviews*. Bouchard, D.F, ed. Cornell University Press, Ithaca, New York. p. 124.
2. Maturana, H.R., and Varela, F.J., 1973. Autopoiesis: The organization of the living. In *Autopoiesis and Cognition*. Maturana, H.R., and Varela, F.J., eds., 1980. D. Reidel, Boston.
3. Fleck, L., 1979. *Genesis and Development of a Scientific Fact*. University of Chicago Press, Chicago.

4. Habermas, J., 1987. *The Philosophical Discourse of Modernity.* Translated by Frederick Lawrence. MIT Press, Cambridge, Mass., p. 137.
5. Leenhardt, M., 1979. *Do Kamo* (translated by Gluati, B.M.). University of Chicago, Chicago, p. 22.
6. Snell, B., 1960. *The Discovery of the Mind* (translated by T.C. Rosenmeyer). Harper Torchbooks, New York, p. 8.
7. Elias, N., 1978. *The Civilizing Process : The History of Manners* (translated by E. Jephcott). Urizen Books, New York, pp. 252–253.
8. Lacan, J., 1977. The mirror stage as formative in the function of the I. In *Écrites: A Selection* (translated by A. Sheridan). New York, W.W. Norton. pp. 1–7.
9. Sagan, D., 1990. What Narcissus saw: The Oceanic "I"/"eye". In *Speculations: The Reality Club* 1. Brockman, J., ed. Prentice Hall Press, NY pp. 245–266.
10. Sagan, D., 1990. *Biospheres: Metamorphosis of Planet Earth.* McGraw-Hill, New York
11. Huxley, J., 1912. *The Individual in the Animal Kingdom.* G.P. Putnam and Sons, New York. p. 125.
12. Dawkins, R., 1982. *The Extended Phenotype: The Gene as the Unit of Expression.* Oxford, W.H. Freeman and Co.
13. Dobell, C., 1913. Observations on the life-history of Cienkowski's "*Arachnula.*" *Arch. Protistenkund.* **31:** 317–353.
14. Wheery, W.B., 1913. Studies on the biology of an amoeba of the limax group. *Vahlkampfia* sp. No. I. *Arch. Protistenkund.* **31:** 77–94.
15. Margulis, L., Enzien, M., and H.I. McKhann, 1990. Revival of Dobell's "chromidia" hypothesis: Chromatin bodies in the amoebomastigote *Paratetramitus jugosus. Biol. Bull.* **178:** 300–304.
16. Guerrero, R., Pedrós-Alió, C., Esteve, I., Mas, J., Chase, D., and L. Margulis., 1987. Predatory prokaryotes: Predation and primary consumption evolved in bactera. *Proc. Nat. Acad. Sci.* **83:** 2138–2142.
17. Pert, C., and Griffiths-Marriott, N., 1988. Bodymind. *Woman of Power* **11:** 22–25.
18. Margulis, L., 1991. Symbiosis in evolution: Origins of cell motility. In *Evolution of Life: Fossils, Molecules and Culture*, Osawa, S. and T. Honjo, eds. Springer-Verlag Tokyo. pp. 305–324.
19. Vetter, R., 1991. Symbiosis and the evolution of novel trophic strategies: Thiotrophic organisms at hydrothermal vents. In *Symbiosis as a Source of Evolutionary Innovation: Speciation and Morphogenesis*, Margulis, L., and R. Fester, eds. MIT Press, Cambridge, Mass. pp. 219–245.
20. Margulis, L., 1981. *Symbiosis in Cell Evolution.* W.H. Freeman & Co., San Francisco.
21. Margulis, L., and D. Sagan, 1986, *Microcosmos: Four Billion Years of Evolution From Our Bacterial Ancestors.* Summit Books, New York, and Touchstone, New York.

References 5, 6, and 7 were cited it Pollner, M., 1987. *Mundane Reason: Reality in Everyday and Sociological Discourse.* Cambridge University Press, Cambridge, England. pp. 136, 143–144, respectively.

INDEX